Lecture Notes in Networks and Systems

Volume 122

The series "Lecture Notes in Networks and Systems" publishes the latest developments in Networks and Systems—quickly, informally and with high quality. Original research reported in proceedings and post-proceedings represents the core of LNNS.

Volumes published in LNNS embrace all aspects and subfields of, as well as new challenges in, Networks and Systems.

The series contains proceedings and edited volumes in systems and networks, spanning the areas of Cyber-Physical Systems, Autonomous Systems, Sensor Networks, Control Systems, Energy Systems, Automotive Systems, Biological Systems, Vehicular Networking and Connected Vehicles, Aerospace Systems, Automation, Manufacturing, Smart Grids, Nonlinear Systems, Power Systems, Robotics, Social Systems, Economic Systems and other. Of particular value to both the contributors and the readership are the short publication timeframe and the world-wide distribution and exposure which enable both a wide and rapid dissemination of research output.

The series covers the theory, applications, and perspectives on the state of the art and future developments relevant to systems and networks, decision making, control, complex processes and related areas, as embedded in the fields of interdisciplinary and applied sciences, engineering, computer science, physics, economics, social, and life sciences, as well as the paradigms and methodologies behind them.

**** Indexing: The books of this series are submitted to ISI Proceedings, SCOPUS, Google Scholar and Springerlink ****

More information about this series at http://www.springer.com/series/15179

Monica Rossi · Matteo Rossini · Sergio Terzi
Editors

Proceedings of the 6th European Lean Educator Conference

ELEC 2019

Springer

Editors
Monica Rossi
Department of Management,
Economics and Industrial Engineering
Politecnico di Milano
Milan, Italy

Matteo Rossini
Department of Management,
Economics and Industrial Engineering
Politecnico di Milano
Milan, Italy

Sergio Terzi
Department of Management,
Economics and Industrial Engineering
Politecnico di Milano
Milan, Italy

ISSN 2367-3370 ISSN 2367-3389 (electronic)
Lecture Notes in Networks and Systems
ISBN 978-3-030-41428-3 ISBN 978-3-030-41429-0 (eBook)
https://doi.org/10.1007/978-3-030-41429-0

This Springer imprint is published by the registered company Springer Nature Switzerland AG
The registered company address is: Gewerbestrasse 11, 6330 Cham, Switzerland

Preface

This publication constitutes the refereed post-conference proceedings of the 6th European Lean Educator Conference 2019, ELEC 2019, held in Milan, Italy, in November 2019.

ELEC is an international conference hosted in Europe that focuses on lean approach, on its forefront development and on the educational process necessary/useful to spread it in both academic and industrial worlds. That is why, ELEC brings together a diverse audience of participants including academics (teachers and researchers), intermediaries (consultants and coaches) and practitioners (managers and employees from industry) in an eclectic mix every year.

The ambitious aim of ELEC is to create the right environment for a fruitful exchange within and between academia and industry by pursuing the mission of shaping the existing generation and growing the next generation of lean educators. How to engage in continuously innovative educational methods is paramount to create more prosperous and fairer societies. And this publication contributes to this aim, from a scientific perspective with practical evidences.

Beyond lean thinking's roots in manufacturing, countless applications have emerged and other fields are experiencing more and more successful lean transformations, from product development and innovation to office and all knowledge work, in services from health care to education and macro-projects in construction and mining, to mention just a few. This means there are so many cross-learning opportunities between the various fields, as well as teaching and learning approaches face a constant evolution towards more effective and efficient ways to understand the potentialities of lean thinking in practice.

ELEC 2019 scope and objective explore the latest academics and industrial contributions to lean education, looking for innovative methods and approaches that allow lean thinking benefits to materialize in practice in as many industries as possible, hence the name given to 2019 conference: *The Lean Educator and Practitioner Mashup.*

In this publication, the editors introduce 41 revised full papers that were carefully reviewed, presented and deeply discussed during 11 parallel sessions at ELEC 2019, around the following topics: lean trainings within university and industry collaborations; lean product and process development; lean and people empowerment; emerging contexts for lean applications; measuring lean performances; lean, green and circular; continuous improvement initiatives; lean thinking in practice; organizational culture in lean journeys; innovative trainings to teach lean management.

Milan, Italy

The editors
Monica Rossi
Matteo Rossini
Sergio Terzi

Contents

A Learning Factory for Remanufacturing: A New Configuration at Valladolid Lean School

Jose A. Pascual, Carina Pimentel, Manuel Mateo, Ignacio Hoyuelos, João Matias and Angel M. Gento

Abstract Collaboration between university and industry is essential in today's changing environment, where limited resources must be used by as many users as possible in an efficient manner. At the University of Valladolid we have a Lean School equipped in collaboration with Renault-Nissan Consulting where training is given to students, workers and professionals: initially in Lean Manufacturing, but with a rapid evolution and improvement towards other paradigms, also integrating the concept of Circular Economy. The main contribution of this paper is the presentation of a training in a learning factory, combining lean manufacturing and circular economy. The assimilation of knowledge by the students with this training process based on role play and simulation of a real process are much higher than those obtained with traditional methods. Furthermore, the new training contributes to the reduction of

J. A. Pascual · A. M. Gento (✉)
Organización de Empresas y CIM, Universidad de Valladolid, Paseo del Cauce 59, 47011 Valladolid, España
e-mail: gento@eii.uva.es

J. A. Pascual
e-mail: pascual@eii.uva.es

C. Pimentel · J. Matias
DEGEIT, Universidade de Aveiro, GOVCOOP, Campus Universitário de Santiago, 3810-193 Aveiro, Portugal
e-mail: carina.pimentel@ua.pt

J. Matias
e-mail: jmatias@ua.pt

C. Pimentel
UNIDEMI, Departamento de Engenharia Mecânica e Industrial, Faculdade de Ciências e Tecnologia, Universidade Nova de Lisboa, Campus de Caparica, 2829-516 Caparica, Portugal

M. Mateo · I. Hoyuelos
Renault-Nissan Consulting, Avda. Madrid, 72, 47008 Valladolid, España
e-mail: manuel.mateo-prieto@renault.com

I. Hoyuelos
e-mail: ignacio.hoyuelos@renault.com

M. Rossi et al. (eds.), *Proceedings of the 6th European Lean Educator Conference*, Lecture Notes in Networks and Systems 122,
https://doi.org/10.1007/978-3-030-41429-0_1

economic and environmental impacts, mainly: reduction of inventory, raw materials, solid wastes, energy and motions.

Keywords Circular economy · Lean manufacturing · University-industry collaboration · Learning factory

1 Introduction

Due to the use of limited resources and the increasing amount of waste in the world, remanufacturing is having a growing push, at least in some sectors such as the automobile industry, where the industry tries to maximize the recovery of the value of products and parts used.

At present, more than 25 tonnes of used vehicle material are recycled (more than 80% of a vehicle by weight can currently be recycled, although the aim is to reach 95% in the next few years), and the automobile industry is also one of the industries that consumes the most recycled material, although there is still a high percentage of vehicles that do not know their destination at the end of their useful life (between 30 and 40%) [1].

Taking into account this worldwide situation and the fact that in the area of Valladolid there are 3 factories of one of the largest automobile manufacturers (Renault), a factory of Iveco, and numerous factories of tier 1 suppliers a learning factory integrating manufacturing and remanufacturing operations is a subject of great interest for the training of students of the University of Valladolid as well as for factories employees' training.

Therefore, and using the learning factory facilities designed jointly by Renault and the University of Valladolid, for the joint training of students and workers in lean manufacturing (applying concepts and tools such as 5S, job definition, waste elimination, line balancing, and quality), a new configuration in the Lean School has been defined, to value the conception of products (and services) focused not only on giving a satisfactory product to the customer, but in the ease to recover the parts that compose it and using them later (after a series of operations and treatments) in the manufacture of new products, thus working in a closed-loop supply chain.

In this way, training participants acquire knowledge on how to better manage materials and energy utilisation and other valuable resources through higher rates of recycling, reuse and remanufacturing. This change greatly affects the activities of different companies and their supply chains in general: reverse logistics of used products, remanufacturing processes of disassembling, component inspection and assembling, and customer demands of remanufactured products. Moreover, they have an open mind and acquire sensitivity to the new paradigm of manufacturing based on Circular Economy.

In this paper we will use the experience acquired with the above presented learning factory to present its fundamental principles, as well as to provide some insights about how it can contribute to the integration of sustainable principles in operations management.

This paper is organised as follows: in Sects. 2 and 3 a theoretical background about Industry and Academia collaboration and the Circular Economy is presented; then, in Sect. 4 a learning factory training integrating lean manufacturing and circular economy is shown; finally, Sect. 5 summarizes the study main results, limitations and future developments.

2 University-Industry Collaboration

In general, university-industry collaboration (UIC) refers to the interaction between universities and companies with the main objective of promoting the exchange of knowledge and technology between them, having UICs a long tradition in several countries of the world [2].

There are many forms of university-industry collaboration. According to Bonarccorsi and Piccaluga [3], six major forms of collaboration can be identified: Personal Informal Relationships, Personal Relationships, Third Party, Formal Targeted Agreements, Formal Non-targeted Agreements and Creation of Focused Structures. In recent years there has been an increase in collaborations between both parties with clearly differentiated objectives. On the universities side, the objectives are several: from an academic point of view, to be able to give their students a formation adapted to the necessities of the environment and to improve the employability of their students, and from the research point of view, to be able to get access to scholarships and funds of the government for applied research and equipment of laboratories, access to real data to test applications to different theories, development and exploitation of patents as well as the publication of papers. On the other side, companies, immersed in such a world that changes so fast technologically, where product life cycles are shorter and where competitors are found all over the world, need: highly qualified human capital, access to research networks and the most modern technologies, as well as national (or regional) incentives to develop these relationships through tax exemptions and grants. And all this, without forgetting social responsibility policies and social pressure to promote innovation and regional and national development [4].

Following this paradigm of worldwide implantation, the University of Valladolid and the Renault-Nissan-Mitsubishi alliance consultancy have been collaborating in several ways for more than 20 years, namely through teaching courses for students developed jointly, teachers and employees taking internships, providing scholarships and awards for work and research. As part of this collaboration, in 2015, Renault Nissan Consulting (RNC) created a laboratory at University of Valladolid for the teaching and development of lean tools for the university students and for employees of Renault and auxiliary companies: the Lean School (LS). Thus, during these 4 years, more than 500 students from the University of Valladolid and nearly 2000 workers from the Renault Group's factories in the Iberian Peninsula have been engaged with the Lean School to acquire or refine their knowledge and competencies.

The development of the learning in this laboratory is carried out through a series of rounds or simulations (productions), in an environment very close to reality, in which the participants solve different real problems in the Lean School production system, through Lean tools, and evaluate them. The main objective of the Lean School is to offer all participants a learning environment where the transmission of specialized theoretical and practical knowledge is based on "learning by doing" [5].

3 Circular Economy

Traditionally, industries have followed a linear pattern of consumption: taking raw materials, transforming them into products and selling them to consumers (who discarded them when they did not need them). In terms of volume, some 65 million tonnes of raw materials entered the economic system in 2010, and is expected to grow to about 82 tonnes in 2020 [6].

Recently, many companies have also realised that this linear system increases their exposure to risks, mainly due to higher resource prices and supply disruptions: there is high volatility in the prices of metals and agricultural products due to shortages or extraction difficulties. To avoid this, some industries have started to use materials extracted from waste products moving towards a circular economy model.

In [7], a circular economy is explained as an economy 'where the value of products, materials and resources is maintained in the economy for as long as possible, and the generation of waste minimised' (Fig. 1).

The European Commission in 2008 [8] and 2011 [9] had already identified critical raw materials and recycling as key aspects for improved resource efficiency by defining in [7] the EU plan for the circular economy.

In the automotive sector, where our industrial partner (Renault) is located, several critical prime materials can be identified, both in conventional vehicles (combustion engine vehicles) and in hybrids (HEVs) and electric vehicles (EVs), from graphite to platinum group metals and rare earth elements (in catalysts).

According to a study by the European Commission [11], 14% of the world's use of graphite is for automotive components. Moreover, the consumption of some metals such as palladium, platinum and rhodium in this sector accounts for more than 70% of total consumption. And, if we take into account the increasing needs for catalysts, it is foreseeable that the consumption of other metals such as nickel and cerium will become critical in the coming years.

Figure 2 shows an illustration of the use and potential use of recycled material in electric vehicles [7]. It is for this reason that the recycling and reuse of components and materials is fundamental in the automotive industry, and that is why Renault, together with the University of Valladolid, aims to raise awareness of its importance by developing cases and training that help to reduce the consumption of raw materials.

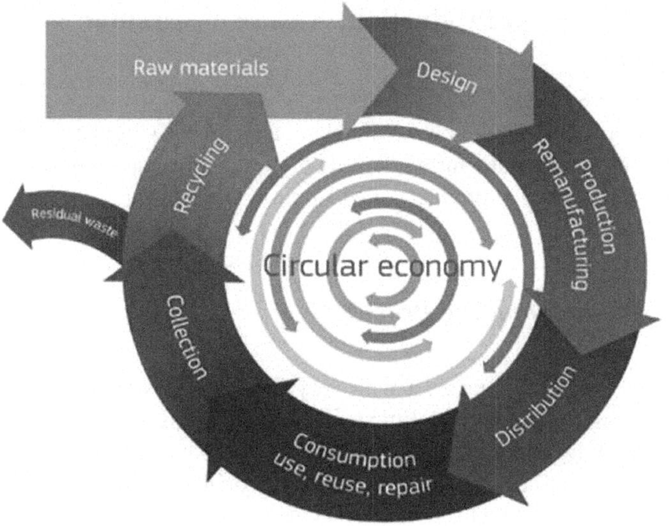

Fig. 1 Conceptual diagram illustrating the circular economy in a simplified way [10]

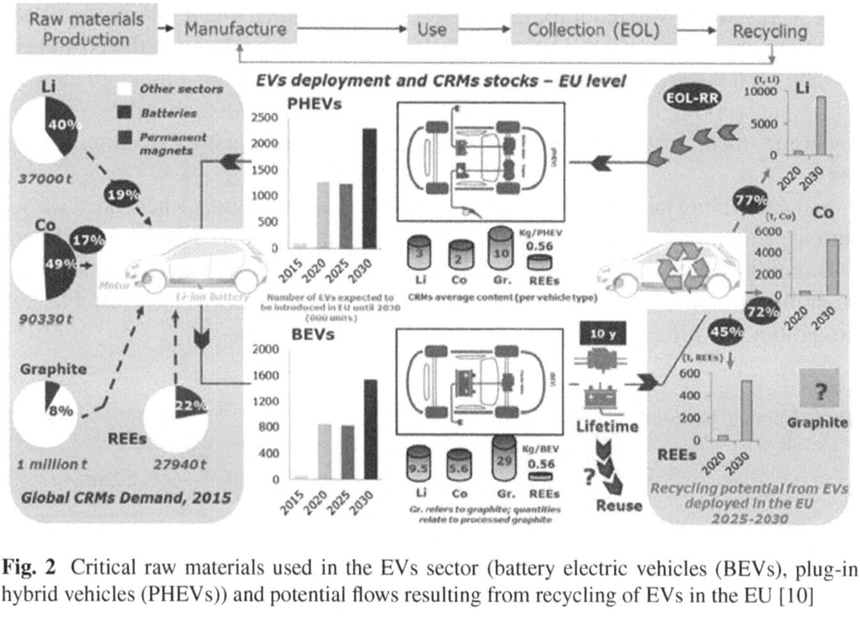

Fig. 2 Critical raw materials used in the EVs sector (battery electric vehicles (BEVs), plug-in hybrid vehicles (PHEVs)) and potential flows resulting from recycling of EVs in the EU [10]

4 Cubiton Product Production: A Case of Circular Economy

In this section an example of how Lean School approaches Circular Economy, integrating it with Lean Manufacturing, will be presented, through the exploration of a training offered in which a product called Cubiton is produced.

4.1 The Product and the Process

The "cubiton" is a simple product with very little complexity in the assembly to facilitate learning for people with little manual dexterity. It is formed by a base and four layers. Each level consists of four different components that fit together (like a puzzle) in different colours (green-yellow-blue-orange). The components of the odd layers are simple, while the components of the even layers include an additional piece with different shapes (hexagon, circle, rectangle, and cross) (Fig. 3).

The assembly of the "cubiton" is carried out in the Lean School in an assembly factory, and the disassembly in a recycling factory. In the assembly factory there are 4 workstations where all the components are assembled and a fifth workstation where the quality control is carried out before sending the cubiton to the customer. Initially each layer/colour is assembled on a different workstation (although there is a different working load between the odd and even layers). Also, in order to reduce downtime, there is work in process between workstations, with a maximum of 9 units.

In the recycling factory (by symmetry with the assembly factory) there are 5 workstations: a first one where the quality control of the products is carried out, and four workstations where the cubitons are disassembled. Initially, each layer/colour is disassembled in a different workstation, although different processes can be established throughout the different stages carried out by the participants.

Fig. 3 Cubiton: base and two levels

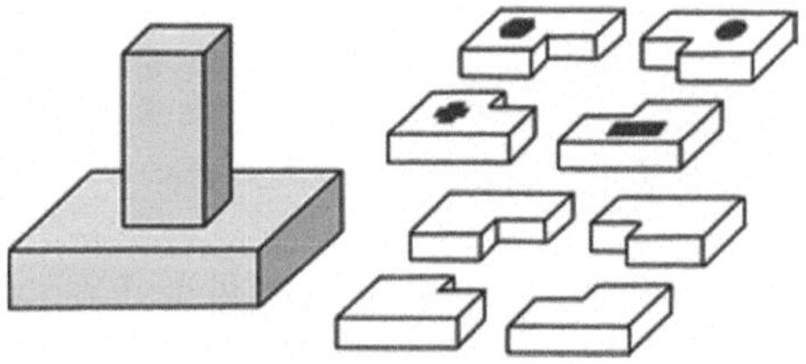

4.2 Learning Methodology

The model for learning process followed in the training is proposed in [12] (Fig. 3). Initially, participants are given an initial session of basic concepts in Circular Economy, Lean Manufacturing and the characteristics of the game are explained. Then, the participants manufacture the product (cubitons) based on their previous knowledge, the behaviour of the rest of the peers and making their own decisions. After manufacturing between 25 and 30 products (depending on the time available), the participants together with the trainers discuss the results obtained and try to identify problems and improvement opportunities. Logically the participants need to do some work outside the laboratory (in their office or at home) to assimilate the concepts.

After the first phase, the scheme proposed along the remaining simulation runs is similar: the phase starts with a theoretical session in which the required concepts are explained. Afterwards, the aim is to guarantee that they are practiced and assimilated during the manufacture of the product with a new configuration resulting from the identification of improvement opportunities in the previous simulation run. This new configuration must be proposed by the participants themselves (guided by the trainers) on the basis of the learning obtained in the previous phase. Normally, three phases are carried out with the same approach.

Along the improvement process the space required to the production system is used as an important improvement measure. Figure 4 shows the values that are typically achieved in this measure after the application of different Lean tools and practices to reduce waste [13]. Some of the lean tools and practices contributing to this achievement are: layout reconfiguration (Fig. 5), pull flow, 5S, standardisation, load balancing, VSM.

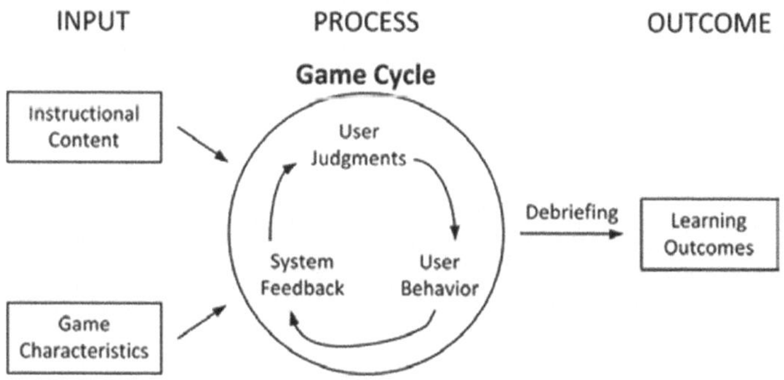

Fig. 4 Cubiton training process (based on [12])

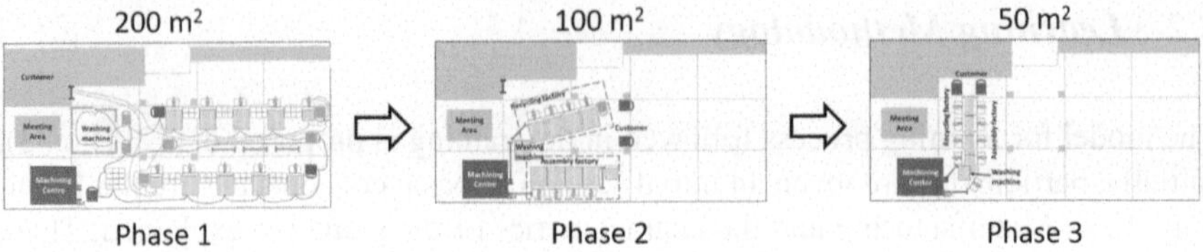

Fig. 5 Lean school layout of the training process

4.3 *Circular Flow*

After the product is assembled in four workstations (A1–A4), and subjected to a quality control in A5, the product must be sent to the customer who, after using it, leaves it at the disposal of the recycling company that subjects it to an inspection process (not all products can be recycled) in workstation R1 and if it is accepted it is disassembled, for the recovery of the components, between workstations R2 and R5. Disassembled components must pass through a washer process previously to be sent back to the assembly factory, while the base is transported to a machining centre to ensure that defects caused during use by the customer are eliminated (Fig. 6).

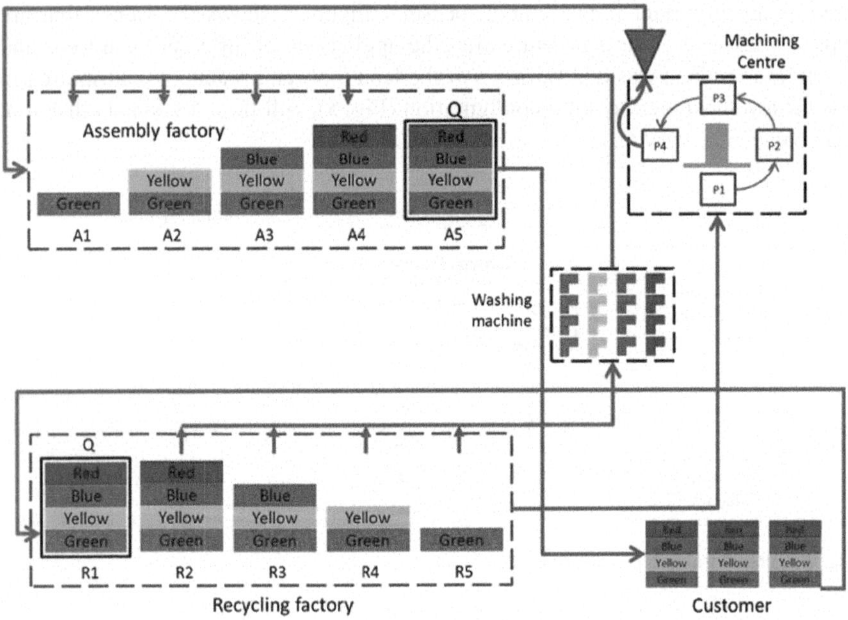

Fig. 6 Flow between the *Assembly factory (top left)* and the *Recycling factory (bottom left)* via the *customer (bottom right)*. After leaving the Recycling factory, the parts must pass through the *Washing process (centre right)* and the bases through the *Machining Centre (top right)*

5 Conclusions

In this paper a learning factory training program was presented, combining Lean Manufacturing concepts with Circular Economy ones, through the integration of an assembly and recycling factory. Through this training two basic principles are acquired by participants: (1) the reduction of motions, waste and inventory allows for the reduction of the quantity of raw materials and energy needed; and (2) the use of the components in a circular flow reduces costs, because to manufacture 25–30 products only are needed components to manufacture 12 products (in the extreme case and in the third phase) because we create a continuous flow with 4 manufacturing workstations and 4 recovering workstations. This amount of circulating components must be calculated by the participants (at all phases of the training) to understand the importance of the circular economy, especially when critical raw materials exist.

Despite the contribution of this paper, there are limitations that should be noted: (1) the use of a unique product and (2) the lack of quantification of the effects of integrating circular economy. So, future research could include increasing complexity and variety in the product. Also, another line of research could be the development of a set of KPIs considering the economic, social and environmental perspectives.

References

1. European Commission (2018) Report on critical raw materials and the circular economy. Commission Staff Working Document, Part 1/3
2. Ankrah S, AL-Tabbaa O (2015) Universities–industry collaboration: a systematic review. Scand J Manag 31:387–408
3. Bonarccorsi A, Piccaluga A (1994) A theoretical framework for the evaluation of university-industry relationships. R&D Manag 24:229–247
4. Rybnicek R, Königsgruber R (2019) What makes industry-university collaboration succeed? A systematic review of the literature. J Bus Econ 89:221–250
5. Pascual J, Hoyuelos I, Mateo M, Gento AM (2019) Lean school: a learning factory for training lean manufacturing in a physical simulation environment. Manag Prod Eng Rev 10(1):4–13
6. Ellen MacArthur Foundation (2013) Towards the circular economy economic and business rationale for an accelerated transition
7. European Commission (2015) Communication from the Commission to the European Parliament, the Council, the European Economic and Social Committee and the Committee of the Regions, closing the loop An EU action plan for the circular economy, COM(2015) 614
8. European Commission (2008) Communication from the Commission to the European Parliament and the Council—the raw materials initiative: meeting our critical needs for growth and jobs in Europe, COM(2008) 699
9. European Commission (2011) Communication from the Commission to the European Parliament, the Council, the European Economic and Social Committee and the Committee of the Regions, tackling the challenges in commodity markets and on raw materials, COM(2011) 25
10. European Commission (2018) Report on critical raw materials and the circular economy. Commission Staff Working Document
11. European Commission (2015) Report on critical raw materials for the EU critical raw materials profiles

12. Garris R, Ahlers R, Driskell JE (2002) Games, motivation, and learning: a research and practice model. Simul Gaming 33:441–467
13. Womack J, Jones D (1996) Lean thinking. Banish waste and create wealth in your corporation. Free Press, Simon & Schuster, Inc., New York, NY

Integrated University-Industry Training: A Collaborative Journey

Manuel Mateo, Ignacio Hoyuelos, Jose A. Pascual and Angel M. Gento

Abstract University-Industry collaboration is fundamental in today's changing and competitive world, as it allows us to take advantage of the strengths of each of them. The University of Valladolid has been collaborating for more than 25 years with Renault-Nissan Consulting in different fields: scholarships, contract research and, more recently, a laboratory adapted to the academic and industrial environment, where Renault students and employees are trained to improve their skills in lean management. This school provides training in production systems concepts and lean manufacturing in an innovative way simulating real manufacturing processes. This work shows one of the latest and most innovative training processes carried out at the Lean School and the results obtained with a small product manufactured with toy building blocks and which is also carried out in companies on demand. The levels of assimilation and retention of knowledge of the students are highly superior to those achieved by traditional methods.

Keywords University-industry collaboration · Industrial engineering · Lean management

M. Mateo · I. Hoyuelos
Renault-Nissan Consulting, Avda. Madrid, 72, 47008 Valladolid, España
e-mail: manuel.mateo-prieto@renault.com

I. Hoyuelos
e-mail: ignacio.hoyuelos@renault.com

J. A. Pascual · A. M. Gento (✉)
Organización de Empresas y CIM, Universidad de Valladolid, Paseo del Cauce 59, 47011 Valladolid, España
e-mail: gento@eii.uva.es

J. A. Pascual
e-mail: pascual@eii.uva.es

M. Rossi et al. (eds.), *Proceedings of the 6th European Lean Educator Conference*, Lecture Notes in Networks and Systems 122,
https://doi.org/10.1007/978-3-030-41429-0_2

1 Introduction

The need for permanent adaptation, of companies and organizations, to the changing and globalized environment, generates in those that achieve it through the adoption of creative solutions, an important competitive advantage. The new solutions, aimed at improving its operation, are directed towards the idea of generating value for its clients, eliminating from the processes everything that does not contribute value, which represents the basis of the Lean philosophy. That is why in the last decades the application of Lean Management in business processes has experienced a very significant growth. Lean Manufacturing is more than a system; it is a global production and management philosophy aimed at generating value at the lowest cost.

The growing implementation of Lean methodology in different sectors (aeronautics, banking, food, distributors, health, …), following the trail of the automotive industry pioneer in its application, has generated a growing demand for trained personnel in the same. Companies have been the first to detect this need, creating their own training environments, and demanding from Universities students trained in Lean methodologies and tools. Universities have thus been forced to adapt the curricula of engineering degrees, so that they include the acquisition of skills inherent to the Lean philosophy and the skills to make use of the tools that are part of the Lean environment. However, from the training point of view, in universities we have been slow in reacting to explain all these concepts in an active way involving students in learning, and not just being mere receivers of documentation. This change is also a demand from students since access to information in the twenty-first century has no similarity with access in past centuries. Considering all the above, the idea arises to implement a learning factory that integrates the main Lean management tools, allowing both students of the University of Valladolid and employees of companies access to experiential training.

To facilitate this change, at the University of Valladolid we have established a strong collaboration with one of the most important car manufacturers (Renault), so that once students have received this training they can do an internship at the company or at one of its Tier 1 suppliers.

2 University-Industry Collaboration

Globalization, the environment in rapid and continuous change, requires rigorous control of costs and markets, which has made both Logistics and the productive improvement of production processes, key tools to improve the competitiveness of organizations and ensure their sustainability.

Industry increasingly demand true specialists in the field of Operations Management, and significantly in the field of Production and Logistics. The top companies, in their search to reach the Excellence in the Operations and thus to obtain competitive advantage, rely on Lean Management.

Table 1 Organizational forms of UIC (based on [2])

Personal informal relationships	Individual consultancy, informal exchange forums and workshops, personal contact with university academic staff or industrial staff
Personal relationships	Student internships, sabbaticals periods for professors, hiring of graduate students, employment of relevant scientists by industry, or use of university or industrial facility
Third party	Institutional consultancy, industrial associations, government agencies
Formal targeted agreements	Contract research, cooperative research projects or training programs for employees
Formal non-targeted agreements	Industrially sponsored R&D in university departments, research grant and donations
Creation of focused structures	Innovation/incubation centres, research, science and technology parks

The "European Higher Education Area" promotes collaboration between universities and companies (UIC), understood as the mutual exchange of knowledge and technology through interaction. Its application allows to know the skills and capacities demanded from the labour market, bringing the training given in the universities closer to the business reality. This has been done for some time in numerous countries [1].

University-industry collaboration can cover different aspects and depths, and according to [2] can be classified into 6 large blocks (Table 1).

Renault-Nissan Consulting (RNC), the consulting firm of the Renault Group, has collaborated with the University of Valladolid for more than 25 years. The relationship began because several of the workers had been students at the University of Valladolid and knew and collaborated selflessly with professors giving some practical session at the university based on their professional experience.

After these first informal contacts, collaboration was increased with internships, company visits for students and the use of some laboratories for testing.

An important collaboration from the formal point of view was the participation of the company in the definition of new degrees within the European Higher Education Area (Bologna Process), a collaboration that has continued over time and has allowed the subjects to adapt to the real needs of companies in the region.

After this collaboration, which continues to be the same over time as the previous ones, collaborations arose with different research groups and participation in national and international research projects.

In the year 2014, this collaboration increased providing RNC and laboratory for the teaching of Lean tools to the students of the University of Valladolid. The collaboration agreement also includes the use by the company of the laboratory for the training of its employees.

This has allowed the University of Valladolid to be the first Spanish university to have a laboratory for teaching Lean tools using the techniques of "learning by doing" widely used in industry: Lean School (LS) [3]. To this end, an old chemistry lip has been transformed into a modern space made up of different workstations, warehouses, shelves, transport elements, etc., which allow participants to carry out active training, forming part of the training process beyond mere observation. In this way, students advance to be at the forefront of their training, as they must use learning methodologies in which the acquisition of skills, abilities and knowledge is done actively, through experimentation, participation, discussion, decision-making, etc.…

The laboratory can be used (in a coordinated way) both for the training of students at the University of Valladolid, as well as for the training of Renault workers, and even on some occasions jointly to favour the exchange of experiences and the incorporation of students in internships in industry. In order to maintain coherence and a standard level, university professors and company consultants participate in the training, thus ensuring a balance between the transmission of theoretical concepts and their applicability in different real situations.

3 The Training Process

In the laboratory (Lean School), the participants in the different formations face a productive process close to reality, in which they can apply the different Lean tools previously explained to them. In the case presented in this work we use the construction of different products, using toy building blocks. The objective of the training, once the main concepts of Lean Management (Pull, Push, Kanban, Kaizen, Continuous Improvement, Standardized Work, Balanced Work, 7 + 2 wastes) [4] have been presented, is to consolidate these concepts and apply different Lean tools.

The training is aimed at both university students and employees of the company or its suppliers. As the knowledge of "Lean Manufacturing" can be very different depending on their experience and/or university degree all participants take a small questionnaire (anonymous) prior to the start of the course in order to adapt the tools on which we focus during the course.

It consists of 3 sessions of 4 h (Fig. 1). Each session begin with a short theoretical introduction (45–60 min) followed by a simulation of a production process where the participants manufacture three types of products in the same productive area: Trailers, Tippers and Cranes. The product mix of customer demand is known, although not the order (40% trailers, 40% dump trucks, 20% cranes). The customer requests with a frequency and a defined delivery time and that do not change during the different productions. Each simulation begins with a short training phase so that the participants become familiar with the activities to be carried out in the different workstations: suppliers, sub-assembly manufacturers, logistics, warehousing, quality control and all of them led by a plant manager. Once 15 products have been manufactured, the participants fill in a questionnaire in which, after a brief reflection,

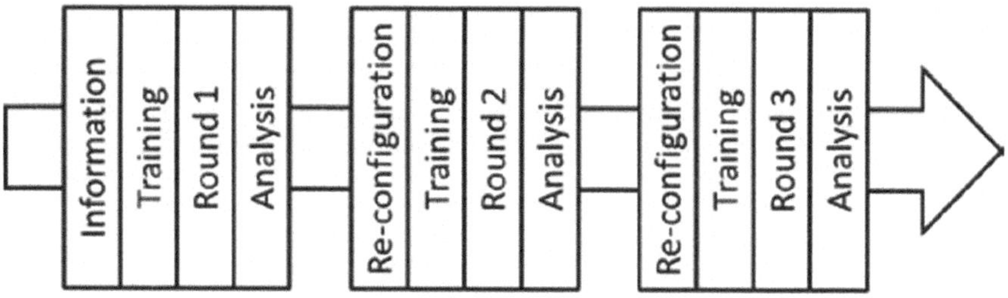

Fig. 1 Lean School training process

they discuss the strengths and weaknesses of the manufacturing process. With these questionnaires, the trainers summarize the results, placing special emphasis on all the problems that have appeared in the manufacturing process from the comments made by the participants and listing some of the Lean tools that will be used in the next manufacturing process to try to eliminate or reduce some of the problems.

Several objectives are intended to be achieved: practice on evolution process of manufacturing in "push" to one in "pull" discovering the benefits, experiment the concept of waste and how to eliminate them, discover the Lean logistics and Just in time concepts, understand the concept of takt time and the need to synchronize our production with it and analyse the concepts of manufacturing capacity and work balance. The analysis of the results in each session shows the practitioners the spectacular improvements that are achieved in all facets of the operational activity.

The three phases are characterized by the following:

1. The first production is carried out following a "push" batch production scheme: the suppliers manufacture components and sub-assemblies, the factory manufactures and logically demands material from the warehouse all 3 by 3.
2. After the first revision of the results (quite unsatisfactory), since only the vehicles were manufactured in the time foreseen for the opening of the factory, and moreover none of the 3 vehicles arrived in the time agreed with the customer, begins to study the balance of jobs (based on their knowledge of operations), improving logistics and reducing the size of lots (2 vehicles).
3. Finally, the third production is carried out through a flow of a single piece, after having continued to improve logistics and warehouse (from 3 people in the warehouse to 1) and balanced workloads over previous productions.

Table 2 shows the most outstanding characteristics of the three phases in which training is developed for 6 fundamental aspects: Lay-out, Production lot size, Transport lot size, Warehouse organization, Line balancing and Over processing.

Table 3 shows the positions occupied by the participants that are defined in phase 1 and how they evolve throughout the different phases. It can be seen that between phase 1 and phase 2 there is no reduction in personnel as the objective is to balance the workloads between the different jobs, define work standards and guarantee the manufacture of quality products (to avoid rework).

Table 2 Main characteristics of the different phases

	First production	Second production	Third production
Lay-out	• Two classrooms • First classroom: factory and two suppliers • Second classroom: warehouses	• Two classrooms • First classroom: factory and two suppliers • Second classroom: ware-houses	• One classroom Factory • Two suppliers and supermarket
Production lot size	3	2	1
Transport lot size	3 (1 between final assembly, quality and customer)	2 (1 between final assembly, quality and customer)	1
Warehouse organization	• Ordered by code number • Codes that are never assembled • There's a lot of stock	• Common Parts located in a sub-zone of warehouse • Small parts (codes 1–15) in workstations • Non used references are eliminated	• Zero stock in warehouse • Small parts (codes 1–15) in workstations • Components groups in supermarket
Line balancing	• Workstations very badly balanced • Much difference of work content among the four workstations of the factory	• Improved Balanced work content of factory workstations • Different distribution of work content of workstations 3 and 4	• Improvement in logistics operations by: – Reduction of work surfaces – Approach of warehouses, suppliers and factory – Kanban
Over processing	• The cockpit must be disassembled to fit another component • The crank must be disassembled to be reassembled on the tipper • Crank crane-chassis • Part attached to the "chassis"	• The three processes of 1st production are eliminated	–

In the different phases, the participants are shown the cycle times in each of the positions depending on the manufactured product. Obviously it is very difficult to make a perfect balance (in the same way that happens in the factories), and they have to learn how to make the balance based on the manufacturing mix (Fig. 2).

Figure 3 shows the results of delivery of vehicles to the customer. It can be seen that of the 3 vehicles delivered to the customer in phase 1 (out of time), 7 were delivered

Table 3 Jobs and functions of the different phases

Organization	Job title	First production	Second production	Third production
Manufacturing	Factory manager	1	1	1
	Wheels operator	1	1	1
	Cockpit operator	1	1	1
	Rear part operator	1	1	1
	Final assembly operator	1	1	1
Quality	Internal quality operator	1	1	0
Logistics	Internal logistics operator	1	1	0
	Warehouse manager	1	1	
	Warehouse operator n°1	1	1	1
	Warehouse operator n°2	1	1	
	Supplier transport	1	1	0
Suppliers	Wheels supplier	1	1	1
	Chassis supplier	1	1	1
	Customer	1	1	1

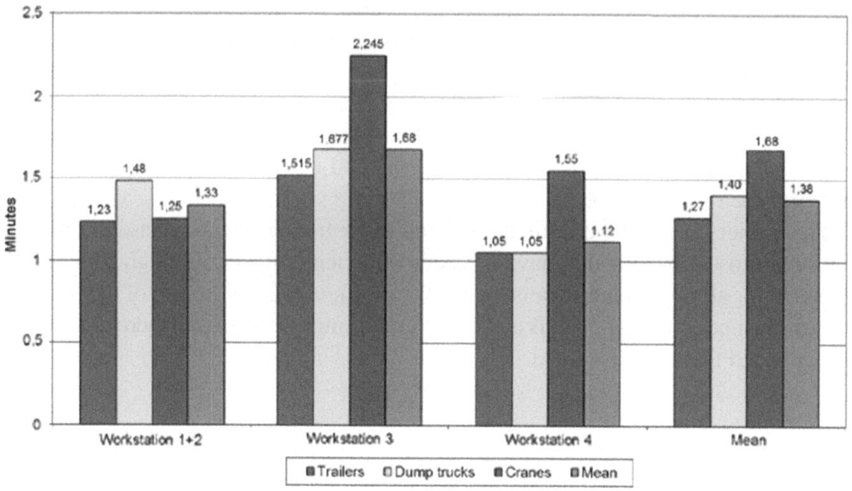

Fig. 2 Cycle time of the 3rd production in the different workstations depending on the product (trailers, dump trucks and cranes) and mean cycle time

Fig. 3 Dashboard with the results of the vehicles produced in the 3 phases

in phase 2 (although only one in time) and in the last phase the 15 requested vehicles were delivered (only 2 out of time and one with quality defects).

4 Main Results for the Learning Process

Participants are given a test of 26 questions divided into three main blocks: concepts (8 questions), principles (8 questions) and tools (10 questions). The test is carried out at the beginning of the training in order to know the level of the participants and to be able to adjust the level and the examples. Afterwards, the test is carried out at the end of each of the productions during the analysis period, although the answers will only be commented by the trainers at the end of the third phase so as not to influence the results (Table 4).

The trainers analyse the results at the end of each intermediate phase in order to reinforce those concepts that have not been sufficiently clear, although it is not an objective for all participants to answer the 26 questions well at the end of the training as they must be clear that there is always room for improvement and there are always concepts and tools to be learned.

Table 4 Percentages of correct answers in the tests

	Q1	Q2	Q3	Q4	Q5	Q6	Q7	Q8	Q9	Q10	Q11	Q12	Q13
Test 1	29	100	79	50	14	7	71	29	7	57	14	36	14
Test 2	100	100	100	36	93	57	79	93	43	71	43	36	29
Test 3	100	100	100	36	93	71	86	86	50	86	100	57	100
Test 4	100	100	100	71	93	71	93	93	86	86	100	79	100
	Q14	Q15	Q16	Q17	Q18	Q19	Q20	Q21	Q22	Q23	Q24	Q25	Q26
Test 1	0	36	36	57	71	86	64	21	14	14	36	14	43
Test 2	0	29	43	64	86	86	57	29	29	21	57	21	71
Test 3	86	86	79	93	100	93	86	21	21	21	43	21	79
Test 4	86	86	100	86	100	93	100	79	100	93	100	64	79

5 Conclusions

At this point, we can identify two major blocks of conclusions. On the one hand, about this particular training, the participants who have carried it out (both students and workers) assimilate the concepts much better than when they are only explained in class in a theoretical way, because they have experienced the different tools proposed and have identified the results. In this sense, the satisfaction of all those who have done this training (both in the university and in business) has been very high not only in terms of learning concepts but also in terms of the teaching format. And the results obtained are very similar, regardless of whether they are students or workers, since the operations to be carried out are very simple (and this is where workers could have a certain advantage).

And the second big block of conclusions refers to the importance of university-industry collaboration where we can identify numerous advantages for both parties. The company identifies students with high growth potential in their factories, as well as being able to test different tools in different environments. On the other hand, the university provides students with training in current topics and adapted to new paradigms, which facilitates their incorporation into companies.

References

1. Ankrah S, AL-Tabbaa O (2015) Universities–industry collaboration: a systematic review. Scand J Manag 31:387–408
2. Bonarccorsi A, Piccaluga A (1994) A theoretical framework for the evaluation of university-industry relationships. R&D Manag 24:229–247
3. Pascual J, Hoyuelos I, Mateo M, Gento AM (2019) Lean school: a learning factory for training lean manufacturing in a physical simulation environment. Manag Prod Eng Rev 10(1):4–13
4. Womack J, Jones D (1996) Lean thinking. Banish waste and create wealth in your corporation. Free Press, Simon & Schuster, Inc., New York, NY

The Power of Six: Relation Between Time and Money in Manufacturing for Segments of the Value Stream

Christoph Roser, Bernd Langer and Jochen Deuse

Abstract A major influence on the cost of a product is the time it takes to make this product. Traditional cost accounting can grasp part of this relation, but misses many critical aspects of having a faster time to the customer. Rajan Suri analyzed this relation empirically. Based on a data set with industrial data he determined an empirical mathematical relation between the turnaround time to the customer (or replenishment time) and the product cost for the entire value stream. This paper modifies the approach by Suri to be applied also to segments of the value stream, creating a relation between the cost within of a segment of a value stream and the time it takes for a part to pass through this segment of a value stream. This allows the estimation of the improvement in cost and the reduction in turnaround time also for sub-segments of the value stream, helping decision makers to better understand the impact of their decisions.

Keywords Lead time · Replenishment time · Cost

1 Introduction

One major factor affecting production cost is time. In this paper we look in particular at the time it takes between the order of a product and the delivery of this product. This time is sometimes also called the replenishment time. As for the cost, we will look at the overall production cost.

Some aspects of this relation can be captured through conventional cost accounting, which is covered in most cost accounting textbooks [1], including factors like cost of capital, storage cost, insurance expenses, etc. The impact of the lead time on cost is often discussed in manufacturing, as for example on ordering cost [2, 3], or more generally procurement cost [4]. However, many elements in the relation between

C. Roser (✉) · B. Langer
Karlsruhe University of Applied Sciences, Karlsruhe, Germany
e-mail: christoph.roser@hs-karlsruhe.de

J. Deuse
TU Dortmund University, Dortmund, Germany

M. Rossi et al. (eds.), *Proceedings of the 6th European Lean Educator Conference*, Lecture Notes in Networks and Systems 122,
https://doi.org/10.1007/978-3-030-41429-0_3

cost and time cannot be captured well by conventional cost accounting. Examples for these are the effort in managing product changes, delays in information flow due to delays in material flow, aging of products, increased need for management oversight and organizational effort, etc. While these are hard to quantify there is a positive relation between reducing the replenishment time and reducing the product cost [5–7]. Unfortunately, in conventional cost accounting any benefit that cannot be calculated is mathematically assumed to be zero.

2 The Power of Six

Rajan Suri detected an empirical relation between production cost and replenishment or turnaround time, which he called the *power of six* [8–10]. There are two elements that the Power of Six brings into relation. The first one is the turnaround time, the second one is product cost. The Power of Six originated with Quick Response Manufacturing (QRM). One key metric is the Manufacturing Critical Path Time (MCT). Since the MCT is used for quite a few different things, it has a very detailed definition. The definition is:

Manufacturing Critical Path Time MCT: The typical amount of calendar time from when a customer submits an order, through the critical path, until the first end item of that order is delivered to the customer [8, 11, 12].

2.1 The Equation of the Power of Six

For our purposes, however, we can simplify this to the average time between receiving an order and delivering the (first part of the) order to the customer. This is based on real time, not working time, and includes for example all off-shifts, weekends, plant holidays, etc. This could be for example the average days between receiving the order and delivery. It is important to take the average, and not a subset of e.g. rush orders, as this may be a biased sample and will be far from the average duration. It is important that this includes the entire value chain. If a customer order is fulfilled merely by grabbing the make-to-stock item from the shelf, this rule of thumb won't work. The cost is simply the total product cost including overhead, materials, and everything else. Assume the goal is to reduce your cost by reducing your turnaround time (or MCT). Let's assume you have the following variables:

- C0 is the current product cost
- C1 is the desired new product cost
- T0 is the current turnaround time
- T1 is the new turnaround time needed.

The relation according to the power of six is shown below:

$$\frac{T_1}{T_0} = \left(\frac{C_1}{C_0}\right)^6 \tag{1}$$

For a desired cost reduction, the required percentage reduction of the turnaround time can be calculated. For example, if the cost should be reduced by 5% (i.e., to 95% of the previous cost), the formula would be

$$\frac{T_1}{T_0} = \left(\frac{C_1}{C_0}\right)^6 = \left(\frac{95\%}{100\%}\right)^6 = 74\% \tag{2}$$

Hence it is estimated that the turnaround time would have to be reduced by around 26% (i.e., to 74% of the original value) to achieve a 5% cost saving, assuming again that only the turnaround time is influenced and other cost saving levers kept constant.

It would also work the other way round starting with the reduction of the turnaround time. As an example, a part is re-shored from China back to the USA. The turnaround time would be reduced by three months shipping time from eight months to five months. The formula can be turned around as shown below by not taking the ratio of the costs to the power of six, but the ratio of the times to the power of 1/6th.

$$\frac{c_1}{c_0} = \left(\frac{T_1}{T_0}\right)^{\frac{1}{6}} = \left(\frac{5}{8}\right)^{\frac{1}{6}} = 92\% \tag{3}$$

Hence, a reduction of the turnaround time by 38–63%, would reduce the cost by around 8–92% of the previous cost. The overall relation is shown in Fig. 1 Please

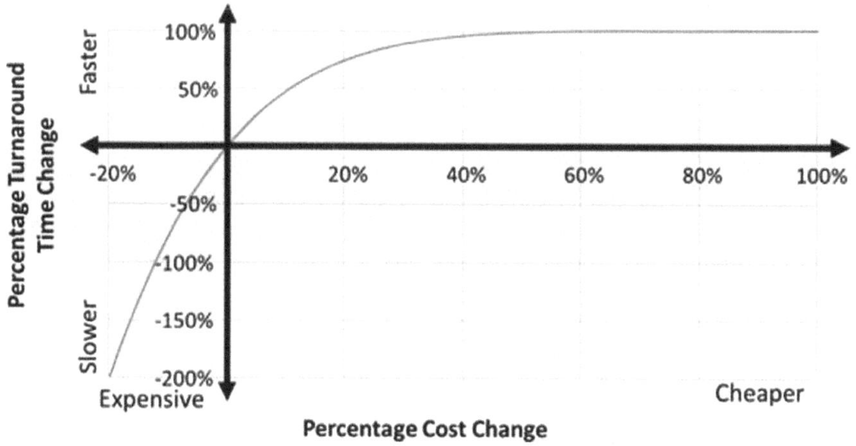

Fig. 1 Relation between cost reduction and turnaround time reduction based on the power of six

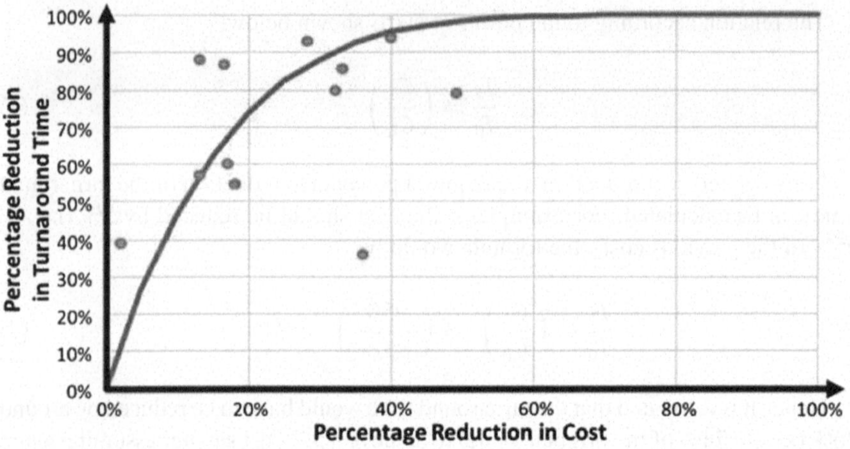

Fig. 2 Comparison of the original data with the power of six relation

note that this is only an empirical estimate, and is not valid for cost reductions above 50% and turnaround reductions above 60% as shown in Fig. 2.

Figure 1 also shows the behavior if the turnaround time is increased and subsequently the cost is increased. For example if production is outsource from the USA to China, the supply chain is extended by three months from three to six month, or a 200% increase in turnaround time. This is calculated in the equation below.

$$\frac{c_1}{c_0} = \left(\frac{T_1}{T_0}\right)^{\frac{1}{6}} = \left(\frac{6}{3}\right)^{\frac{1}{6}} = 112\% \tag{4}$$

Doubling the turnaround time would increase the cost to 112%, an increase of 12%. Therefore, production in China would have to be at least 10.7% cheaper just to break even.[1]

2.2 *Accuracy of the Power of Six*

It is important to remember that the power of six is a rough rule of thumb. Figure 2 shows the original data from Tubino and Suri [12]. While the power of six fits the data well, there are definitely fluctuations and outliers in the data.

[1]These numbers are different because the way percentages work. An increase of 100€ by 12% would get 112€. Decreasing 112€ by 12% would get 98.56€. Decreasing it by 10.7% would give 100€ again.

3 Power of Six for Segments of Value Chain

One limitation of the original Power of Six rule is that it applies to the entire value stream. However, the entire value stream is often difficult to access and data may not be available for the entire value stream, especially if the value stream crosses multiple different enterprises. The original source itself looks only at the value stream under control of a single enterprise [12], even though this is not stated explicitly. Nevertheless, by extrapolation the power of six can be adjusted for segments of the value stream within the limits of the accuracy of the method. In many instances, an improvement is done only within a segment of the value stream, and the personnel responsible for the improvement have access only to the data within that segment of the value stream. For example, assume the system consists of a series of Kanban loops as shown in Fig. 3, of which only one is improved. The power of six can be adapted to analyze these segments as shown in the next sections.

3.1 Mathematically Correct Approach in Using the Power of Six

If only a segment of the value stream is improved, and if this segment is on the critical path, the share of the improvements with respect to the entire turnaround time would have to be estimated to get an estimate of the improvement of the entire cost. Figure 4 shows such an example. Segment B in the value stream is improved, reducing the turnaround time for this segment from six days to four days, improving it by two days. Since the entire turnaround time across the value stream is twenty days, the overall improvement is still only two days, and the new turnaround time comes down to eighteen days.

Hence the overall reduction of the turnaround time was by 10–90% of the original value, and according to the power of six the cost should go down by around 1.74–98.26% of the original value as shown below.

$$\frac{c_1}{c_0} = \left(\frac{T_1}{T_0}\right)^{\frac{1}{6}} = \left(\frac{18}{20}\right)^{\frac{1}{6}} = 98\% \tag{5}$$

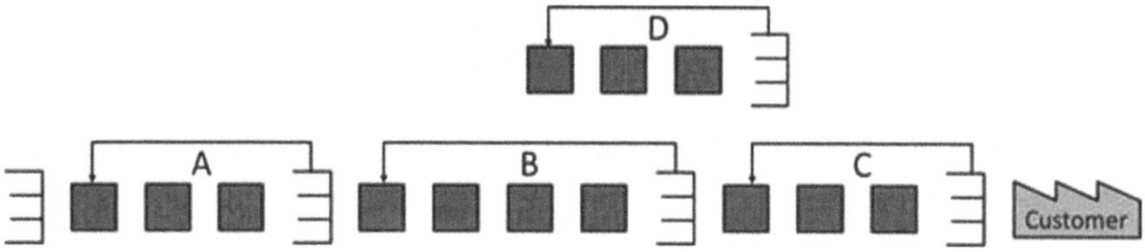

Fig. 3 Example of segments of a value stream

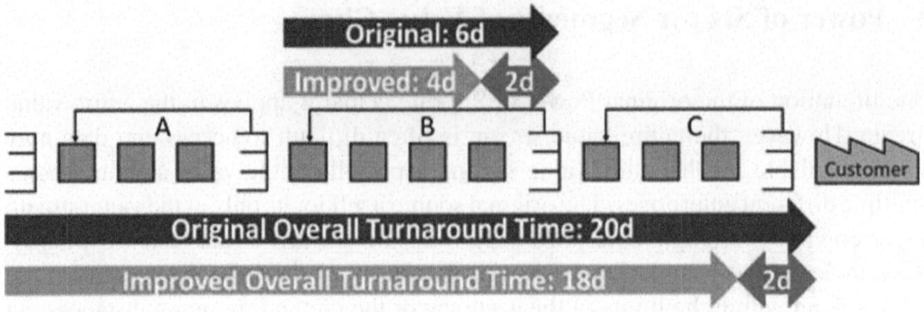

Fig. 4 Example for improving a segment of the value stream

This approach would work, although it still requires an understanding of the turnaround time and cost for the entire value stream, which may sometimes be tricky. It also does not work for subsegments that are not along the critical path, as for example segment D in Fig. 3.

3.2 *A Simplified Estimation*

The power of six also works for subsegments, where you relate the turnaround time for that segment T0,S and T0,S and its impact on the value add within this segment C0,S and C1,S. The equation and its reverse form would look as follows:

$$\frac{T_{1,S}}{T_{0,S}} = \left(\frac{c_{1,S}}{C_{0,S}}\right)^{6} \tag{6}$$

$$\frac{C_{1,S}}{C_{0,S}} = \left(\frac{T_{1,S}}{T_{0,S}}\right)^{\frac{1}{6}} \tag{7}$$

Applying these equations to the example from the mathematically correct approach from above where we reduced the turnaround time for that segment from 6 to 4 days. This would reduce the cost in this segment by 6.53–93.47% as shown below.

$$\frac{c_{1,S}}{c_{0,S}} = \left(\frac{T_{1,S}}{T_{0,S}}\right)^{\frac{1}{6}} = \left(\frac{4}{6}\right)^{\frac{1}{6}} = 93\% \tag{8}$$

Assuming a proportional relationship of the cost of the segments to the time in the segment and hence assuming that this segment that has six of the twenty days turnaround time has also 6/20th of the cost, the estimate of the overall cost improvement as shown below. Please note that to understand the savings of the segment in relation to the savings of the entire line we again would need information about the entire line, hence the equation below may not always be feasible. The equation above, however, is all that is necessary to estimate the cost–time relation for segments of value streams.

$$1 - \frac{c_{1,S}}{C_{0,S}} = \left[1 - \left(\frac{T_{1,S}}{T_{0,S}}\right)^{\frac{1}{6}}\right] \cdot \frac{T_{1,S}}{T_1} = 1.96\% \tag{9}$$

While the mathematical approach above gives an improvement of 1.74%, the simplified approach gives 1.96%. Within them accuracy of the method this is close enough to be valid. We tested this also for different situations, and the error between the correct and the practical approach is marginal unless you reduce the turnaround time in the segment by more than 70%. However, the original power of six is only valid for changes of the cost of less than 50% and turnaround times of 60%, hence it falls within the accuracy of the original method. An additional benefit is that this segment based approach allows the calculation of the improvements of segments not on the critical path. In this case you merely calculate the improvement for the value add based on this part of the segment.

4 Summary

The power of six is a useful rule of thumb for the relation between the product cost and the turnaround time. This paper adapted the rule for segments of the value stream, showing the relation between the turnaround time within the segment of the value stream and the value add of this segment of the value stream. While an accurate cost calculation is usually preferred, in many cases the necessary data and the complexity of an accurate calculation make this impractical. A simple estimation based on the modified power or six approach allows faster and hence less expensive decision making at the expense of a slightly decreased accuracy.

Acknowledgements We would like to thank Rajan Suri for his input.

References

1. Bhimani A, T. Horngren C, Foster G, Datar S (2012) Management and cost accounting
2. Ouyang L-Y, Chen C-K, Chang H-C (1999) Lead time and ordering cost reductions in continuous review inventory systems with partial backorders. J Oper Res Soc 50:1272–1279. https://doi.org/10.1057/palgrave.jors.2600840
3. Chang H-C, Ouyang L-Y, Wu K-S, Ho C-H (2006) Integrated vendor–buyer cooperative inventory models with controllable lead time and ordering cost reduction. Eur J Oper Res 170:481–495. https://doi.org/10.1016/j.ejor.2004.06.029
4. Chandra C, Grabis J (2008) Inventory management with variable lead-time dependent procurement cost. Omega 36:877–887. https://doi.org/10.1016/j.omega.2006.04.009

5. Hartley JR (2017) Concurrent engineering: shortening lead times, raising quality, and lowering costs. Routledge
6. Clark KB, Fujimoto T (1989) Lead time in automobile product development explaining the Japanese advantage. J Eng Tech Manage 6:25–58. https://doi.org/10.1016/0923-4748(89)90013-1
7. Fujimoto T (1999) The evolution of manufacturing systems at Toyota, 1st edn. Oxford University Press, Oxford, New York
8. Suri R (1998) Quick response manufacturing: a companywide approach to reducing lead times. Taylor & Francis Inc, Portland, OR
9. Suri R, de Treville S (1992) Rapid modeling: the use of queueing models to support time-based competitive manufacturing. In: Fandel G, Gulledge TR, Jones A (eds) Operations research in production planning and control. Springer, Hagen, Germany, pp 21–30
10. Suri R (2014) MCT quick reference guide. Suri Consulting and Seminars, LLC
11. Suri R (2010) It's about time: the competitive advantage of quick response manufacturing, 1st edn. Productivity Press, New York
12. Tubino F, Suri R (2000) What kind of "numbers" can a company expect after implementing quick response manufacturing? Empirical data from several projects on lead time reduction. In: Quick response manufacturing 2000 conference proceedings

Teaching Lean with Virtual Reality: Gemba VR

Torbjørn H. Netland, Rafael Lorenz and Julian Senoner

Abstract Lean production is best taught on the factory floor. Yet, in higher education, it is almost exclusively taught in classrooms. We want to keep and proliferate the learning experience of exploring a real factory's "Gemba" and, at the same time, to remove limitations to factory visits. Due to recent developments in virtual reality (VR) technologies, VR offers excellent opportunities to achieve this. In this paper, we present an innovative way to teach lean production with VR. We show how we implemented a solution to let students be immersed in the factories of Toyota, ABB, and other world-renowned companies without having to travel. We also report on our experiences and provide other teachers the information needed to adopt "Gemba VR" in their own teaching.

Keywords Teaching lean · Lean production · Virtual reality · Toyota · Gemba

1 Introduction

Training in lean production is most effective at the place it is supposed to be implemented, that is, the "Gemba." The term is Japanese, meaning "the actual place" [1, 2]; for lean production, that means the factory floor. Yet almost all lean training and instruction in the higher education sector takes place in classrooms. There are obvious reasons for this, but in this paper we show that—due to recent advancements in virtual reality (VR)—teachers can bring the Gemba to the classroom. We demonstrate and discuss the incorporation of online virtual environments from real factories in course designs.

T. H. Netland (✉) · R. Lorenz · J. Senoner
Department of Management, Technology, and Economics, Chair of Production and Operations Management, ETH Zurich, Weinbergstrasse 56/58, 8092 Zürich, Switzerland
e-mail: tnetland@ethz.ch

R. Lorenz
e-mail: rafaellorenz@ethz.ch

J. Senoner
e-mail: jsenoner@ethz.ch

M. Rossi et al. (eds.), *Proceedings of the 6th European Lean Educator Conference*, Lecture Notes in Networks and Systems 122,
https://doi.org/10.1007/978-3-030-41429-0_4

Field trips are an effective way of teaching lean in higher education [3], but field trips are generally inefficient. For example, it can be difficult to gain access to factories, and entering all the areas relevant to the class may not be permitted. Factory visits require considerable resources to coordinate and organize, especially for class sizes exceeding 25 students. It can also be difficult to fit factory visits into busy study and teaching schedules. Because field trips are difficult to plan and organize, they are rare learning opportunities for students. What if—rather than the students going to the field—the field could be brought to the students?

We want to keep and proliferate the learning experience of exploring a real factory environment and, at the same time, remove limitations to field visits. Recent developments in VR technologies offer excellent opportunities to do so. In this paper, we present a teaching innovation that takes advantage of VR to teach lean production. In the next section, we present the technology and concept of "Gemba VR." After that, we draw on our experiences from using this concept in three different courses at ETH Zurich to discuss opportunities and challenges. We also provide advice for teachers who would like to teach with VR.

2 Introducing Gemba VR

There is agreement in the pedagogy literature that active learning is superior to passive learning [4, 5]. For example, the "70-20-10 rule" of learning suggests that managers learn 70% from on-the-job experiences and challenges, 20% from what they hear from other people, and only 10% from classroom training [6]. In light of these findings, it is disappointing that the vast amount of training in lean production takes place in classrooms. While this is obviously true for higher education, it is unfortunately also true for many training programs in companies. This problem is exaggerated when it comes to training in lean production, as it is hard to teach accurately in classrooms. While the introduction to tools and techniques can be covered [7], it is not easy to convey the complexity involved in applying them to real factory setting. It is even harder to teach the important cultural and behavioral elements of lean in the classroom [8].

VR can help mitigate some of the drawbacks of traditional classroom teaching in higher education [9]. VR is an artificial environment presented to the user so that the user experiences it as a close-to-real environment [10, 11]. The VR environment can be an artificial creation, a digital copy of the real world, or a combination of both. The form of VR that we refer to in this paper is 3-dimensional images and videos that can be explored with a computer screen or wearable VR headsets. By offering the students VRs of real factories, they can actively explore the virtual content guided by assignment questions. We call this idea "Gemba VR." We take advantage of available technologies to show how Gemba VR can be implemented cost-effectively and effortlessly at scale.

Gemba VR consists of hardware, software, and a task assignment, all of which we introduce below.

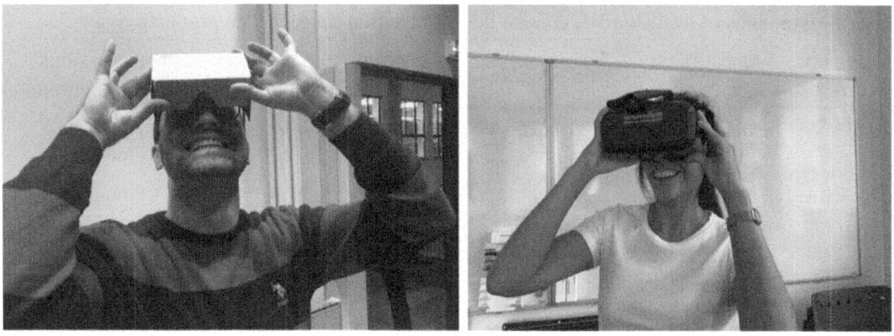

Fig. 1 Teaching staff using a VR cardboard headset (left) and a plastic headset (right)

2.1 Hardware

The hardware consists of students' own smartphones and a VR headset. It is anticipated that students have their own smartphones that they can use for this purpose. Figure 1 shows teaching staff using two different VR headsets: The person on the left uses a cardboard headset that cost $3, and the person on the right uses a plastic headset that cost $30. We have used both in Gemba VR. The cardboard version is cheap enough to hand out freely to all students, while the plastic headsets are reused by other students. More professional VR equipment (e.g., HTC Vive, Oculus Rift) also exist, but these technologies are difficult to scale due to use requirements (e.g., space and marking) and costs.

2.2 Software

We use software that is freely available online or in mobile applications (apps). Table 1 provides a list of exemplar software that can be used. YouTube contains a number of 360° videos of factories. In addition, more and more companies are launching VR apps in which users can explore factories freely at their own pace. For example, since 2017, the ABB Group has offered VR environments of some of their factories through an app available in major app stores. The app consists of 360° pictures and videos from five factories blended with virtual instructions and information. In the app, students can visit a number of predefined areas in the factories in any order they like. In our course, we have chosen to use the ABB app and a Toyota YouTube video, as the richness of this software fit well with our specific learning objectives.

360° videos and VR apps have pros and cons. Among the pros are that the apps are easier to navigate and allow the user to visit different locations in the factory at his or her own speed. They can also be operated seamlessly with most VR headsets. Among the cons are that the virtual rooms are often static or only show short videos on repeat. The apps are therefore generally unsuitable for showing the material flow

Table 1 Examples of available VR content that can be used to teach lean production

Company	Location(s)	Title	Format	Duration (min)	Access links
ABB	CH (2), DE, FIN (2)	ABB 360 VR tours	App	N/A	AppStore GooglePlay
AkzoNobel	UK	Ashington VR	App	N/A	GooglePlay
Toyota	FRA	Toyota VR/360° factory tour	360 video	2.56	YouTube
GE	US	Inside a gas turbine Factory	360 video	3.50	YouTube
Tesla	US	See where Tesla makes its cars	360 video	2.06	YouTube
Hyundai	KOR	Explore VR plant	360 video	4.43	YouTube
Niftylift	UK	VR factory tour From Above	360 video	2.01	YouTube
TVS	IND	TVS 360 factory tour	360 video	2.07	YouTube
Würth	DE	Würth Elektronik factory tour	360 video	2.28	YouTube

in the factory. The videos (e.g., the Toyota video listed in Table 1), however, are excellent for showing material flow. The drawback of videos is that they are often played at a high speed, which makes it hard to absorb all their details. This issue can be mitigated by pausing the videos or manually setting the play speed at a fraction of the standard (e.g., in YouTube, set the "playback speed" to 0.25). While a slower play speed makes it much easier to follow, any audio material will be practically unusable and should be muted.

By using both VR apps and 360° videos in our courses, we take advantage of the strengths of both. Figure 2 shows two examples from the apps we use in class. The left picture is a snapshot of a running 360° video that follows the assembly of a forklift from start to end in a Toyota factory. The right picture is a repeated 360° video that shows how the operator uses a machine in an ABB electronics assembly factory. In the right picture, the information icon (the red "i" in a white circle) contains blended information that can be clicked and read. In both the video and the app, the user can explore the assembly halls by turning the headsets in 360°, or if navigated on a screen, the user can use the navigation panel to look around.

Fig. 2 Snapshots from VR environments in Toyota (left) and ABB (right)

2.3 *Assignment*

We implement Gemba VR as a work assignment in the courses. The assignment requires students to solve questions related to the learning objectives of the course. To answer the questions, the students have to visit the virtual environments. We ask the students to form groups of 3–5 people to discuss the content, but the assignment could also be given as individual coursework. Every student is encouraged to first solve the questions individually before meeting and discussing in their groups. The assignment lasts about three weeks. The student groups must hand in group reports answering all the assignment questions and also prepare short presentations containing their answers to the questions. We dedicate a class to present and discuss randomly drawn group reports.

Table 2 lists the questions we ask in the 2019 fall semester course. We separate three different types of questions (seek-and-find, explore-and-think, and compare-and-analyze), which increase in difficulty as the students progress with the assignment. Other teachers can and should adapt the assignment questions to fit their own course's learning objectives.

3 Experiences from Teaching with Gemba VR

Our experience with teaching lean with VR is that students generally appreciate the innovation. In particular, they report that they enjoy the opportunity to be immersed in a real factory. Figure 3 shows a group of Master of Business Administration (MBA) students solving questions with the use of VR. One of these students summarized the assignment as follows: "The VR experience was tremendous. It was a great way to get a good insight into production facilities and to start analyzing the situation." However, the VR technology and the way we integrated it had several drawbacks. In this discussion, we summarize the benefits and drawbacks that we encountered when integrating VR into teaching lean production.

Table 2 Assignment questions

Task type	Question
Seek-and-find questions	1. For the ABB Baden factory and the Toyota factory, search for the following lean methods, take screenshot of where you find them, and note their location [and time]: a. Andon b. Shadow board c. Kaizen board d. Kanban e. Two-bin system f. Poka yoke g. One-point-lesson 2. Briefly explain each method from 1a–g as it is used in the Toyota factory 3. Visit the assembly line of the Toyota factory. Search for information that you can use to calculate the takt time of the factory. What is the takt time of the forklift assembly line? What is cycle time and how does it relate to the takt time? 4. What are muri, mura, and muda? Please find and explain one example of each in the ABB factories
Explore-and-think questions	5. In the five ABB factories, did you notice any lean methods, tools, or techniques not covered in Question 1a–g? If so, please name the method(s) and where you found it (them). (Limit the response to maximum five methods) 6. Consider all five ABB factories and give an assessment of the levels of 5S. Do the same for the Toyota factory
Compare-and-analyze questions	7. Compare the Toyota assembly plant to the ABB Baden assembly line. What are the main differences in terms of operational characteristics? How should lean implementation differ across these settings? 8. Compare the ABB Lenzburg factory to the Toyota assembly plant. What lean practices from the Toyota plant do you think are applicable in the semiconductor fabrication plant in Lenzburg?

3.1 Benefits

With VR, students can explore factory environments whenever and wherever they like. One student reported, "It was good that you could discover the factory [at] your own speed." In contrast to factory tours, students can revisit the VR environment for any purpose, such as discussing content in groups. VR also allows the students to virtually visit multiple production sites in different locations and compare their differences without needing to travel. They can access areas and views that might be unavailable during a field trip, such as clean rooms and close-up views of machinery. VR also offers the possibility to blend different types of information, such as real images and videos of factory operations with overlaid digital information, to support students' learning experiences and learning outcomes.

Fig. 3 Student group solving an assignment task with VR

One of the biggest advantages of VR is that it enables inquiry-based learning. One student remarked, "I think it's because it's more like an active discovery than like a passive [one]." While the questions guided students to seek certain answers, they also needed to use their own curiosity and intelligence when visiting the virtual environment and seeking the answers. One student asserted, "Since you have to look around, you start thinking... hey, so what am I actually looking at, or what should I be looking at?"

Furthermore, the use of a modern technology, such as VR, is exciting for many students. This enthusiasm can improve their learning motivation. "Oh cool," commented one student. "We're doing something using VR. You know it's kind of a buzz word—it automatically excites you a little bit." Although the technology had a visible boost on student motivation, we did not find this excitement to last beyond the first explorations of the VR app.

3.2 Drawbacks

We also learned that the current state of technology has several drawbacks. For example, there is no way to ask questions during a virtual factory tour. It is also very hard to take notes or discuss with peers when using full-immersion VR glasses. Vision-impaired students using spectacles, found the VR glasses to be inconvenient. Also, in very few cases, we had students with older, incompatible smartphones or who experienced the app crashing. Some students who had experienced higher-end VR viewers, such as the Oculus Rift, were disappointed by the limited possibilities and low resolution offered by the cardboard variant.

The most serious drawback was that many students experienced motion sickness when using the VR app. To avoid this, several students reported that they stopped using the cardboard viewers and instead looked directly at the smartphone screen or used the computer. This workaround reduces the level of immersion, but it also affords the possibility to tour the virtual facility with others and take notes simultaneously.

4 Conclusions

VR can increase the effectiveness and efficiency of teaching lean production. In this paper, we showed how we integrated VR into classrooms for this purpose. We call this "Gemba VR." We also described all the details needed for other teachers to design similar assignments. In our experience, VR had mostly positive effects on students' learning experiences, as the immersion it offers can certainly improve learning outcomes. However, we also found that the current state of VR technology has some limitations and drawbacks that inhibit its use to teach all aspects of lean production.

A typical concern for teachers is that new technologies are expensive to access and difficult to master. In our experiences with the type of VR applications we used, those concerns do not hold water. Students can use technologies they already possess (smartphones), factory apps are offered for free, VR headsets are cheap, and their use is self-explanatory. A completely free alternative is to ask students to explore the 360° videos on YouTube. With the rising improvement and availability of VR over the next few years, there is good reason to believe that its applications to lean teaching in higher education will only widen. In the near future, we hope to see an increasing availability of VR apps, such as the one provided by ABB.

Acknowledgements The authors are grateful for the financial support from Innovedum, the teaching innovation fund at ETH Zurich. We also want to thank the many companies that openly publish VR content from their factories online. Finally, we are indebted to the more than 200 students who have tested VR in our classes and provided informal and formal feedback on its applicability and opportunities.

References

1. Bicheno J (2001) Kaizen and kaikaku. Manufacturing operations and supply chain management: the LEAN approach, pp 175–184
2. Mann D (2014) Creating a lean culture: tools to sustain lean conversions. CRC Press
3. Candido JP, Murman EM, McManus H (2007) Active learning strategies for teaching lean thinking. In: 3rd international CDIO conference, Cambridge, MA, USA. www.cdio.org
4. Furlan A, Galeazzo A, Paggiaro A (2019) Organizational and perceived learning in the workplace: a multilevel perspective on employees' problem solving. Organ Sci 30(2):280–297
5. Scholten K, Dubois A (2017) Advancing the skill set of SCM graduates—an active learning approach. Int J Oper Prod Manag 37(11):1683–1699
6. McCall MW, Lombardo MM, Morrison AM (1988) Lessons of experience: how successful executives develop on the job. Simon and Schuster

7. Bicheno J (2004) The new lean toolbox: towards fast, flexible flow. Production and inventory control, systems and industrial engineering books
8. Netland TH (2016) Critical success factors for implementing lean production: the effect of contingencies. Int J Prod Res 54(8):2433–2448
9. Dede C (2009) Immersive interfaces for engagement and learning. Science 323(5910):66–69
10. Sherman WR, Craig AB (2002) Understanding virtual reality: interface, application, and design. Elsevier
11. Burdea GC, Coiffet P (2003) Virtual reality technology. Wiley

Continuous Improvement; One Recognizable Approach, One Language and Plenty of Results

Simone van der Donk, Janet Tabak and Carien van Horne

Abstract This paper describes the first results of the Continuous Improvement program at Saxion University of Applied Sciences. The main objective of this research is to characterize the successful elements of the program at the Saxion University of Applied Sciences. The program has trained 1610 out of 2700 employees of Saxion over the past two years. A successful Continuous Improvement program for higher educational organizations should focus on three focus points: leadership, teams and projects. To implement the program there is a need for trained coaches to support the units and the program should be part of the policy of the educational institution to guaranty the sustainability. Further research needs to be done under staff and management to measure the effectiveness of applying the Continuous Improvement Program. Institutions of Higher Education can use the outcomes of our research for stimulating interventions in Continuous Improvement.

Keywords Continuous Improvement · Higher Education · Lean · Design Research

1 Introduction

Saxion as a Continuous Improvement organization. One of the defining characteristics of a Continuous Improvement organization is that staff and management analyse their work processes on a daily basis to see how these can be improved, made smarter or be done differently to create added value from the perspective of students and the professional field. A Continuous Improvement organization will help to realize the True North goals of Saxion for 2018–2025.

S. van der Donk (✉) · J. Tabak · C. van Horne
Saxion University of Applied Sciences, Executive Committee, Handelskade 75, 7417 DH Deventer, The Netherlands
e-mail: s.e.vanderdonk@saxion.nl

J. Tabak
e-mail: j.tabak@saxion.nl

C. van Horne
e-mail: c.h.m.vanhorne@saxion.nl

M. Rossi et al. (eds.), *Proceedings of the 6th European Lean Educator Conference*, Lecture Notes in Networks and Systems 122,
https://doi.org/10.1007/978-3-030-41429-0_5

The process to becoming a Continuous Improvement organization consists of a Continuous Improvement transition which is supported by three focus points: Leadership, Teams and Projects. These three focus points will be integrated in the participating units during this Continuous Improvement transition. The motive for Continuous Improvement at Saxion is that Saxion wants to become the best University of Applied Sciences in the Netherlands.

2 Theoretical Framework

As the main research question we define: what are the characteristics of a successful Continuous Improvement program for higher educational institutions? To define this research questions, four sub-research questions are formulated: (1) What is the need for a Continuous Improvement program in Higher Education? (2) What is a logical design for a Continuous Improvement program in Higher Education? (3) Is the prototype of the Continuous Improvement program usable for staff and management, to analyze their working processes daily? (4) What is the relevance and sustainability of the final prototype of the Continuous Improvement program? In other words, will staff and management apply this program in their daily work after the 'wave'?

The ultimate goal requires an effective organization that meets the needs of students and employers and that is continuously committed to improvement and adjustment. But this also demands a different way of working and working together. Continuous improvement provides the building blocks needed to achieve this goal [6].

The Saxion Continuous Improvement programme is based on various management philosophies. Lean forms the foundation but is supplemented by elements taken from Scrum. Lean and Scrum have more similarities than differences. Their difference can predominantly be found in their origins, as Lean was developed by Toyota in Japan and Scrum is the result of software development in the USA. These two methods also have a slightly different approach in which Lean is focused on increasing client value by preventing waste while Scrum is aimed at increasing client value by quickly delivering subproducts in short bursts. Despite these minor differences, both philosophies are aimed at increasing client value, short cyclical developments and learning, visual management, personal leadership for each staff member and working in an action-oriented way. For our Saxion Continuous Improvement approach it is not necessarily important which philosophy, method or tool we use, but whether it matches the Saxion context and helps us realize the Saxion Continuous Improvement organization. Therefore program is not only based on scientific literature of continuous improvement or lean. Fields like change management, leadership, project management, program management, team cooperation, coaching and kaizen are also embedded. Out of all these fields the three models had the largest share in the development of the program: (1) Kotter's 8 step change model [3] (2) Managing Successful Programmes (MSP®) [5] and

(3) Wouter Hart: Verdraaide organisaties – Terug naar de bedoeling [1] (Twisted organizations—Back to the intent. No English translation available).

3 Methods

The method that is used for this study is the design research method. The characteristics of this type of study is to design and develop an intervention as a solution to a complex educational problem. In addition it develops our knowledge about the characteristics of these interventions and the processes to design and develop them. This is done by a cyclical iterated process of analysis, design, evaluation and revision until the appropriate balance between the intended (purpose of the continuous improvement program) and realization (results of the program) has been achieved. The process of this design research method is visualized in Fig. 1 [4].

A review of the literature and projects about continuous improvement and lean in higher education has resulted in a blueprint for the Saxion Continuous Improvement program and forms the context of the analysis.

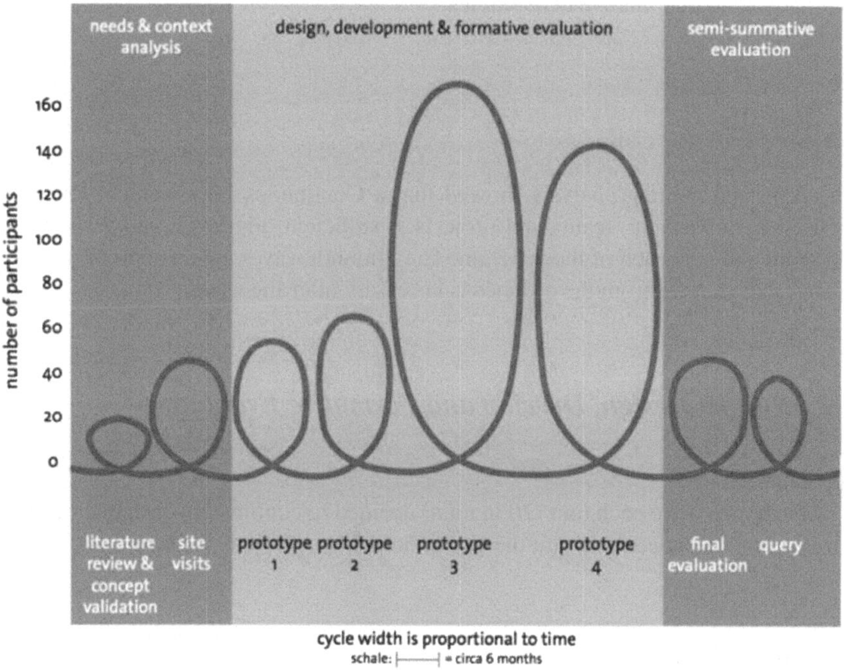

Fig. 1 Design research method [4]

The basic concept of the design was to divide the organization into sections: a 'wave'. Each wave consists of four or five departments who are committed to accelerate the change.

In wave 1 the first version of the Continuous Improvement program has been executed with 385 employees and was revised based on the first formative evaluation. In wave 2 de adapted version of the program was executed with 402 employees and revised based on the second formative evaluation. In wave 3 (n = 726) the third version and in wave 4 (n = 97) the fourth version was executed. Every wave took about 5 months. In these five months the prototype of the program was implemented, evaluated and revised. Each unit is coached by two coaches. The coaching team consists of 9 coaches especially trained to implement this program. An evaluation tool has been developed to measure the consistency and the effectiveness of the prototype. All the directors of the units that participated in the wave joined together in an expert group to evaluate the prototype in a qualitative way. The outcomes of these experts judgments are used to revise the prototype when a wave has completed.

After the fourth wave the program team evaluated whether target users (directors, team leaders, project managers and teachers) can work with the Continuous Improvement program (actual practicality) and are willing to apply it in their unit (relevance & sustainability). Also, whether the program is effective in terms of qualitative and quantitative deliverables.

4 Results

The need and content analysis showed that a Continuous Improvement Program focused on leadership, teams and projects is sufficient. Figures 2 and 3 show the blueprint and approach of the program. The 5-month waves are an appropriate way to start, but support from the coaches is necessary after the wave.

4.1 Results Design, Develop and Formative Evaluation

See Table 1.

After every wave each unit (20 in total) decided to continue the program (100%) None of the units decide to quit the Continuous Improvement program.

5 Conclusion/Discussion

The conclusion that can be drawn is that a successful Continuous Improvement program for higher educational organizations should focus on leadership, teams and projects. To implement the program there is a need for trained coaches to support

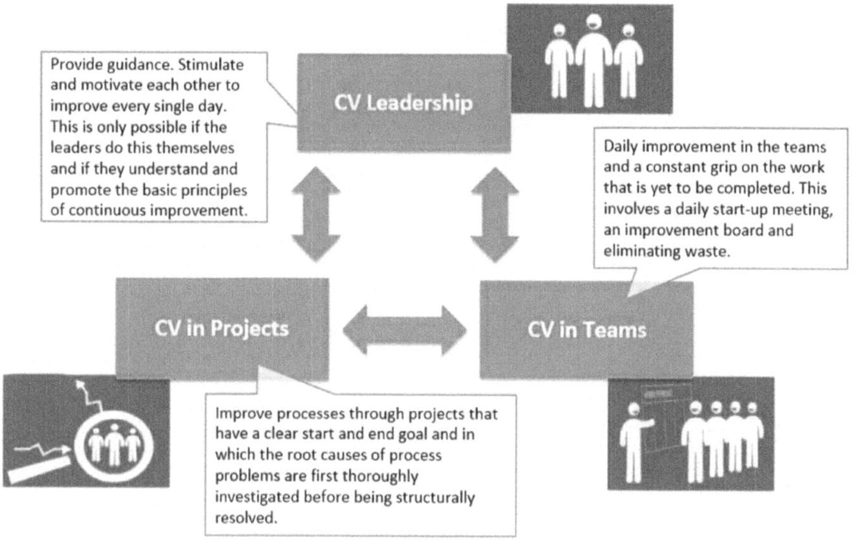

Fig. 2 Blueprint Continuous Improvement Program Saxion [2]

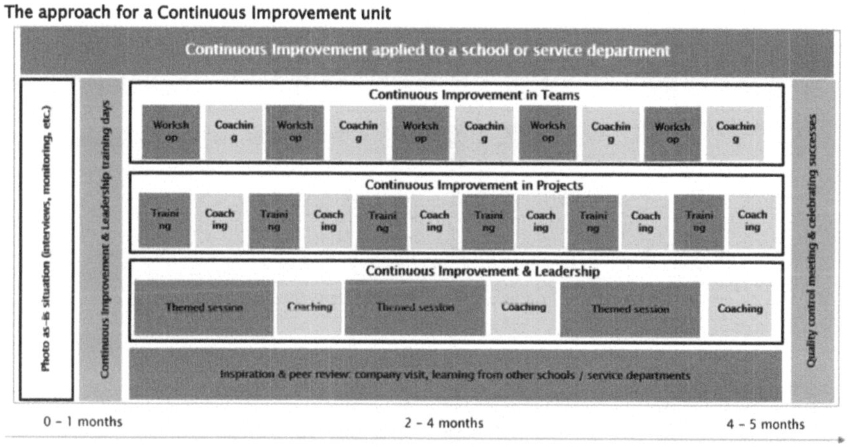

Fig. 3 The approach for the Continuous Improvement unit [2]

the units and the program should be part of the policy of the educational institution to guaranty the sustainability. The content of the program should be customized to the needs of the unit therefore the usability of the program should be evaluated and revised frequently. If a Continuous Improvement program fulfills these characteristics the staff and management will recognize the relevance and usability according to the outcomes of this research.

Table 1 Results design, develop and formative evaluation

	Name unit and number participants	Deliverables	Evaluation	Revision
Wave 1	HRM Department (n = 33) International Business School (n = 48) School of People & Society (n = 153) Saxion Part-time School (n = 28) School of Business, Building & Technology (n = 123)	5 GB[a] 20 workshops 2 GB 20 workshops 5 GB 20 workshops 2 GB 24 workshops 4 GB 32 workshops	Executive Board should start with a True North to which other True Norths contribute 5 months is too short for complete transition Directors are not well informed what to expect from the program Usability mark = 8 on 1–10 scale Green Belt training not usable for Saxion	Start with Executive board with Saxion True North as outline Follow-up program outlines developed Modification information brochure Modification Green Belt training to needs of users
Wave 2	Marketing & Communications (n = 57) School of Life Science, Engineering & Design (n = 115) Timetabling and Assessment (n = 50) School of Applied Psychology & Human Resource Management (n = 95) School of Governance, Law & Urban Development (n = 63) Offices and Departments (n = 20) Executive Board (n = 2)	4 GB 25 workshops 5 GB 26 workshops 10 GB 6 workshops 7 GB 33 workshops 6 GB 10 workshops 2 workshops 2 workshops	Sharing results between directors, not only an expert meeting on process Evaluation tool "voortgangsmeter" (progress indicator) Insufficiently usable for leaders and projects Workshop TPI (to determine KPI's) for teams was chaotic Lack of project leaders Usability mark = 8 on 1–10 scale	Adding in-depth meetings for sharing experiences between directors adding evaluation Meter for projects and leadership Adjusting workshop TPI based on needs teams Obliged participating (mechanical use) Project manager to Green Belt training during wave

(continued)

Table 1 (continued)

	Name unit and number participants	Deliverables	Evaluation	Revision
Wave 3	Financial Control Department (n = 40) Facility Services Organization (n = 50) School of Creative Technology (n = 225) School of Health (n = 268) School of Finance & Accounting (n = 25) School of Commerce & Entrepreneurship (n = 118)	3 GB 5 GB 4 workshops 4 GB 8 workshops 20 workshops 5 GB 5 workshops 5 GB 25 workshops	Workshop Kaizen simulation case not applicable for education Continuous Improvement stands alone Usability mark = 8 on 1–10 scale need for extra workshops More focus on deliverables/outcomes waves	Adjusting workshop to 'bring your own case' simulation Work sessions with HR, communication and business control counselors Development of three new workshops Developing template for results
Wave 4	ICT& Education Department (n = 32) (part of FSO) International office (n = 20) Department Pedagogic and Education (n = 20) Program Smart Solution Semester (n = 25) (part of School LSED)	5 workshops 4 GB 3 workshops 3 workshops 1 GB	Different units deliver different output Unclear what to expect from coach Usability mark = 8 on 1–10 scale Hot topic as stress and wellbeing employees not in the program More focus on sustainability of the program	Developing a template for results program Developing FAQ for directors, project leaders and team leaders Sharing best practices and tools to reduce stress Adjusting Continuous Improvement in Saxion policy
Total	Approximately 1685 employees (staff and management) participate in the program	77 active GB 293 workshops		

[a]GB = number of employees that is trained for their Green Belt LCS-certification to become a project manager. 28 GB who are in training in spring 2019 are not counted

The cyclical iterated process of analysis, design, evaluation and revision that the design research method is stating, is consistent with de Plan Do Check Act method in the continuous improvement philosophy. Therefor the design research method is an applicable research method that can be used in future research to design continuous improvement programs in Higher Education.

Further research needs to be done under staff and management to measure the effectiveness of applying continuous improvement to their work to contribute to the strategy of Saxion to become the best University of Applied Sciences of the Netherlands.

References

1. Hart W (2012) Verdraaide organisaties - Terug naar de bedoeling. Vakmedianet Management B.V, Deventer
2. Janmaat K (2018) Information about the program. Continuous Improvement. Saxion, Enschede
3. Kotter J (1996) Leading change. Harvard Business Review Press, Brighton
4. Mckenny S, Reeves TC (2001) Conducting educational design research. Taylor & Francis Ltd., Abingdon
5. Sowden R (2011) Managing successfull programmes, 2011st edn. The Cabinet Office, London
6. Stam M, Janmaat K (2018) Interview programme management. Continuous Improvement (Vrerink L, Interviewer), November 29

i-FAB: Teaching How Industry 4.0 Supports Lean Manufacturing

Violetta Giada Cannas, Maria Pia Ciano, Giovanni Pirovano, Rossella Pozzi and Tommaso Rossi

Abstract The link between Industry 4.0 (I40) and lean manufacturing has recently gained significant popularity in both academia and industry. The implementation of I40 has been proved to be beneficial for lean programs, supporting lean practices and increasing the flexibility of lean. In this context, the present paper introduces i-FAB, a learning factory developed by Università Carlo Cattaneo (LIUC) to demonstrate the benefit of the adoption of I40 technologies in a lean managed assembly system. The paper provides details on the i-FAB lean tools, I40 technologies and the training modules developed for Industrial Engineering and Management students and executive learning programs, showing empirical evidence of the benefits linked to the implementation of I40 technologies in a lean managed assembly system.

Keywords Industry 4.0 · Learning factory · Lean manufacturing · Assembly

1 Introduction

The link between Industry 4.0 (I40) and lean manufacturing has recently gained significant popularity in both academia and industry. The literature claims that the implementation of I40 does not moderate the effect of lean manufacturing but rather helps to increase the maturity of the firm's lean program and further optimizes lean

V. G. Cannas (✉) · M. P. Ciano · G. Pirovano · R. Pozzi · T. Rossi
School of Industrial Engineering, Università Carlo Cattaneo—LIUC, Castellanza, Italy
e-mail: vcannas@liuc.it

M. P. Ciano
e-mail: mciano@liuc.it

G. Pirovano
e-mail: gpirovano@liuc.it

R. Pozzi
e-mail: rpozzi@liuc.it

T. Rossi
e-mail: trossi@liuc.it

M. Rossi et al. (eds.), *Proceedings of the 6th European Lean Educator Conference*, Lecture Notes in Networks and Systems 122,
https://doi.org/10.1007/978-3-030-41429-0_6

systems to deal with higher complexity [1]. The integration between lean manufacturing and I40 technologies is denoted as Lean Automation [2]. Among the examples of Lean Automation that can be found in the literature, the digitalization of the Kanban systems, the robot-based solutions to support employees' within Chaku Chaku lines, the augmented reality systems to assist employees in autonomous maintenance, the internet of things (IoT) technologies for real time data collection, to monitor the status and locations of materials and support the management of the shop floor [3–5]. In this context, the present paper introduces i-FAB, a learning factory developed by Università Carlo Cattaneo (LIUC) to demonstrate the benefit of the adoption of I40 technologies in a lean managed assembly system.

Learning factories are defined as systems that include elements of learning or teaching as well as a production environment [6]. Usually the production environment is a simulated, reality-conform one, and processes and technologies implemented in the learning factory are based on real industrial applications [7]. Simulating a production environment, i.e., reproducing the complexity of the shop floor, provides two main benefits that have led to its extensive use to complement traditional lectures: directly experience in a risk-free way which effects any decision or behavior produces when applied; receiving an instant feedback about the quality of the decisions stimulates curiosity and discovery learning [8]. Several learning factories dealing with lean concepts have been developed and discussed by literature (e.g. the KartFactory described by [8], the Mini-factory by [9]). In addition, with the advent of I40, several learning factories have been developed as research, teaching and training platforms with regard to this new industrial paradigm and presented by literature. Among them, the I40 Learning Factory at TU Wien. Whereas, a limited number of studies describes learning factories dealing with lean manufacturing and I40 applications, despite their importance in teaching participants how to systematically conduct the digital transformation [10]. To fill this gap, this paper provides details on the implemented lean tools and I40 technologies in the i-FAB learning factory and describes the training modules developed for Industrial Engineering and Management students and executive learning programs.

2 The i-FAB Learning Factory

The i-FAB environment is a learning factory in the narrow sense, as a real production department within an approximate area of 230 m^2, where foosballs are assembled and disassembled with real factory tools implementing the concepts of lean production and I40. The factory is located in a dedicated building in its campus in Castellanza. i-FAB (www.liuc.it/la-ricerca-in-liuc/ifab/) was set up in 2016 and continuously upgraded in collaboration with leading technology and industry partners (Bosch-Rexroth, Bossard, Comau, Grassi, Harting, Incas, Omron-Adept, Rivetta Sistemi).

The motivation that led to choose foosball as the output of the process is related to its dimensions (130 × 74 × 85 cm) and weight (approximately 40 kg) that could increase the similarity between the simulated processes and most of actual ones, making the training closer to a real world working experience [8] and let trainees better understand the benefits of I40 technologies. The 582 components (of 25 different typologies), their difference in size and weight and the possibility to introduce customized parts causes complexity in the materials management.

2.1 The i-FAB Lean Production Principles

The process is based on lean production principles: safety, process stability, total quality, just in time and continuous improvement. Safety is fundamental, and trainees must wear personal protective equipment, such as safety shoes and gloves, when actively working in the process. The stability of the system is pursued through the application of lean tools and practices such as 5S, standard work and visual management. Total quality is chased through the implementation of poka-yoke systems. Just in time production is ensured by the avoidance of batches, i.e. the process works one-piece flow, and pull materials management, obtained by Kanban system and supermarkets along the line, providing a minimum inventory coverage in time, and line balancing through Yamazumi chart. With the aim of ensuring continuous flow, according to lean principles, assembly and disassembly activities are organized in line, instead of job shop.

2.2 The Industry 4.0 Technologies Implemented in i-FAB

In i-FAB seven I40 enabling technologies are implemented: Internet of Things (IoT), Horizontal and Vertical Integration, Big Data, Data Analytics, Collaborative Robots, Additive Manufacturing, Simulation and Augmented Reality/Virtual Reality. In the following, details on the application of each technology to the assembly system are provided.

Products, materials and operators are connected by IoT systems. The trolleys are equipped with RFId transponders that make each item recognizable and traceable. An RFId reader in each assembly station detects the presence of a trolley (i.e. item) and communicates the Manufacturing Execution System (MES) which item is in the station. All information about the characteristics of each foosball are stored in a SQL database, communicating with MES via middleware. Visual instructions describing how to correctly perform each job element of the assembly activities to be completed in the station are projected on screen and the MES updates the projected instructions as assembly operators signal the conclusion of a job element. MES gathers then data on the duration and which operators perform the assembly activities and store them in the database. Together with the assembly/disassembly instructions, the operators are assisted in picking the right components from the supermarket by a poka-yoke pick/drop to light system developed together with INCAS. The lights update coherently with the activity progress registered by the MES. The application of IoT in i-FAB is aimed also at improving the activities performed by line operators. A Real Time Locating System is used in i-FAB to track operators moving around the workstation with the aim to have the corresponding spaghetti chart automatically drawn.

Integration plays an important role in materials management, sustaining the pull management system. To ease the management of small components, that are more difficult to manage through traditional visual methods, so that the level of inventory is always coherent with the planned coverage of the supermarket, they are stored in smart bins equipped with weight sensors that constantly check stock levels and, when the minimum stock level is reached, a signal is sent to the warehouse. Here a screen depicts the status of stock level in the supermarkets and the information related to the signal can be visualized. A red ben (low inventory) triggers the supply of a defined amount of material by the logistic operator.

The MES gathers data also on breakdowns of assembly tools and materials shortages that affect line efficiency and, together with other data from the MES and the ones gathered by sensors on the bins and on operators, are stored in a database. The amount of data to be handled is huge, as the collection frequency is up to one record each 30 ms. All data are collected and combined by the Bosch Rexroth system integrator, synthetizing all relevant information about the assembly process. Data are collected also by the R developed MES and depicted with descriptive purposes (see the application at https://ifabanalysis.shinyapps.io/centrale/). The MES shows in real time for each assembly station the time performance (i.e. the completion time gathered from the field) of the activities over time. Equipment breakdowns and materials shortages are described by the overall frequency of occurrence and duration.

Two examples of Collaborative Robots are provided: a self-navigating autonomous indoor vehicle and collaborative. The self-navigating robot dynamically moves material from one drop/pick up point to another, intelligently navigating around people and unplanned obstacles. The followed route is not fixed, and the next point the autonomous vehicle moves to can be easily selected by the operator. A 6 axes open-source robot works together with operators. The robot is devoted to a precision positioning activity (i.e. centering the scorekeeper so that the screws holes of the scorekeeper are aligned with those on the foosball), unburdening the operator of such a time-consuming activity and reducing the possibility of error and repositioning, i.e. time losses. Moreover, as the robot is equipped with an open-source software, students and operators are given the possibility of programming it, easing possible robot activity changes that could be required for further assembly process improvement.

The development of ad hoc tools, such as the robot hand or poka-yokes, is one of the opportunities offered by Additive Manufacturing technologies, besides the production of unique or small series. In i-FAB the benefits coming from the possibility of design, additively build a tool for process improvement, and apply it to the process in few hours can be experienced. The additive manufacturing activities are performed through the exploitation of three 3D printers.

Discrete event simulation is aimed at providing the digital representation of the physical i-FAB so that all changes in the process, from layout to materials management and line balancing, can be tested in advance, before their implementation in the process. The Enterprise Dynamics simulation model is fed with data gathered from the process and stored in the database, such as the cycle time of each activity. Desired changes to the physical process can be applied to the simulated one and predict the future behavior of the system. When unbalanced situations are detected by the data gathered and depicted by the MES, simulation can be implemented as it is a suitable tool to evaluate and compare line balancing alternatives. Enterprise Dynamics simulation program is available in i-FAB for testing the performance (e.g. respect of calculated takt-time) of alternatives and evaluate the respect of constraints (e.g. maximum amount of materials in the supermarket). Such application is an initial step towards the development of a digital twin of i-FAB, that would automatically acquire data, process them, define scenarios to be tested, according to implemented optimization methods or algorithms, and calculate the performance of each design.

Augmented Reality/Virtual Reality application in i-FAB consists in the application of an immersive virtual reality tool (Oculus Rift) aimed at enhancing the benefits to process improvement offered by simulation. The possibility of experiencing the future system with a 360° perspective gives the decision maker the possibility of understanding how the space would be in the future. In the context of assembly lines, as i-FAB is, experiencing the dimensions of space occupied by the material needed to feed the line according to a particular line balancing is worth to understand whether it could cause problems when implemented or not. Similarly, a different layout or number of assembly stations can involve issues related to space requirements. Virtual reality allows all stakeholders (i.e. operators and managers) facing doubts related to physical constraints of the systems.

2.3 The Learning Module

Trainees work in teams of 12–15 for two subsequent sessions of 8 h. The aim is creating in participants familiarity with lean and I40 technologies, awareness of the benefits that their application is able to provide to the run, management and design of production systems. Trainees work actively on the process chasing safety, quality, productivity, on time delivery and cost targets and work to reach them. The training module continuously alternates work sessions in the factory, debriefing, and class room sessions. During the work sessions in the i-FAB factory the team runs the factory aiming at achieving the set goals in 30 min runs. Each participant is assigned a role to play in the factory, that changes every work session to avoid learning factors. At the end of the run, the achieved performances are discussed and, supported by i-FAB instructors, what could help in improving the performance is acknowledged. The class sessions are devoted to examining the I40 enabling technologies with the aim to understand which better fit the lean context. Working sessions begins when the team, supported by i-FAB instructors, implement the selected tools and technologies before starting another 30 min run of the factory.

Before starting the first run of i-FAB instructions to run the production system are given to trainees. The initial run of i-FAB turns in no ability to reach the target performance goals. Through the first discussion supported by i-FAB instructors, the team agrees upon general instability of the production system as primary cause of the mismatch of target and actual performance. Instability is tackled through the implementation in the factory of lean tools such as 5s, visual management and standard work that are introduced by i-FAB instructors in a class session. Line unbalance is then tackled through the Yamazumi chart, that the trainees develop timing trials of activities. As the available time for data gathering is limited, the Yamazumi is based on few timings. Thus, from the discussion the need of a tool that continuously assesses the workload of each station usually emerges from the discussion with the aim to balance the line based on more accurate data. In the class session instructors present IoT devices and MES, that are then going to be used to gather data from the system.

In the second run, the lean tools applied to the systems lead to performance enhancement and the possibility of further improvements usually emerges from discussion. Due to the data gathering performed in the second run, the trainees are able to exploit descriptive analytics and the need of a re-balancing of the line emerges. Discrete event simulation is exploited to identify the best line balancing in short time. Moreover, trainees experience the use of a virtual reality model corresponding to the obtained line configuration and exploit this technology to validate the line layout from the analysis of the automatic spaghetti chart obtained from RTLS. Collaborative robots are introduced by the instructors to help operators in performing assembly and logistics activities in the third run. In this last run, the MES provides visual and dynamic instructions to trainees based on the item they have to assemble/disassemble. By the end of the third run, due to the implementation of lean and Industry 4.0 tools and technologies, the team is frequently able to achieve the target performances.

Table 1 synthetizes the three runs, providing details on the applied lean techniques and practices, applied industry 4.0 technologies, objectives and time.

3 Evidence from the Learning Modules

Collected data refer to production performance measures: (i) quality, measured through the number of defects, (ii) productivity, measured through the number of good products, (iii) throughput time, the time to complete the entire process for a single foosball, (iv) percentage of time spent in value-added activities and (v) cost, measured through the proxy of the steps per each good product, according to the Toyota's motto "*one second, one step, one yen*". The i-FAB instructors record performance (i), (ii) and (iii). Performance (iv) is obtained through the observation by

Table 1 The learning modules

	Applied lean techniques and practices	Applied Industry 4.0 technologies	Run objectives	Given time
Run 1	–	–	Recognize the need of standardization to bring stability	30′
Run 2	5s and visual management Standard work Yamazumi chart	IoT devices (RTLS) MES (both to gather data)	Recognize the need to re-balancing the line	30′
Run 3	5s and visual management Standard work	Simulation and virtual reality MES Collaborative robot	Experience the benefits of the adoption of both paradigms	30′

time-method operators during the first run and adding IoT and MES during the second and third runs. Pedometers worn by the trainees record the steps to measure performance (v). During the academic year 2018/19, 75 students divided into five groups experienced the above-described learning modules in i-FAB.

The evidence from this first set of learning modules proves that i-FAB allows the trainees to experience the benefits of lean and then the role of I40 as its booster. As the boxplots that represent data (Fig. 1) show, performance measures improve every run. In particular, lean adoption drastically reduces the number of defects in three out of five cases while sustaining the results of one and zero defects. Such improvement affected the increase in the number of compliant products as well, going from an

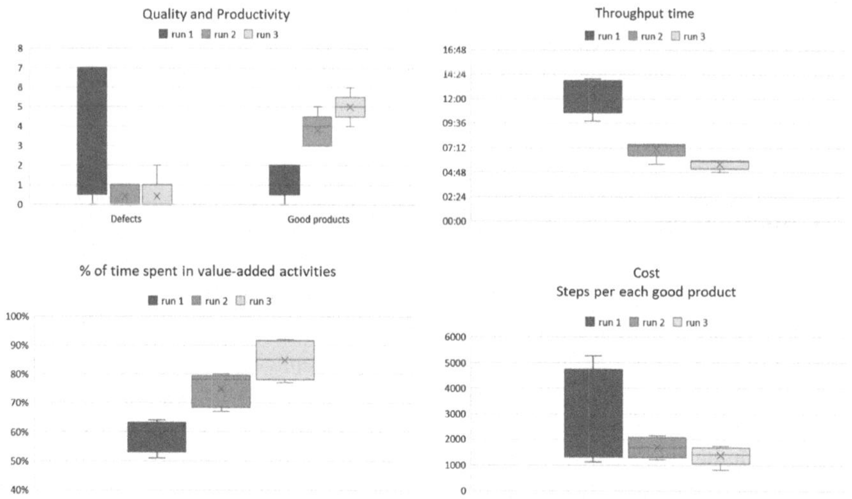

Fig. 1 Performance achieved in the three runs charactering the learning modules

average value of 1.2 in the first run to an average value of 3.8 in the second run. The results in terms of quality achieved in the second run were still maintained during the third run, indeed the average number of defects was the same.

From the gathered data, productivity benefitted of both lean and I40 tools, indeed the dynamic instructions provided by the MES and the better line setting, allowed to increase the average productivity by 1.2 compliant product. Data also reflects that balancing and layout improvement based on data gathered through IoT and MES and tested via simulation is able to boost the lean benefits on motion and time reduction.

Lean adoptions, such as line balancing, reduce the average throughput time from 12′26″ in run 1 to 7′04″ in run 2, but I40-driven balancing and layout improvement led the average throughput time to 5′35″ in the third run, providing further reduction. The boxplots also show a reduction in variability of results, reflecting a more stable and standardized context.

In the same way, the ultimate result of such I40-driven solutions is the impact on the lean goal to eliminate non-value-added activities: data reflect the increase by 13% of the average percentage of time spent in value-added activities in the third run compared to the second one.

Moreover, as 5s and visual management has helped in reducing the movements of trainees, layout reconfiguration based on I40 technologies and the adoption of collaborative logistics robots reduced the average cost per each good product by 301 steps compared to the second run.

Finally, trainee's opinion on the learning action he/she took part to was gathered by means of a questionnaire containing four items, each presented as a single statement which can be endorsed on a uniform four-point Likert scale: (i) correspondence with declared objectives; (ii) relevance of the course contents considering the learning need; (iii) suitability of teaching resources; (iv) overall course evaluation. 45 questionnaires were completed by trainees. The course was positively evaluated from trainees: (i) average 3.80 and 0.40 standard deviation; (ii) average 3.73 and 0.44 standard deviation; (iii) average 3.82 and 0.39 standard deviation; (iv) average 3.84 and 0.37 standard deviation. These results enforce the positive performance results gathered from the i-FAB runs.

4 Conclusions

The link between I40 and lean manufacturing has recently gained significant popularity in both academia and industry, but a limited number of studies describes learning factories dealing with lean manufacturing and I40 applications. The present work aims to fill this gap presenting the i-FAB learning factory and the developed learning module. Moreover, this work contributes to the research on the link between lean and I40 with the evidence from the learning modules. The data gathered proves that i-FAB allows the trainees to experience the benefits of lean manufacturing tools and practices adoption and the role of I40 as its booster.

The research is limited by the number of studied groups of trainees. Increasing the number of tests with groups of trainees to statistically demonstrate the suggestion from the presented analyses represents a future step of research.

References

1. Buer S-V, Strandhagen JO, Chan FTS (2018) The link between Industry 4.0 and lean manufacturing: mapping current research and establishing a research agenda. Int J Prod Res 56:2924–2940
2. Tortorella GL, Fettermann D (2018) Implementation of industry 4.0 and lean production in Brazilian manufacturing companies. Int J Prod Res 56:2975–2987
3. Kolberg D, Zühlke D (2015) Lean automation enabled by industry 4.0 technologies. IFAC-PapersOnLine 48:1870–1875
4. Sanders A, Elangeswaran C, Wulfsberg JP (2016) Industry 4.0 implies lean manufacturing: research activities in industry 4.0 function as enablers for lean manufacturing. J Ind Eng Manag 9:811–833
5. Cannas VG, Pero M, Pozzi R, Rossi T (2018) An empirical application of lean management techniques to support ETO design and production planning. IFAC-PapersOnLine 51:134–139. https://doi.org/10.1016/j.ifacol.2018.08.247
6. Wagner U, AlGeddawy T, ElMaraghy H, MŸller E (2012) The state-of-the-art and prospects of learning factories. Procedia CiRP 3:109–114
7. Abele E, Metternich J, Tisch M, Chryssolouris G, Sihn W, ElMaraghy H, Hummel V, Ranz F (2015) Learning factories for research, education, and training. Procedia CiRp 32:1–6
8. Pozzi R, Noè C, Rossi T (2015) Experimenting 'learn by doing' and 'learn by failing'. Eur J Eng Educ 40:68–80
9. Matt DT, Rauch E, Dallasega P (2014) Mini-factory—a learning factory concept for students and small and medium sized enterprises. Procedia CiRP 17:178–183
10. Küsters D, Praß N, Gloy Y-S (2017) Textile learning factory 4.0—preparing Germany's textile industry for the digital future. Procedia Manuf 9:214–221

Creating Employee 'Pull' for Improvement: Rapid, Mass Engagement for Sustained Lean

Frank Devine and John Bicheno

Abstract The paper describes an approach to engagement called Rapid Mass Engagement (RME) that has been used at a quarter of Shingo Prize winning sites in Europe between 2010 and 2017. The approach has been developed over the past 20 years with an ongoing process of experimentation. Particular features include the involvement of ALL employees, a series of diagnostic and problem-solving not merely consultative meetings, and the development of an employee-created organizational culture, locally developed by employees and codified in their own words. This 'bottom-up' dominated approach is in contrast with many top-down approaches, but helps to facilitate lean by enhancing leader standard work, policy deployment and continuous improvement.

Keywords Employee · Engagement · Improvement · Human resources

1 Introduction

Gallup's 2017 report 'State of the Global Workforce' stated that '85% of employees worldwide are not engaged or are actively disengaged in their job', but 'in the best managed companies' as many as 70% of employees are engaged. Moreover, Gallup found that 'Business or work units that score in the top quartile of their organization in employee engagement have nearly double the odds of success (based on a composite of financial, customer, retention, safety, quality, shrinkage and absenteeism metrics) when compared with those in the bottom quartile'. In a recent article from 'The Insider' (HBR, 17 May 2019), Buckingham and Goodall report, after a huge US survey, that 'engagement averages a paltry 16%', but doubles when 'what really

F. Devine
Accelerated Improvement, Coventry, UK
e-mail: acceleratedimp@aol.com

J. Bicheno (✉)
University of Buckingham, Buckingham, UK
e-mail: john.bicheno@buckingham.ac.uk

M. Rossi et al. (eds.), *Proceedings of the 6th European Lean Educator Conference*, Lecture Notes in Networks and Systems 122,
https://doi.org/10.1007/978-3-030-41429-0_7

engages is their experience on a team' and further improves when the team is in a trusting environment.

Clearly, then, high engagement should be an important aim for any Lean-aspiring organisation. This paper aims to

- Outline a new and promising approach to engagement in Lean Transformation called Rapid Mass Engagement (or RME) that has achieved success in productivity gains and in assisting a number of organisations to win Shingo Prizes—gold, silver, or bronze. From 2010–17, 25% of all Shingo awards in Europe went to organisations deploying the approach outlined in this paper. The implementation is an ongoing experiment involving collaboration between an organisation, a consultant and, in some cases, a university.
- Contrast RME with other approaches to engagement in Lean.
- Focus on one particular stage of RME whereby an entire workforce creates its own culture of continuous improvement. This culture is codified in a set of behavioral standards.
- Alert practitioners to the risks of when the process has not achieved its objectives.

2 Literature Survey: Lean and Engagement

Academic interest in employee engagement and work engagement has risen sharply over the past 20 years. The engagement literature is huge and growing exponentially [19, 20]. The significant impact of engagement on employee performance has also been researched. Many are written from a Human Resource perspective and appear to orient towards engagement being an HR or top-down concern [7]. It would appear that many academic papers in the area are concerned with definition, with measurement, and with conceptual frameworks.

Differences in the definition of work engagement and employee engagement remain undecided. The most often used definition of work engagement in the scientific literature is '… a positive, fulfilling, work-related state of mind that is characterized by vigor, dedication and absorption' [21].

The importance of culture and engagement is now also prominent in Lean transformation [1, 9–11, 14]. Engagement is often cited as being an important consideration in, for example, 5S, Hoshin Kanri, suggestion schemes, problem solving, and team development. Of course, 'Respect for People' is one of the two pillars of the Toyota 'House of Lean' as is 'Capability Development' in the Lean Enterprise Institute model.

Here we briefly mention just three of several routes to engagement. The Shingo Prize [18] includes 10 principles of which two ('respect every individual' and 'lead with humility') are strongly aligned with engagement and three ('embrace scientific thinking', 'think systemically', 'create constancy of purpose') are indirectly aligned. In the 'Toyota Way Fieldbook' [10] the responsibilities of team members, team leaders, and group leaders are detailed. This makes clear that engagement is not simply an edict from neither the top, nor a stand-alone bottom-up activity but requires active and ongoing involvement through all levels. TWI (Training Within Industry) is a set of concepts—Job Instruction, Job Methods, Job Relations—known as the 'three legged stool' [8], that had a major influence on early development of the Toyota Production System and is still used today, in modified form, for 'developing competent and able people' [11]. Here we note the similarity of TWI Job Relations to engagement—'the foundation for good relations' particularly 'people must be treated as individuals'. More recently, TWI has increasingly been linked with 'Kata' [15]—to judge by several 'TWI Kata' conferences in Europe and USA in 2018 and 2019.

3 Rapid Mass Engagement: Development and Approach

Rapid Mass Engagement (RME) has been in development since 1990. The 6-stage culture change process outlined below is the result of the 23-year cycle of PDSA-type experimentation via hypothesis-based application and improvement that continues to this day. The methodology tests hypotheses in real work situations with large numbers of employees over long periods across multiple organisations with repeated feedback loops and improvements. Examples include hypotheses that recognition must be timely and that recognition can be enhanced by attention to body language. These hypotheses were tested separately to avoid contagion and predicted that these approaches to recognition would increase employee survey results concerning recognition. In both cases, applied across large sample sizes (to increase statistical reliability) and for a longer time period (to avoid measurement failing to capture slow moving improvements) the hypotheses did not produce the predicted results and, in the case of body language, produced worse results than the control groups! In contrast, many organisations have measured significant increases in their recognition scores after the application of RME's approach to recognition. Examples include Rolls Royce, Coca-Cola, Johnson & Johnson, Bacardi-Martini and GKN.

Many Lean transformations have in the past failed due to insufficient attention to the 'people' aspects [5]. Today, to judge by the huge volume of articles—a small fraction of which have been mentioned above—the need for engagement is well established, perhaps even over-emphasized. However, in such conventional Lean transformation, even though employees 'have a say' or are involved or consulted, the power to make the final decisions remains with management alone [2]. RME takes a different route, with the fundamental belief that true engagement must begin with employees themselves, reflecting their own beliefs and removing obstacles that prevent full participation. To do otherwise is merely 'pseudo engagement'.

Throughout the RME process employees are involved in adult-to-adult conversations and make decisions not merely react to management decisions. Employees never 'ask management' or 'make representations'; they make many operational decisions themselves and actively prioritise. As a result, employees are not the passive recipients of 'engagement'—rather they act on their system of work in such a way that they become actively engaged.

In this sense, management doesn't engage employees; management *creates a process whereby employees become engaged* and then work to sustain the new system thus created [2, 3]. RME is therefore a radical alternative to the traditional top down approach to engagement and enablement. In this approach:

- *All* the employees on a local site who constitute an interdependent system make decisions and are not merely 'involved' or 'consulted'.
- Employees create their Behavioural Standards in the language chosen by them, not in managerial or academic language.
- Employees agree, by consensus, not negotiation or compromise, with the local senior management team, a jointly-owned and prioritized change plan to overcome obstacles to achieving the site's Higher Purpose.
- RME aims to undermine any legacies of negative and limiting assumptions and aims to create rapid momentum and sustainability from the bottom up.

By involving all employees the process ensures the *width* of ownership necessary for a new culture to withstand the kind of early challenges that can undermine it before it grows strong enough to sustain itself. *Depth* of ownership is achieved by the more intense experience of *collective, joint decision making* (known as Consensus Day). Both width and depth are sustained and leveraged by continuous improvement outputs and from joint decision-making.

The overall process of RME is illustrated in Fig. 1.

The phases are as follows:

1. The process starts with *employees* rapidly creating a new and competitive culture. A Joint Decision-Making event called Consensus Day agrees, by consensus not negotiation, with the senior leadership, a change plan to both enable (by removing obstacles to the organisation's Higher Purpose) and engages employees via the creation of a new, employee-owned continuous improvement culture.

   The leadership approach is called the Cathedral or Higher Purpose Model. A core driver is that work is not just about earning wages but also can be harnessed to create jobs and sustain communities. When employees see the *genuine* focus on changing their experience at work, skepticism reduces and engagement deepens and widens. The engagement is deepened by the intensive nature of the process—e.g. 'Consensus Day' at Boston Scientific and Seagate both involved 60–90 employees making joint decisions with their Senior Team about key business issues by consensus over 24 h of contact time. The mass nature of the process widens engagement as all employees create their own culture.
2. To enable and sustain an initially fragile new culture, standards of leadership *outputs* have to be consistent and high. This makes it difficult for opponents of the new culture to point to examples of individual managers who are not both operating at a high-performance level and *modeling and referencing* the new culture in their day-to-day activities.

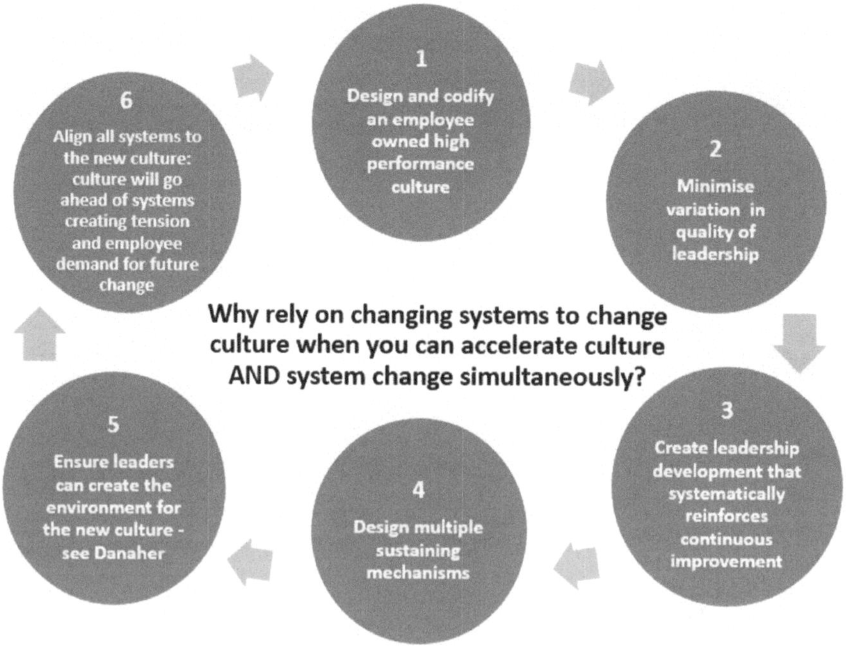

Fig. 1 Process overview [2]

3. The new fragile culture needs to be sustained by process change as well as behavioural change and this combination is designed to be mutually reinforcing in nature. The leadership approach thus leverages improvement science.
4. To sustain the new culture, multiple sustaining mechanisms are designed and implemented to avoid natural degradation over time and to *make the new culture independent from the energy and commitment of the original leadership group.* As an example the specific role of the internal facilitator group includes acting as a permanent 'conscience' of the new culture and helps ensure all new policies and procedures are measured against it.
5. Creating an environment where the new culture is reinforced every day. As an example, Boston Scientific have a standard internal workshop called 'Creating the Environment' whereby the front-line leader and the team are taken through a process of agreeing how to make the new culture a reality not just words on the wall. This includes ensuring that leaders can be challenged without consequence.
6. Once the new culture is created it will expose systems that are inconsistent with it thus creating the tension and pull to improve and align these systems.

Throughout the process, including the leadership development and continuous improvement aspects, existing systems and procedures are examined to identify potential conflicts and barriers—both social and technical. This is a specific design feature and aligns with 'Socio-Technical' design and 'Quality of Work Life' [22]. The 'socio' aspects include issues such as the effect of changes on employees' social

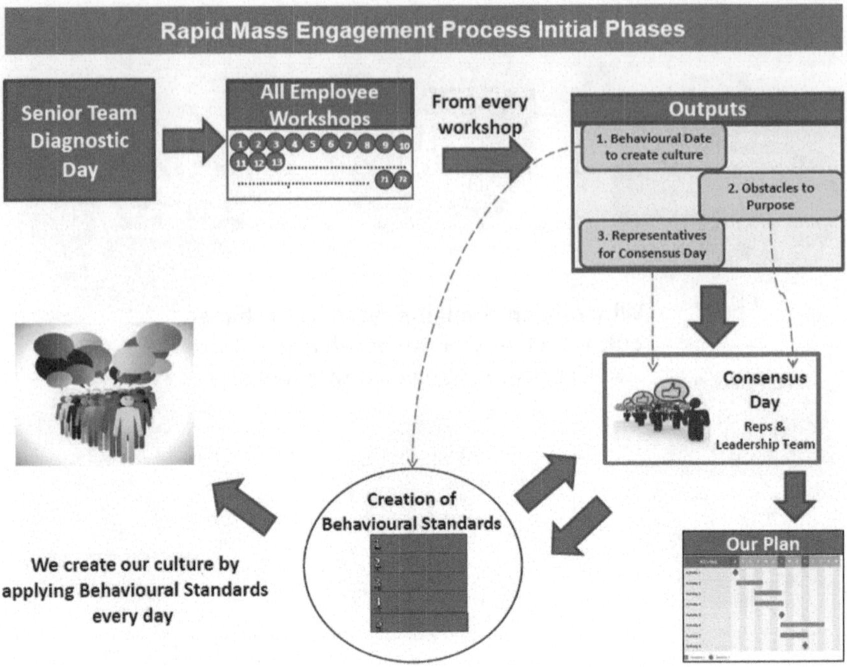

Fig. 2 Early stages of RME

standing and self-image and which are sometimes missed even when employees are involved in substantial technical changes.

In this process, all employees diagnose the key obstacles to achieving the organisations' Higher Purpose and agree the nature of the new culture needed to overcome such obstacles. The early stages of RME are illustrated in Fig. 2.

4 Five Points of Comparison with Conventional Approaches to Engagement

1. RME creates engagement and pull for Lean from the bottom up not top-down. There is a similarity with lean policy deployment in as far as values are interpreted 'top down', but an important difference in that Behavioural Standards are employee-created and 'bottom up'.
2. The Behavioural Standards methodology (see below) of codifying the employee-created culture is more behaviourally specific and locally meaningful than top-down corporate values and thus increases accountability.
3. The improvement methodology is direct, measuring results in real time in real work situations with permanently repeating feedback loops and improvement

opportunities. This is designed to overcome limitations that often frustrate managers and HR staff. An example is having to rely solely on indirect forms of measurement such as surveys and corporate KPIs that have to be interpreted and thus suffer from lack of follow-up opportunities [16].
4. Behavioural Standards are directly aligned with the lean concept of 'leader standard work' in as they specifically encourage and systematise the 'gemba' dialog that should be present [12].
5. Continuous improvement is facilitated by the process of decentralization and rapid, mass engagement and by encouraging Deming's famous point 'Drive out fear' to take hold, as described at length by Liker and Hoseus [11], Liker and Meier [10] and by the Richardson and Richardson [14].

5 Behavioral Standards as Local Differentiation

Behavioural Standards are designed as differentiators, i.e. locally decided cultures agreed by the employees in a particular location and owned by those employees. They *supplement, but don't replace*, the integration effects of well-designed and communicated Corporate Values. They are a deliberate form of differentiation, creating an approach that respects the local culture and language of employees whilst nevertheless being compatible with stated corporate values, 'mission statements' or 'credos' [13].

At Boston Scientific, Ireland a team of 20 volunteers worked for 3 days with Devine distilling 1400 behaviours into a small, agreed set. A series of 3-h workshop for groups of up to 34 employees at a time was conducted covering all 3000 employees at the site.

To quote a senior manager from Boston Scientific 'Ideal results need ideal behaviours, which are driven by system design and people beliefs, both of which are informed by principles' [17].

In another example, 1200 Seagate employees agreed a recent set of RME Behavioural Standards in February 2019, as shown in Fig. 3.

6 Risks and Examples of When the Process Has Not Achieved Its Objectives

Such a high impact process cannot be risk-free. Sub-optimisation is predicted in the following circumstances.

1. Lack of complete ownership of the process by **all** members of the senior team.
2. The promotion of a senior leader at an early stage of the process before the culture is strong enough to sustain itself.

Our Behavioural Standards

- We assume positive intent and are happy for others.
- We show respect by listening, seeking to understand and being there for each other.
- We appreciate each other.... regardless of our roles.
- We encourage different ideas....and give them space to grow.
- We involve people early and trust them to do the right thing.
- We get ahead by taking responsibility and getting it done!
- We challenge in ourselves and in others:
  - Seeing failure instead of learning
  - Negative Limiting Assumptions
  - Complacency
  - Back biting
  - Any breach of our standards

Fig. 3 Seagate behavioural standards

3. Lack of support from corporate leaders.
4. Any gaps re the experience in the process itself and the quality of the facilitator, and the related issue of underestimating the depth of knowledge needed to understand the process well enough to produce a powerful Consensus Day and change plan.
5. Not being willing to match the ambition of the business objectives with the ambition of the engagement process necessary to achieve it.
6. Not taking the time necessary to understand the process and do due diligence re the independent facilitator/consultant.
7. Not ensuring that leaders are trained specifically to reinforce and sustain such a High Performance, non-hierarchical culture.
8. Not integrating the engagement process with the approach to continuous improvement.

7 Typical Results

- Boston more than doubled its output on the same footprint within 5 years.
- Rolls Royce increased output to such an extent that a new Test Bed facility (£30–50 m) was not needed and the employees concerned moved from bottom, with frequent IR issues, to top in the corporate engagement survey.

- At Coca Cola, Investors in People found that RME produced 'the greatest transformation in employee attitudes ever measured' across IIP's extensive database.

Tangible results were studied in two MSc Dissertations of the implementation at DePuy. In summary, they found

- A 73% increase in productivity
- A 45% reduction in absenteeism
- A 300% increase in ideas implemented per person
- A 30% increase in engagement scores and a 34% increase in Q-12 engagement scores [6, 23, 24].

The successful implementation of the RME in both Shingo (DePuy) and non-Shingo (Boston Scientific) sites indicates that the application of the Shingo Principles rather than pursuit of the Shingo Prize per se, was important. The Shingo Principles thus acted as an intellectual underpinning of both the need to engage with employees in a deep and meaningful way and to have a systematic approach to leadership development designed to do so [4].

An external endorsement is: "The best example we have seen of an organisation that truly embraces the cultural aspects of the Toyota Way to deliver sustainable results" (Kevin Robinson Assistant General Manager Toyota Manufacturing UK, 2010).

8 Conclusions and Experiences

During the experimental process of developing RME, learning included:

The process should be Rapid. For Engagement to be meaningful, *it has to change the working experience of employees*, and the speed with which action is taken is extremely meaningful to employees, partly because of the powerful contrast to their normal experience. Speed also signals powerful leadership intent and seriousness.

The process should involve everyone *in the system to be optimised*. It should involve the 'culture' being self developed, and hence owned, from bottom-up rather than being imposed 'top down'.

'Mass' means engaging all employees not a sub-set of them.

Making changes via a pilot process to test the process carries risks of 'Not invented here' attitudes leading to a rejection of the 'foreign body' by those not involved.

Experience over 23 years and at over 20 sites indicates that RME is a promising approach to improvement, particularly at a time when employees are increasingly better educated (20% of Seagate engineers have PHDs), more demanding, less passive, and less willing to accept top-down 'command and control'.

References

1. Bortolotti T, Boscari S, Danese P (2015) Successful lean implementation: organizational culture and soft lean practices. Int J Prod Econ 160:182–201. https://doi.org/10.1016/j.ijpe.2014.10.013
2. Devine F (2016) When employees create their own high performance culture: the rapid, mass engagement process. Lean Manage J, September
3. Devine F (2016) The rapid, mass engagement process—part two. Lean Manage J, November
4. Devine F (2016) Demystifying leadership-setting leaders up for success. Lean Manage J
5. Emiliani ML, Stec DJ (2005) Leaders lost in transformation. Leadership Organ Dev J 26(5):370–387. https://doi.org/10.1108/01437730510607862
6. Garvey P (2015) Engaging an organisation in operational excellence: a case study in mass engagement. MSc dissertation, University of Buckingham
7. Garrad L, Chamorro-Premuzic T (2016) The dark side of high employee engagement. HBS (accessed)
8. Graupp P, Wrona R (2016) The TWI workbook, 2nd edn. CRC Press, New York
9. Hopp W (2018) Positive lean: merging the science of efficiency with the psychology of work. Int J Prod Res 56(1–2):398–413. https://doi.org/10.1080/00207543.2017.1387301
10. Liker JK, Meier DP (2007) Toyota talent. McGraw Hill, New York
11. Liker JK, Hoseus M (2008) Toyota culture. McGraw Hill, New York
12. Mann D (2010) Creating a lean culture. CRC Press, New York
13. Quinn RE, Thakor A (2018) Creating a purpose driven organisation. Harvard Business Review, July–August, pp 78–85
14. Richardson T, Richardson E (2017) The Toyota engagement equation. McGraw Hill, New York
15. Rother M, Aulinger G (2018) Toyota culture. McGraw Hill, New York
16. Shah R, Ward P (2007) Defining and developing measures of lean production. J Oper Manage 25:785–805
17. Shields C (2019) Boston Scientific Galway culture journey. Presentation made at enterprise Ireland conference, Croke Park, May 2019
18. Shingo Institute (2019) The Shingo model. https://shingo.org/model. Accessed June 2019
19. Shuck MB (2011) Four emerging perspectives of employee engagement: an integrative literature review. Hum Resour Dev Rev 10(3):304–328
20. Shuck MB, Rocco TS, Albornoz CA (2010) Exploring employee engagement from the employee perspective: implications for HRD. J Eur Ind Training 35(4)
21. Schaufeli WB (2018) Work engagement in Europe. Org Dyn 47:99–106
22. Trist E (1981) The evolution of socio-technical systems: a conceptual framework and an action research program. In: Van de Ven A (ed) Perspectives on organizational design and behaviour. Wiley, London. Accessed July 2019
23. Twomey W (2011) Beneath the waterline—a study of the effect of applying leader standard work to the social aspects of a manufacturing system on the performance of the system. MSc dissertation, Lean Enterprise Research Centre, Cardiff University
24. Whyte P (2011) Exploring the use of systems thinking to understand and plan change. MSc dissertation, Lean Enterprise Research Centre, Cardiff University

Lean: Quo Vadis?

Richard Schonberger and John Bicheno

Abstract Inventory turn performance is, or should be, seen as an important measure of lean, since it relates directly to lead time and is a concomitant of many lean activities. Inventory turns, based on published data, has been studied for decades. Among more recent of those studies are findings showing long-term decreases in inventory turns in several industries and in several countries, including some prominent names in lean. This is disturbing. There may be good reasons for declining performance in inventory turns but questions as to their root causes should be of concern to lean practitioners and educators. After establishing hard-data indicators, some tentative thoughts are proposed. Hence, Lean Quo Vadis: 'Where are you marching?'

Keywords Lean · Inventory · Decline · People · Methodology

1 Introduction

Clearly, 'lean', its predecessor 'just in time', and its well-known exemplar, the Toyota production system, have had a huge impact. The number of hits on Google relating to lean now numbers over 1.4 million. Annual conferences relating specifically to lean, such as Shingo Prize, AME (Association of Manufacturing Excellence), and indeed the ELEC conferences, report hundreds of success stories. Yet, and yet, the number of reported failings and fade-outs of lean also continues to grow [1, 8]. Of course, there are numerous reasons for lean's shortcomings. This paper makes tentative suggestions as to those reasons—one being that inventory performance may be overlooked, ignored or deliberately downplayed when the facts are uncomfortable to lean enthusiasts (or to those with vested interests?). Both authors of this paper have

R. Schonberger
University of Nebraska, Lincoln, USA
e-mail: sainc17@centurylink.net

J. Bicheno (✉)
University of Buckingham, Buckingham MK18 1EG, UK
e-mail: john.bicheno@buckingham.ac.uk

M. Rossi et al. (eds.), *Proceedings of the 6th European Lean Educator Conference*, Lecture Notes in Networks and Systems 122,
https://doi.org/10.1007/978-3-030-41429-0_8

been strong proponents of lean, but contend that some practices under the name of lean have led it away from achieving its potential.

Inventory is, of course, one of the ‘seven wastes’. As Ohno [13, 19–20] put it, ‘The greatest waste of all is inventory.’ Further, Little’s Law [12] establishes the relationship between inventory, throughput rate and lead time. The central role of lead time is, in turn, established through Ohno’s famous dictum ‘All we are trying to do is to reduce the time from order to cash’ [13].

Since the early 1990s, Richard Schonberger [15] has traced inventory turn performance based on annual reports amongst publicly-traded companies, as *one* measure, albeit arguably the single most important measure, of their progress with lean or world class. The contention is that many lean initiatives should eventually work through to an increase in inventory turns. By contrast, following Little’s Law, a decline in inventory turnover presupposes a lengthening in end-to-end lead-time, and thus, via Ohno, a decline in a central aim of lean. Moreover, lead-time is a critical customer-focused metric, especially relevant, enterprise-wide, in this ‘want-it-now’ competitive climate.

Changes in inventory turn performance have been measured by Schonberger using a scoring system that allocates points between +2 and −1 for long-term trends in inventory turnover. For instance, a score of +2 is given for sustained growth for 10 years and −1 is given for sustained decline. Note that it is inventory turnover *trends* that are tracked and enumerated rather than absolute values. Key findings of the inventory studies have previously been reported [17–19]. The studies have now been updated to the present time.

Samples of inventory trend performance are given in Figs. 1 and 2. The inventory turn data is derived from cost of sales and inventory holdings, both of which are

Rank-Ordered Industries	Score	Number of Firms
Electric	0.46	115
Metalworking/machining	0.39	325
Retail	0.34	173
Electronics	0.30	363
Liquid/gas/powder/grains	0.22	399
Plastic/rubber/glass/ceramic	0.21	257
Vehicular components	0.20	124
Food/beverage/tobacco	0.20	155
Autos, light trucks, bikes	0.15	37
Pharmaceuticals	0.11	90

Fig. 1 Rank-ordered industries (these 10 sampled from full list of 32 sectors) based on long-term trends in inventory turnover—as of 2018

Rank-Ordered Regions	Score	Sample Size
Scandinavia	0.29	60
Canada/Mexico/Israel	0.27	88
Romance (Southern Europe)	0.21	64
Germany/Austria	0.18	54
United States	0.16	875
Grand Average	0.14	1,194
Asiana/So. Africa	0.14	111
Benelux/Ireland	0.12	37
United Kingdom	0.07	92
Japan	-0.09	132

Fig. 2 Rank-ordered regions based on long-term trends in inventory turnover—as of 2018

published and externally audited. Inventory turn data can therefore be independently replicated.

Figure 1 is an abbreviated portrayal of the full data set of 32 industry sectors, most of those excluded having overly small n's (e.g., furniture and paper) or overlapping character (e.g., a semiconductor sector is also included under electronics, and chemicals are included under liquid/gas/powder/grains). (The full dataset of 32 sectors is given in Schonberger [18], though with older data.) Figure 1 shows stark differences in long-term inventory scores for the 10-industry sample. Mean score for the bottom six sectors is 0.182, whilst for the top four is, at 37.2, two times greater.

The scores and rankings have changed a good deal over decades of their compilation, although pharma has stayed at or near the bottom all along. As for top rankings, metal-working/machining has long placed well. On the other hand, electronics once ranked at or near the top in inventory-reduction trends. That rank dates back to the JIT era of the 1980s, in which companies such as Hewlett-Packard, IBM, Intel, and (the former) Control Data were among the globe's most adept at implementing cells, one-piece flow, supplier development, and so on. Before long, however, much of electronics manufacturing had been outsourced to Southeast Asia. Best practices in supply-chain management have not made up for quick, low-inventory production that those companies had mastered in the West under JIT. Moreover, best JIT practices (e.g., cellular organization of production) failed to accompany the off-shoring of electronics: most off-shored electronics production reverted to assembly lines and the batch-and-queue mode.

Figure 2 scores and ranks global regions on the long-term inventory scale as of the year 2018. But is that also reflective of earlier-year results? The answer is no. Figure 3 shows the region-by-region scores and rankings as of several data points

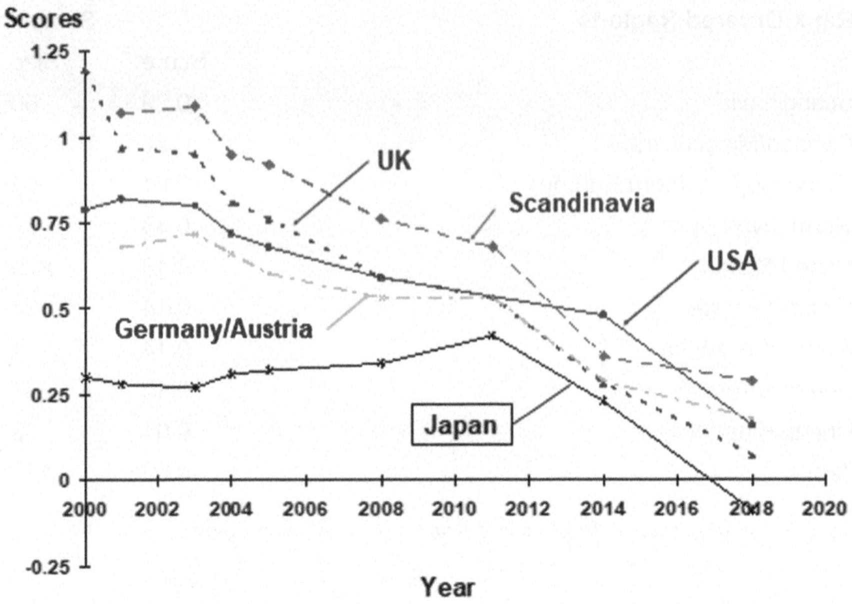

Fig. 3 Long-term trends in inventory turns: scores for five global regions (this figure is an update of a similar graph shown in Schonberger [13])

beginning in 2000—a trend-in-a-trend graph. (To keep the graph reasonably uncluttered it includes the five most continuous regions but not the four other regions, the complex makeup of which is rather random.) Notably, all five regions show steep declines from 2000 to 2018. That Japan is bottommost among the five could have any number of explanations. One that suggests itself is that the Japanese nation's economic malaise, now in its third decade, may have been accompanied by, even partially triggered by, strategic shifts toward bottom-line objectives, with lean and related initiatives, including quality, losing emphasis. That Japan's score has gone negative is shocking, in that the scoring system was originally formulated based on a likelihood that companies should be able to attain mean scores near to 1.0.

Using this measurement system, some companies have indeed shown impressive results over 16 years. Included in this group are Ingersoll Rand, Unilever, and Novo Nordisk. Noteworthy, however, are disappointing trends by major automotive organizations, including Toyota, Volkswagen, and General Motors, and in various other industries such as pharma. Moreover, inventory performance on a corporate level indicate that many have worsened irrespective of the location of headquarters.

We may note that the trends are grouped by industry and by global region. (It should be remembered that the scores in the tables do not represent absolute values; rather they arise from assigning points to long-term trends in turns.) It is also the case that a worsening in overall trends in inventory turns is frequently accompanied by an improvement in work-in-process inventory—a plant may have higher turns but its supply chain may be, and often is [19], chock-a-block with finished inventory, offsetting the plant-level gains.

2 Why Do Inventory Turns Decline?

Of course, there may be many reasons for a decline in inventory turn performance—some good (or justified responses), some bad (or unwarranted). Further, some reasons may be self initiated actions (or inactions) within the control of managers, whilst other responses may be forced by external events. It is important to distinguish between these. Some examples are given in Fig. 4.

Some critics of inventory turn trends as an important lean indicator do not seem to have made these distinctions. Instead they attack the measure as being inadequate or invalid in that it fails to consider attractive visual appearance (housekeeping, 5S, display boards, and the like) within a site, or market factors; or they may simply have a myopic viewpoint that rejects any criticism of lean. The authors of this paper believe that such viewpoints undermine progress of lean—'there is none so blind as those who will not see'. (For instance, Kahneman [9], details several cognitive biases, such as the sunk cost fallacy and confirmation bias.) To ignore the inventory trends is simply self-denial and not in line with the questioning approaches that were a feature of the early days of TPS and lean.

Spector [14] looking at pharma, is enthusiastic about turns ('a reliable indicator') but says that inventory reduction is beneficial only if such reduction derives from process improvement. If inventory reduction comes without process improvements, then stock-outs and customer dissatisfaction will far outweigh any benefit. We could attach a corollary to this: Process improvement should result in inventory reduction, because process failings of most kinds are buffered by inventory.

Cost, quality and delivery (QCD) are often taken as desirable measures of operations management. Internally these are satisfactory provided that no manipulation

	Internal	External
Poor Reasons (Failures?)	Managerial/leadership Inappropriate tools Shop floor absence 'Cooking the books' Inappropriate KPI's	Declining supplier performance Corporate reporting pressure Wage costs Failure to respond to tech changes
Good Reasons (Strategic and Unavoidable?)	Decrease in WIP but greater RMI/FGI Capacity tradeoff Service improvement Corporate growth Multiple locations Product complexity	Competition Risk aversion Safety Financial regulation

Fig. 4 Reasons for worsening inventory trend performance

(or massaging) takes place. Graban [6] supports this, but adds employee satisfaction. He states that a single measure is a 'huge error' (we agree), but then says that 'Low inventory is not the primary goal - profitability is'. (We disagree with the Goldratt-inspired 'goal' of 'making money today and more tomorrow'. Setting aside dishonest and unethical practice, marketing and price over-recovery rather than operations changes can hugely influence both the lean and non-lean enterprise.) Graban also warns that outsourcing is not the way to go, even though turns increase. Problems of definition and data collection mean that these are unsuitable for inter-company comparative trends. Eroglu and Hofer [5] hypothesize that there is a concave relationship between performance and inventory turns. This is a worthy contention, if it is understood that the optimal point is a continual moving target.

Hopp [8] believes that there are three types of buffer—inventory, time, and capacity. These, he says, should be explicitly traded off—for instance increasing capacity against decreasing inventory, or reducing customer wait time against increasing inventory. Well and good. However, lean has long attempted to avoid tradeoffs (for example by reducing changeover time rather than seeking the optimal 'economic order quantity' or by failsafing so as to obviate 'acceptable quality limits'). By not *first* seeking to avoid tradeoffs is simply non-lean practice. As shown by the Kingman equation [2] the only way to reduce queues (and hence inventory) in a given layout is by reducing arrival or process variation, by decreasing utilisation, or by reducing process time. There are two points here: One is that the causes of variation are both internally and externally generated, some of which is certainly avoidable by concerted action. A second is that both utilisation and process time have waste elements that can be reduced or eliminated. All of these actions help to avoid inappropriate tradeoffs that will lead often to worse inventory turns, and certainly to decreased competitiveness.

Aside from the internally derived inventory turn metric, well-rounded performance measurement generally includes various externally generated measures—for instance JD Powers on quality, and Glassdoor on employee satisfaction. The Shingo Prize points system is comprehensive and externally measured but is complex and more suitable for internal assessment. Less complex systems such as Goodson's 'Read a Plant Fast' [7], and the AME Lean Sensei Self Assessment tool have the advantage of being quick assessments covering a range of aspects but are subject to the 'tick box' temptation. All measurement systems are subject to 'gaming'—in the case of inventory turn trends deliberate running up or running down to massage the financials'.

Of course, any lean (or operations based) organisation in a changing environment needs continually to receive new energy, or it will decline following entropy. Likewise, the competence of managers and consultants, including their focus, plays an important role. Other aspects include appropriate and inappropriate practices, complexity, expectations, organization, commitment, and measures.

3 Pathways

When an organization sees fit to make lead time reduction a primary objective, a manager may have a choice between a 'people prioritized' route, a 'methodology prioritized' route, and a balanced route. The differences are illustrated in Fig. 5. (We prefer methodologies as a more descriptive and specific term than what are sometimes labelled as 'tools'): The people route includes a preference for 'culture change', empowerment, engagement, daily team meetings, structured walkarounds, and an attitude that 'the tools are the easy stuff'. This could be seen as a 'bottom up', incremental, approach, led by managers and with strong HR emphasis. The methodologies route includes a preference for classic lean 'engineering': kanban, changeover reduction, cell layout, ergonomics, process control, and product design.

Based on Schonberger's recent (2018) text—looking at 101 mini case studies, and Bicheno's experience of spending a week at over 65 organizations, and a day at many others—some tentative conclusions—and speculations—can be made:

In our experiences (and in published material found on the web) there would seem to have been a steadily growing trend towards the people route and backing away from early lean or JIT/engineering-oriented methodologies. (Here, recall Ohno's statement that TPS is 'profit-based industrial engineering'). Yet, there are dangers in taking either the people or the methodologies routes. 'Going for methodologies' whilst downplaying 'people' carries sustainability risk—what Emiliani [4] describes as 'fake lean'. A balance is required, as found decades ago by Trist [23] who, in 'socio technical systems' warned against a myopic view of technology, and from Deming [3], who said 'A bad system will beat a good person every time', thus warning about a myopic view of the role of people.

Amidst the flood of publications that continue to reference our (historical) lean exemplar, Toyota', it is noteworthy how much of that has been written on leadership, culture, engagement, talent, and coaching—which could or should be seen as general management rather than lean concepts; in contrast little in the Toyota writings concerns advanced engineering practices.

Of course, many companies continue to develop impressive engineering innovations, and some may qualify as being under the lean 'umbrella'. One is what Toyota calls 'New Global Architecture', or TNGA, which in essence is modular design - one of the hallmarks of design for manufacture and assembly. In its development of TNGA Toyota had benchmarked Volkswagen's so-called MQB modular platform [10]. Both VW and Toyota may have benefited from studying truck maker Scania's highly effective modular design methods [11, 115–140]. For its part DFMA is a

		'Methodology Prioritised'	
		Low	High
'People Prioritised'	High	'Overweight'	'Fit'
	Low	'Under nourished'	"Skinny'

Fig. 5 Three routes to lean implementation

potent lean methodology in that it greatly reduces numbers and variety of parts, thereby 'to greatly reduce throughput times, flows, and defects' [15, 16]. Toyota has long emphasized 'autonomation' rather than automation—a particularly pertinent view in the light of 'Industry 4.0'—again an aspect that has received relatively little attention.

By downplaying methodologies in the hope that inventory and lead time improvement will eventually work through carries the risk of top management impatience, abandonment and searches for a 'new' approach—perhaps agile (or 'leagile'?), six sigma (or 'lean six sigma';), demand driven MRP, quick response manufacturing [22], or 'Industry 4.0'. Our hypothesis is that the dismal inventory turn performance in many industries is as a result of too strong an emphasis on either of these routes.

4 Impacts

Further understanding of lean methodologies has to do with their impacts—kind and extent—on inventory For instance, kanban implemented between stores and work centers and from work center to work center has localized impact—on the work-in-process component of inventory. On the other hand, kanban from suppliers to assemblers and beyond toward final customers has breadth of impact, including inventories of purchased materials and on finished-goods inventories. Figure 6 shows these differences, plus the still greater dimension of endurance over the long term.

Localized, internal	Internal kanban Cells Quick changeover Fail-safing Cross-training/job rotation 5S TPM Total quality.
Encompassing external value chain	External kanban Product/customer focused productive units Concurrent production Right-sized equipment Supplier development Supplier-managed inventory Cross-docking Continuous replenishment
Enduring	Re-focus lean on flexibly quick customer response Tighten up lean concepts/terminology (e.g., toward the term, flow) Cultural implantation

Fig. 6 Impacts of various lean practices or methodologies on lean performance: a typology

Caveat: Here we distinguish between lean 'analysis' (i.e., second-order impact) tools and 'action' or 'intervention' oriented lean practices (i.e. first-order). The former includes tools such as value stream mapping, A3, lean accounting, and waste classification. The latter includes the 'classic' lean methods, such as cell design, setup reduction, kanban, fail-safing, and TPM. Though there has been an apparent trend in growth of the analysis tools, the action methodologies yield actual changes in operational processes. For these reasons of impact on inventory, the analysis tools are omitted from Fig. 6.

The lean methods having strong but localized, internal impacts include lean's stalwarts. Cells, for one, generally have maximal impact on internal inventory in that they can greatly shrink space among work centres and between work stations, space otherwise holding WIP inventory. Another, 5S, shrinks spaces where inventories tend to accumulate, whilst also enabling quick changeover. TPM and total quality activities reduce needs for inventory as buffers against equipment malfunctions and quality issues.

The second set of lean practices in the typology has broadening impacts that encompass inventories and lead times in external value chains to final customers. These include kanban linkages with suppliers and with downstream value-chain entities. Concurrent production with multiples of smaller-scale (non-monument) equipment allows multiple products to be produced at the same time, rather than serially in large lots [20]. Supplier development, supplier-managed inventory, cross-docking and continuous replenishment are lean practices that reduce inventories in supply chains.

For lean not to fade (as is common) calls for interventions with still greater breadth. Rather than the heads-down approaches of maximizing efficiency, enduring lean adopts heads-up emphasis on flexibly quick response; lean, so presented, taught, and measured—in customer terms—gains stature in marketing and throughout the executive suite. Add to that an effort to encapsulate lean, minus many of lean's add-on tools and terminology—analogous to a single, well-tested pill rather than a collection of over-the-counter or home-spun remedies. Thirdly, enduring lean needs the implantation of a culture of continuous experimentation in process improvement and knowledge transfer, plus deliberate measures for recording and re-hashing what goes well and what doesn't; without this, companies will tend, at each juncture of new management or dramatic competitive change, to forget the past and go on to 'reinvent the wheel.'

5 Conclusions

Although there are notable exceptions, inventory turn performance in many lean and lean-aspiring organisations have worsened, often throughout most of the 2000s. This statement, based on hard data from companies' audited financial statements, appears not to be widely recognized. More particularly, though many researchers have been claiming that lean has been fairing poorly—often using the definitive term 'failures'—few understand that worsening trends in inventory turns underpin, to a large extent, this disappointing state of lean. Without that awareness, among both practitioners and academics, lean will eventually lose its credibility.

A tentative hypothesis has been offered in this paper as to reasons for poor inventory turns and related declines in lean performance. It is that in recent years there has been an imbalance between 'people prioritized' routes to lean and 'methodology-prioritized' routes. The early days of JIT production gave emphasis to 'engineering-oriented' flow methodologies. Over time, under lean terminology, the emphasis appears to have shifted a good deal toward 'people' and less on 'flow'. Has the swing gone too far?

Also introduced in this paper is a typology of lean practices or methods and their impacts on lean performance. Certain of the methods (e.g., cells, 5S, and quick changeover) yield high impacts of a localized and internal lean nature, including shortened lead times and low work-in-process inventories. A second set of methodologies (e.g., right-sized equipment, external kanban, and continuous replenishment) generate results beyond operations: in encompassing external value chains, they shortened customer lead times, whilst shrinking raw-material and finished-goods inventories. The third level in the typology aims at 'enduring' lean—gained through a strategic refocusing of lean on flexibly quick customer response (as opposed localized efficiencies); and accompanies by a discarding of increasingly obtuse and confusing lean terminologies, and toward the simplicity of the purity of the term, flow. So, Quo Vadis: 'Where are we marching?'.

References

1. Belekoukias I, Garza-Reyes JA, Kumar V (2014) The impact of lean methods and tools on the operational performance of manufacturing organisations. Int J Prod Res 52(18):5346–5366
2. Bicheno J (2018) Towards reducing queues: Muri, Mura, Muda. In: Dinis-Carvalho J, Alves AC, Costa N, Lima RM, Sousa RM (2018) Lean educator's role in lean development. In: Proceedings of the fifth European lean educator conference (ELEC2018), Braga, Portugal, pp 141–150
3. Deming WE (1993) Attributed to Deming at a seminar in Phoenix Arizona. https://blog.deming.org/?s=beat+a+good+person
4. Emiliani ML (2008) Real lean: the keys to sustaining lean management, vol 3. Center for Lean Business Management, Wethersfield, CT
5. Eroglu C, Hofer C (2011) Lean, leaner, too lean? The inventory-performance link revisited. J Oper Manage 29(4):356–369. https://doi.org/10.1016/j.jom.2010.05.002
6. Graban M (2012) Lean cannot be measured by inventory alone. Graban's Lean Blog, Nov 2007. Last updated 1 Mar 2012
7. Goodson E (2002) Read a plant fast. Harv Bus Rev 80(5):105–113
8. Hopp WJ (2018) Positive lean: merging the science of efficiency with the psychology of work. Int J Prod Res 56(1–2):398–413
9. Kahneman D (2011) Thinking, fast and slow. Allen Lane, London
10. Kubota Y (2015) Toyota adopts more modular design. Wall Street Journal (27 March):B3
11. Johnson HT, Bröms A (2000) Design to order (Chap. 4). In: Profit beyond measure. Free Press, New York
12. Little JDC (2011) Little's law as viewed on its 50th anniversary. Oper Res 59(3):536–549

13. Ohno T (2013) Taiichi Ohno's workplace management. McGraw hill, New York
14. Spector R (2010) How lean is pharma? A 10-year progress report; (2018) Lean laggards; exploring the state of lean in pharma, Jan 2018. https://www.pharmamanufacturing.com/articles/2018/109.html
15. Schonbergr RJ (2018) Flow manufacturing—what went right, what went wrong: 101 mini-case studies that reveal lean's successes and failures. Routledge/Productivity Press, New York. https://doi.org/10.1080/09537287.2019.1699971
16. Schonbergr RJ (2018) Disintegration of lean manufacturing and lean management. Bus Horiz 62/3(May–June):359–371
17. Schonberger RJ (2016) Cycles of lean: findings from the leanness studies—part 2. Manage Account Q 18/1(Fall):18–27
18. Schonberger RJ (2016) Cycles of lean: findings from the leanness studies—part 1. Manage Account Q 17/4(Summer):21–33
19. Schonberger RJ (2008) Best practices in lean six sigma process improvement: a deeper look … with telling evidence from the leanness studies. Wiley, New York
20. Schonberger RJ, Brown KA (2017) Missing link in competitive manufacturing research and practice: customer-responsive concurrent production. J Oper Manage 49(51):83–87
21. Shingo Institute (2019) Shingo model handbook. https://lean.nh.gov/documents/Shingo%20Model%20Handbook.pdf
22. Suri R (2010) It's about time: the competitive advantage of quick response manufacturing. Productivity Press, New York
23. Trist EL (1978) On socio-technical systems. In: Pasmore WA, Sherwood JJ (eds) Socio-technical systems. University Associates, La Jolla, CA

JELA: An Alternative Approach to Industrial Engineering and Management student's Lean Management Education

Sara Fernandes, Rute Cunha, Diamantino Torres and Carina Pimentel

Abstract This paper presents an Industrial Engineering and Management (IEM) students' union, JELA, that was born at University of Aveiro (UA), Portugal. JELA is a junior lean company, with the purpose to provide its members with an opportunity to acquire valuable knowledge and competencies during their graduation in the Lean Management (LM) system, in cooperation with the IEM-UA teachers. The objective of this paper is to explore how JELA and such a kind of students' union, fully committed to the development of the LM, can contribute to the IEM student's education and individual learning, and to the integration and development of lean concepts and tools in small and medium companies. To this end, the paper is organized around three specific objectives: (1) to explain how this group emerged and what is its mission; (2) to show how this group contributes to IEM-UA students' independent learning during their graduation; and (3) to describe this group role in the approximation of IEM-UA students with the labour market.

Keywords Lean management system · Industrial engineering and management · Students' unions · Junior lean company

S. Fernandes · R. Cunha
University of Aveiro, DEGEIT, JELA, Campus Universitário de Santiago, 3810-193 Aveiro, Portugal
e-mail: fernandesara@outlook.com

R. Cunha
e-mail: rutecunha13@gmail.com

D. Torres
University of Aveiro, GOVCOPP, DEGEIT, ESAN, Campus Universitário de Santiago, 3810-193 Aveiro, Portugal
e-mail: djtorres@ua.pt

C. Pimentel (✉)
University of Aveiro, GOVCOPP, DEGEIT, Campus Universitário de Santiago, 3810-193 Aveiro, Portugal
e-mail: carina.pimentel@ua.pt

Faculdade de Ciências e Tecnologia, Universidade Nova de Lisboa, UNIDEMI, Quinta da Torre, 2829-516 Caparica, Portugal

M. Rossi et al. (eds.), *Proceedings of the 6th European Lean Educator Conference*, Lecture Notes in Networks and Systems 122,
https://doi.org/10.1007/978-3-030-41429-0_9

1 Introduction

The industrial activities, globally, are facing exponential growth that, in turn, are associated to a significant increase of the competitiveness in different areas and sectors. The Industrial profits positively stimulate the development of infrastructures and advanced manufacturing mechanisms, suited to the desired economic prosperity. Therefore, technology stands out as having a major role in most economies that want to build strong foundations to maintain their competitive capacity against competition. The digital era has transformed static rules into flexible and volatile rules, allowing real-time monitoring of all operations and the involvement of all stakeholders in the business chain, ensuring its sustainability.

According to a study conducted by Deloitte in 2016 [1], which included more than 500 Chief Executive Officers and business managers linked to the industrial production area, it is evidenced the importance of the advanced technologies in the maintenance of the future competitiveness, highlighting the Internet of Things (IoT), smart products and intelligent factories. Therefore, fortify that the concept of competitiveness is indissociable from professional qualification, being mandatory to manage effectively and continuously the binomial education-training to provide human resources with essential skills for insertion and adaptation to the labour market. There is a notable concern on the side of the companies to offer training actions to their workers to enhance their knowledge and performance. On the other hand, the labour market seeks professionals with social and behavioral skills. The development of soft skills is essential for a better professional performance. Candidates with corporate experience, adaptability to different contexts and realities, ease of teamwork, motivated and with proven time management skills are more valued by companies. So, the role of educational institutions should be to create practical and enriching opportunities for students. In addition, students should seek to engage in extracurricular activities that highlight their value to the business world.

Lean management (LM) is an established production system in the organizational context, being part of the culture and business strategies of a huge amount of companies all over the world [2]. Thus, for a young graduate in the Industrial Engineering and Management (IEM) field to have knowledge and experience in LM is an important feature when entering the labour market [3]. Universities play a central role in the preparation of the young IEM graduates, namely by teaching them the LM knowledge and giving them the LM competencies they must acquire to succeed. This may be achieved in many ways, although typically students are mostly exposed to this knowledge and competencies through the inclusion of LM topics in the course curriculum, through company field trips or through learning factories, promoted by the university staff or in cooperation with companies. However, other alternatives may be used to approach students learning and development regarding LM along their University path. In this paper the experience of a students' union created to promote the development and sharing of LM knowledge and competencies through a voluntary association is presented.

The Júnior Empresa Lean de Aveiro or Junior Lean Company of Aveiro—JELA—is a non-profit entity, managed by university students of IEM-UA whose main purpose is to provide services in the area of LM. JELA presents a management philosophy like real companies seeking to integrate its members and collaborators around its own corporate principles, being oriented by statutes and regulations directed to its activity.

This paper is intended to present JELA and to explore its role to the IEM students' education and to small and medium companies from Portugal.

The paper is organized into four sections. In Sect. 2 the basic concepts in Experiential Learning are introduced, as well as Students' Unions contribution to independent learning; Sect. 3 is dedicated to the description of the Junior Lean Company of Aveiro; while in Sect. 4 the main conclusions of the paper are highlighted.

2 Theoretical Background

2.1 Experiential Learning

Almost 2400 years ago, Aristotle wrote "for the things we have to learn before we can do them, we learn by doing them". More recently, William Glasser applying his "choice theory" defended that each person learns 10% reading, 20% listening, 30% watching and 80% experiencing. Reviewing different authors and experts on experiential learning, although all define it in a different way, there is a clear common sense about its definition. Kolb [4] defends that experiential learning is a "continuous process grounded in experience", a synthesis of "experience, perception, cognition and behavior". Merriam et al. [5] says "learning from experience involves adults connecting what they have learned from current experiences to those in the past as well to possible future situations". In the words of Lewis and Williams [6] "in its simplest form, experiential learning means learning from experience or learning by doing. Experiential education first immerses learners in an experience and then encourages reflection about the experience to develop new skills, new attitudes, or new ways of thinking".

Eyler and Halteman [7] present a study where a comparison was made between students learning in classes on legislative politics and students learning in internships at a state legislature. This study found that although both groups did equally well on a traditional test of facts, when challenged to develop a strategy for enacting policy, the internship students presented much better results than the classroom-based students.

Several references enumerate different principles and/or characteristics that should be present in order to define an activity or method as experiential. Some of the most important ones are referred by Chapman et al. [8] and by the Association for Experiential Education [9].

The way how learning happens can vary significantly. Merriam et al. [5] defend that experiential learning can be guided and/or self-directed. That the learner can diverge his/her degree of participation in the experience as it occurs. That it can be presented in group sessions or one by one. That it can happen in different environments that are simulated, drawn from in-field examples, or take place in real working environment.

In addition, the best model of experiential learning to apply should take into consideration whether the goal is advisor training or advisor development.

Based on the study developed by Duslak and McGill [10] some of the most well-known models of experiential learning are:

- Transcription Analysis—approach based on writing the entirety of an advising session providing unique opportunities for self-observation, reflection and cooperative discussion [10].
- Shadowing—refers to observing another advisor perform his/her job tasks. One advantage of shadowing is that is a more passive form of learning allowing the learner to direct their efforts toward focusing on points of the interaction of which he/she wishes to focus [11].
- Case Studies/Role Plays—Case studies afford the opportunity to create scenarios that exemplify a specific problem situation or present the use of a specific approach. Steele [12] suggests a triad model of examining cases in which three individuals recreate a scenario by assuming the roles of advisor, advisee, and observer.
- Clinical Observation/Cognitive Apprenticeships—It is common in other related fields for professionals to undergo clinical observation. Activities "in the field" are observed by a peer or superior for subsequent analysis. With signed consent of the student, advising sessions can be audio or video recorded and then viewed, analyzed and discussed [10].

2.2 Students' Union Contribution to Independent Learning

Students' unions have a long history and tradition in Higher Education. In Portugal, the oldest students' union dates back 1887 and was established at University of Coimbra. Nowadays, in every Higher Education Institution (HEI) several broad and specific students' unions coexist. Despite that fact, the role of students' unions seems to have remained largely unexplored within academic research [13].

The development of professional competencies in undergraduate students is not an easy task as the students do not have a high level of theoretical and practical experience. Thus, independent learning, involving the application of competency-based approach, such as individual approach, subject-subject interaction and activity-based approach is one of the possible ways to solve the problem [14]. There is not an universal definition to independent learning. In this study we follow the definition of Candy [15] that considers it as "a process, a method and a philosophy of education, in which a student acquires knowledge by his or her own efforts and develops the ability for inquiry and critical evaluation; it includes freedom of choice in determining those objectives, within the limits of a given project or program and with the aid of a faculty adviser; it requires freedom of process to carry out the objectives; it places increased educational responsibility on the student for the achieving of objectives and for the value of the goals".

In a recent study Hockings et al. [16] conclude that "students feel that teachers and course teams could play a more proactive role in creating opportunities for peer support for independent learning" and offer that "such support could potentially be offered in collaboration with staff and officers of the students' union as trainers and coordinators of student-led academic support and development". In the same line, Thomas et al. [14] stress that students' unions play an important role to helping fellow students take the required steps towards effective independent learning in their university lives.

3 JELA—Junior Lean Company of Aveiro

3.1 Creation and Evolution of JELA

At the end of December 2015, the students of the degree course in Industrial Engineering and Management of University of Aveiro came up with the need to create something that deepened their practical knowledge about Lean Management. Being that, this was filled with the creation of JELA, Júnior Empresa Lean de Aveiro.

Before JELA, students had a Lean Department that was part of the IEM students' union, also from UA. This department was responsible for providing knowledge and experience in the area of Lean, whether for visits to companies in the area, or for events organized as lectures and training. At that time, lean was an issue that was in vogue, expanding, so that the entire business and industrial community knew or wanted to know this added value. So, the students wanted to create something that differentiated them from the other students of other universities. With the help of the teachers and with the students' wishes, JELA was created.

In its first year of existence, the focus was to understand what would be the best strategy of action coming from JELA, that is, to create structures, departments, select the target market, create communication channels, image and mainly to perceive the correct approach to reach clients. In addition, it was already in this first year that JELA began to feel the difficulties, such as recruiting members, motivating them and retaining them. Since JELA belongs to the IEM students' union of Aveiro, it can only recruit IEM students from UA, which limits the polyvalence of the team.

Since its creation, JELA began to develop its own path, being at this moment, in its 4th year of existence. Therefore, since all its main objectives are already established, nowadays JELA is making efforts towards obtaining projects to achieve the knowledge in practice and the satisfaction of its members.

The participation of the students in this entity symbolizes an added value. Besides putting in practice all the knowledge obtained in the classes, JELA increases their skills to work in teams, organization, time management, and communication skills, among others. All these characteristics make students into more dynamic people who subsequently excel in the world of work.

3.2 JELA Job

JELA, as Junior Company, provides services in the field of Lean Management. Its members solve case studies/optimization problems that its clients present, being these small and medium companies from Portugal and mostly from the Aveiro district. JELA is typically involved in improvement projects that require the use of several lean tools, such as time measurement, 5S, SMED, Visual Management, Value Stream Mapping and Standard Work. In addition, JELA is also careful to accompany its clients and to help the development of their human resources by providing training on concepts that are necessary for the proper functioning of the companies in which they are involved. In Table 1 some examples of the projects developed by JELA are briefly presented. Besides JELA involvement in practical projects, its team is also committed to research. As an example, a research is being developed in partnership with a group of Brazilian students from a HEI, with the support of IEM teachers from the University of Aveiro and from the Brazilian HEI. The project is entitled "Evaluation of the Degree of Adherence to Lean Manufacturing Principles in Small and Medium Enterprises: A Comparative Study between the Regions of Aveiro and Sorocaba". So far, the team main tasks were a literature review, field research and survey development and application.

Regarding the structure of JELA, currently, it has a functional structure being divided into Presidency, Operations Management Department, Commercial Department and Human Resources Department. The Presidency has a president, a vice president and a vice president for finance. All departments have a director who has

Table 1 JELA practical projects—some examples

Company sector	Project description
Pipe industry	In this project JELA's members mostly applied Visual Management tool. Firstly, some problems were identified, such as clutter and lack of security. To contour them, the team developed a color coding to identify different workspaces and associate them with specific activities in the entire shop floor, with well-marked traffic aisles as well as storage areas. In addition, several modifications were suggested to the forklift trucks to improve workers' safety.
Winery	This project encompassed the implementation of 5S. The aim was to organize a storage section of capsules used in bottling. Firstly, the team sought to organize and classify the inventory. Through this process capsules that would no longer be used were identified and removed from the site. Subsequently, a new layout was implemented, organized according to the product type and rotation. Additionally, a methodology for periodic inventory updating was created. There was also a concern to train employees, to familiarize them with the Lean philosophy.
Paper industry	This project, entitled "Commitment to Land at the Top Management Level" was mostly devoted to the application of Leader's Standard Work and Gemba Walk with Kamishibai.

the responsibility of coordinating their team. The Operations Management Department has nine members, and the Human Resources Department has three members, such as the Commercial Department.

The Presidency is responsible for all strategic planning, team coordination, how to ensure a healthy situation for the organization. The Operations Management Department is responsible for meeting the needs of its clients, ensuring all projects. The Commercial Department has two strands, the commercial side where it is responsible for obtaining clients, sponsors, partners and the marketing side, where members sell the image of JELA managing all their online platforms, as all necessary physical marketing. Finally, the main objective of the Human Resources Department is to obtain members and retain them, always guaranteeing their satisfaction. In addition, JELA also act in the training of its members, having the responsibility to ascertain the gaps that exist in their knowledge so that, in this way, they can surpass them through training actions (typically lectured by professionals in the lean management field), visits to companies and conviviality. The department of Human Resources still has as main task the organization of the Lean Conference. This event serves students and professionals and provides two days of experience in the Lean topic, which includes lectures given by multinational professionals, gemba walks and workshops where each participant puts his/her knowledge to test. Usually the Lean Conference has between one hundred and two hundred participants and is held at UA. Also, JELA belongs to the Association of Industrial Engineering and Management of Aveiro (AEGIA), where all its members can get to know and experience other organizational realities. In addition to JELA and AEGIA there are also ESTIEM Local Group Aveiro, and in tune, these three entities function as one where their main attention is to help students from UA to enrich their path during their Integrated master's degree in industrial engineering and Management.

3.3 JELA Contribution to IEM-UA Students' Education During Their Graduation

Lean Management is strongly established in different organizational contexts, being part of the culture of several companies around the world. JELA has contributed over the last few years to the development of knowledge and competencies, essential for a young graduate in IEM, to be successful in the labour market. The enrolment into a Junior Company is, undoubtedly, fundamental for personal and professional growth. In a way, the involvement of students in JELA contributes to overcoming one of the biggest barriers to entry into the labour market, the lack of "professional" experience.

During the academic degree, the students are exposed to varied experiences that allow them to acquire skills essential for their insertion in the labour market. However, JELA emerges as an extra opportunity for development and growth, allowing students to establish a relationship between theory and practice in an organizational context, enabling them to improve working mechanisms and strategic thinking. Among the range of activities carried out by JELA, the support of the IEM-UA teachers proves to be fundamental to sustain the students' work and help them achieving positive results in the projects, allowing them to have a more technical and experienced view on market trends.

In JELA the IEM students can contact the companies and establish relations of mutual benefits with the organizations for which they provide consulting services. On the one hand, allowing micro, small and medium enterprises in the region easy access to a set of services and application of LM and, on the other hand, enabling its members to be closer to business reality and to build an extremely valuable network of contacts.

In addition, integration into JELA allows the development of professionals with a high sense of commitment and proactivity. It also allows the contact with different contexts and, consequently, helps in identifying the most appropriate areas for each profile. Regarding personal development, members have opportunities to develop their brainstorming capabilities, decision-making, leadership, negotiation skills, among others.

In short, JELA is a mechanism for personal and professional development, guarantees an intensification of student's relationships with the companies during their graduation and guarantees to its members a positive differential of experience and maturity. The possibility to actively be in contact with business reality and the world of business provides a factual insight into real difficulties and the stimulation of the problem-solving capabilities, among others, which results in a better consolidation of the concepts and theoretical knowledge acquired at the university.

4 Conclusion

Recent literature highlights that independent learning is poorly understood by university students and that students' unions, such the one presented in this paper, can have an impact in its development. In this paper it is showed how individual learning about lean management system can be fostered through IEM students' experiential learning acquired through clinical observation during projects developed in companies.

Acknowledgements This work was supported by the GOVCOPP research unit (UID/CPO/04058/2019), funded by national funds through FCT.

References

1. Center for Industry Insights (2016) Global manufacturing competitiveness index. Technical report
2. Sony M (2018) Industry 4.0 and lean management: a proposed integration model and research propositions. Prod Manuf Res 6(1):416–432
3. Rosenkrantz PR (2011) Transformational leadership 101: what all industrial engineering graduates should know about the six stages of quality management system implementation. In: Proceedings of the 2011 ASEE annual conference & exposition, pp 22.1544.1–22.1544.16
4. Kolb D (1984) Experiential learning: experience as the source of learning and development. Prentice-Hall, Englewood Cliffs, NJ
5. Merriam SB, Caffarella RS, Baumgartner LM (2007) Learning in adulthood: a comprehensive guide, 3rd edn. Jossey-Bass, San Francisco, CA
6. Lewis LH, Williams CJ (1994) Experiential learning: a new approach. In: Jackson L, Caffarella RS (eds) Jossey-Bass, San Francisco, pp 5–16
7. Eyler J, Halteman B (1981) The impact of a legislative internship on students' political skill and sophistication. Teach Polit Sci 9(1):27–34
8. Chapman S, McPhee P, Proudman B (1995) What is experiential education? In: Warren K (ed) The theory of experiential education. Kendall Hunt Publishing Company, Dubuque, pp 235–248
9. Association for Experiential Education. https://www.aee.org/. Visited on 23 July 2019
10. Duslak MP, McGill CM (2014) Stepping out of the workshop: the case for experiential learning in advisor training and development
11. Duane Hoover J, Giambatista R, Belkin L (2012) Eyes on, hands on: vicarious observational learning as an enhancement of direct experience. Acad Manage Learn Educ 11(4):591–608
12. Steele G (2003) A research-based approach to working with undecided students: a case study illustration. NACADA J 23(1–2):10–20
13. Brooks R, Byford K, Sela K (2015) The changing role of students' unions within contemporary higher education. J Educ Policy 30(2):165–181
14. Thomas L, Hockings C, Ottaway J, Jones R (2015) Independent learning: student perspectives and experiences
15. Candy P (1991) Self-direction for lifelong learning: a comprehensive guide to theory and practice. Jossey-Bass Publishers, San Francisco, CA
16. Hockings C, Thomas L, Ottaway J, Jones R (2018) Independent learning—what we do when you're not there. Teach High Educ 23(2):145–161

Literature Mapping of the Use of Games for Learning Correlating with Lean: A Systematic Review

Andréa de Freitas Avelar and Michele Tereza Marques Carvalho

Abstract Organizations of all kinds have faced new challenges in which achieving the shortest possible cycle time, boost productivity, cost reduction and increase quality are crucial to survive globalization. To this end, it is necessary to transform production processes from the qualification of employees to the Lean philosophy. However, traditional training demands high cost, huge allocation of resources and have not enhanced the learning processes efficiently. In this sense, ICT has emerged to fill the gap of the inefficiency of conventional training. For this reason, this work sought to identify and categorize the different initiatives of Lean training adopted in the companies and their benefits based on Serious Games as a very promising tool. Hence, the major contribution of this work is to bring to the light of the academic and professional universe the need for the use of technological and digital alternatives for the Lean qualification of people through games as learning resources.

Keywords Lean · Serious Games · Games-based learning

1 Introduction

Over the past decades, virtually, all kinds of organizations have been using Lean as a change agent of managerial realities in order to optimize results and make better use of human resources. The interest has been growing as enterprises, engaged in a great variety of sectors, have faced new challenges in which achieving the shortest possible cycle time is imperative to survive globalization. However, all Lean initiatives need to be grounded on clearly defined purposes to deliver value to the customer, enabling efficiency and flexibility. It is no surprise that these competitive

A. de Freitas Avelar (✉) · M. T. Marques Carvalho (✉)
Department of Civil and Environment Engineering, Universidade de Brasília, Campus Darcy Ribeiro, Brasilia 70830-303, Brazil
e-mail: andreadefreitasavelar@gmail.com

M. T. Marques Carvalho
e-mail: micheletereza@gmail.com

M. Rossi et al. (eds.), *Proceedings of the 6th European Lean Educator Conference*, Lecture Notes in Networks and Systems 122,
https://doi.org/10.1007/978-3-030-41429-0_10

conditions have required changes in processes and in workplace organization, clearly evident in companies seeking greater competitiveness [1].

Lean is, therefore, a body of knowledge whose essence is the ability to continuously eliminate waste and solve problems systematically. This implies reshaping the way companies lead, manage, and empower employees through their full engagement in the workplace what can envision opportunities for improvement and sustainable earnings as stated by Gallardo et al. [2] who highlight the "importance of teamwork" in Lean process, as well as Tiagy et al. [3] when referring that Lean thinking must be part of the organizational culture or just "tools may not be sufficient to reach the expected results."

Nevertheless, Lean initiatives need to broaden its focus on production and logistics drawing special attention to the implementation of an integrated training and a qualification system, which should be understandable, interesting, effective and at the same time, low cost and show greater flexibility, since they demand high costs, huge allocation of resources and have not enhanced the learning processes efficiently, and consequently, they have hindered substantial changes in the production processes.

Thus, it can be said that the successful usage of Lean permeates education and production processes in which academia and companies share a common problem: how can they ensure efficient qualifying people for Lean production?

Parallel to this scenario, it is important to take into account that the worker's profile has been changing globally, especially the generation Y (born between 1980 and 1997), whose characteristics are looking for independence, transparency, instant gratification and social connection and which are likely to represent 75% of the global workforce by 2030 [4].

In this context, the need for collaboration between academia and industry become evident, where the core of the problem is to produce knowledge through research, to disseminate knowledge through education as well as to use and apply knowledge through innovation, configuring the "knowledge triangle" [5]. Universities and industrial training facilities are confronted with the challenge of identifying future job profiles and correlated competence requirements, and need to adapt and improve their concepts and educational methods, notably innovative learning environments, which should be able to respond the challenges mentioned in an interdisciplinary manner.

Therefore, new techniques and experiences have been developed and shared continuously, allowing learning to be faster and more effective.

In this sense, information and communication technologies emerge to fill the gap of the inefficiency of conventional training, mainly those which ensure a continuous and real learning experience such as "Serious Games". Their statistics have shown that 78% are more effective for reinforcing learning, 14% have a considerable potential to develop behavioral skills, in addition, they reduce training costs by 62% compared to conventional learning methodologies [6], providing leverage to intensify learning.

For this reason, the present article sought to identify and categorize the different initiatives of Lean training adopted in companies and their benefits. Hence, the major contribution of this work is to demonstrate the importance, efficiency and necessity of the use of technological and digital alternatives for the Lean qualification of people through games as learning resources to the academic and professional universe.

2 Theoretical Framework

Breaking this educational barrier within the community of Lean educators, special attention is being devoted to the innovation in the training of people within the industry and to the professional training of engineering students using simulated production environments because there is a consensus that education for Lean must contain key elements of hands-on experience, that is, it is in the application of the Lean principles studied in academia, with its tools and methods, that one has the consolidation of knowledge [7].

Since traditional teaching methods show limited effectiveness in developing employee and student competencies and skills for current and future manufacturing environments [8].

It should be noted that manufacturing itself has advanced rapidly in the development of new technologies, tools and techniques related to production, thus, workers and engineers will need new lifelong learning schemes to accompany these advances, with industrial education as one of the main drivers for the construction of new generations of "knowledge employees" in production activity [9].

Therefore, the educational paradigm of manufacturing production, as well as Lean production, needs to be reviewed, in which modern training concepts, industrial learning and knowledge transfer strategies that can contribute to improve production performance are required [9–11], ensuring the acquisition of skills and competences by stakeholders. These concepts suggest that education for production and for Lean production cannot be treated efficiently only in the classroom [11–13], and new approaches to learning are necessary, especially for: allowing training in realistic manufacturing environments; modernizing the learning process and bringing it closer to industrial practice; levering age industrial practice through the adoption of new manufacturing knowledge and technology; and boosting innovation in the manufacturing by improving the capabilities of workers to solving problem, creativity and systems thinking capability [14].

In recent years, learning for research and education in environments close to industrial reality has become an innovative and effective alternative to meet these challenges and the "teaching and learning factories", through "Serious Games" and their variations and multiple compositions, such as simulations and gamification, are training alternatives for Lean production.

2.1 Learning, Knowledge, Skill and Competence

"Learning involves acquiring and modifying knowledge, skills, strategies, beliefs, attitudes and behaviours. People learn cognitive, linguistic, motor, and social skills, and these can take many forms" [15]. Due to learning, the changes implemented in the organisms can be relatively permanent [16]. According to Kirkpatrick, learning is "[…] acquired knowledge, improved skills or attitudes changed due to training" [17].

When it comes to "learning", it is inherent the conception of "knowledge", which is the "outcome of the assimilation of information, […], the body of facts, principles, theories and practices that is related to a field of work or study" [18].

And the ability to "apply knowledge and use know-how to complete tasks and solve problems", is called "skills", which can be classified as "cognitive (involving the use of logical, intuitive and creative thinking or practical involving manual dexterity and the use of methods, materials, tools and instruments)" [18].

Bloom et al. [19] consider that "ability" is linked to the "individual who can find information and appropriate techniques from his previous experience to deal with new problems and situations". This requires an understanding of the new situation; knowledge that can be easily used; and also the easy to discern the appropriate relationships between the past experience and the new situation.

Erpenbeck and Rosenstiel [20] consider "competence" to be the use of knowledge and skills, in a given context, to perform creative and self-organized actions in open and complex situations, in which competencies can be divided into: social-communicative (communication, cooperation, conflict management and leadership skills); professional and method (application and acquisition of knowledge, as well as analytical, methodical and problem-solving skills); personal (reflectivity, adaptability and unfolding of motivation); and activity and action (skills to work independently, implement plans and perseverance) [20].

2.1.1 Types of Work-Related Learning

According Dehnbostel [21], there are three types of work-related learning: (i) the work-connected learning, (ii) the work-integrated learning (iii) the work-oriented learning. Table 1 [22] shows different learning approaches related to existing and innovative work.

2.1.2 Experiential Learning and Active Learning

Experiential learning "is the process by which knowledge is created through the transformation of experience" [23]. It is a process in which knowledge is acquired through the following stages: (a) a concrete experience, (b) observation and reflection, (c) abstract conceptualization and (d) active experimentation [23]. Learning methods

Table 1 Approaches to skills development related to work [22]

	Work-based learning	Work-connected learning	Work-oriented learning
Established approaches	• Training workshops • Training centers • Practical learning companies • Learning factory • Self-learning programs	• Quality circles • Circle seminar • Learning station	• Training station • Guiding text method • Instruction • Informal learning by doing in a real work process

such as project-based learning, simulations and case studies are beneficial for the development of cognitive skills especially when combined with other methods of active learning [24].

Active learning focuses on involving students in the analysis, application, manipulation and evaluation of ideas. Thus, the understanding of concepts is promoted, as opposed to the mere reproduction of information [24]. It "involves students in doing things and thinking about the things they are doing," as defined by Bonwell and Eison [25] and is stimulated by methods such as image-based instruction, writing, cooperative learning, discussion, role-playing, games and peer teaching.

Action-oriented learning is a subtype of active learning, focusing on the active integration of the learner through own actions and is intended to improve conceptual knowledge, allowing the understanding of cause and effect relationships as a prerequisite for problem solving.

The appropriate learning environment is characterized by a high degree of fidelity to the real work context, for example, simulations, role plays, virtual reality and learning factories [8].

2.2 *Learning Factories*

Established in 1994, the concept of learning factory is constituted from an educational philosophy oriented to the action of teaching and learning. It combines traditional theoretical sessions with practical and informal learning experiences. Thus, it aims to provide a greater anchorage of transferred knowledge, optimizing the learning effect and the ability to apply this knowledge in real cases [26, 27].

In the last decade, learning factories have been implemented more and more intensively in Europe [28–30] and, also, have been constituted of many different forms of facilities that vary in size, scope, function and sophistication, in order to enhance the learning experience in various areas of application [28]. The complete value chain of real products, from raw materials throughout machining and assembly to shipping, are mapped, whether in physical or virtual reality. Over the past few years, several other learning factories have been established with other focuses and various physical manifestations.

Thus, among several types of existing learning factories and classified by their central theme, it is worth highlighting three, for pertinence to the theme of this research. They are:

(1) Learning factories for production process improvement, which deal with Lean methods and principles, such as value chain analysis and design, just-in-time, line balancing, problem solving or task optimization.
(2) Learning factories for industry 4.0, designed to foster understanding and education for Industry 4.0 which requires complex interdisciplinary connections and the ability to adapt and design all facets of cyber-physical production systems, among others, to manufacturing, information technology and communication sciences such as [31–36].
(3) Digital and virtual learning factories, in which due to the trend of digitalisation of production facilities and processes, have gained increasing attention in the context of production education [37–40] because they constitute a digital and virtual environment to provide added value for the education of future production, being used for training in similar research fields of conventional physical learning factories, mapping all the processes, products and resources of a real learning factory into a digital model. It is an integrated computer and information technology (IT) environment, as well as data models that represent the link between various IT tools.

Virtual environments are used with the motivation to increase the quality of teaching [41] and are considered an important strategic means to enable education in the manufacturing domain [42]. In the literature, various training approaches using virtual production environments can be identified [42–51].

2.3 Serious Games

Games-based learning are structured ways of playing, designed with the purpose of facilitating learning, the acquisition of knowledge, skills, values, beliefs or habits. They can help the user to understand certain issues or concepts, or support the user to learn or improve skills, involving the learner's own actions [52].

Approaches considering elements of the game with a serious purpose can be classified into Gamification and Serious Games [53].

Serious Games are defined as games with an "explicit and carefully thought out educational purpose" [54]. They "are not intended to be played primarily for fun. This does not mean that serious games are not or should not be fun". Gamification is the use of game design elements in contexts that are not games. Serious games and gamification are also used to increase student motivation.

They consist of a simulated environment, where the player is the main player and it is intended that he performs tasks, processes and draws conclusions, in order to increase his productivity and knowledge.

In this context, still, it has that the information and sensations experienced, through this type of games, allow the player to redefine their perception, concentration and attention in order to facilitate changes in the behavior of practical learning.

Serious games are declared to have the following performance indicators: rule and objective, sensory stimuli, control, challenge and interactivity [54].

3 Research Method

The research project adopted for this study is a "systematic" literature review because it is a scientific investigation through a rigorous and explicit process to identify, select, collect data, analyze and describe the relevant contributions to the research. It is a review made with planning and gathering of original studies, synthesizing the results of multiple primary investigations through strategies that limit biases and random errors [55]. The approach of the review is based on the Systematic Search Flow (SSF) method, developed by Ferenhof and Fernandes [56] composed of four phases and eight activities, being structured as illustrated in Fig. 1.

Based on this understanding, we proceeded to develop the work as described below:

Phase I—Definition of the research protocol: the search strategy was defined through the keywords "Lean Thinking" AND "Serious Games" AND "Games-based learning". On April 30, 2019, the search for the publications of the last 5 years began, on the basis of the ScienceDirect, with 102 results found. Then, for the organization of the bibliographic portfolio, it was observed the segmentation of these findings into "review articles", "research articles", "encyclopedia", "book chapters", "conference abstracts", "news" and "other". Then, the readings of interest were identified,

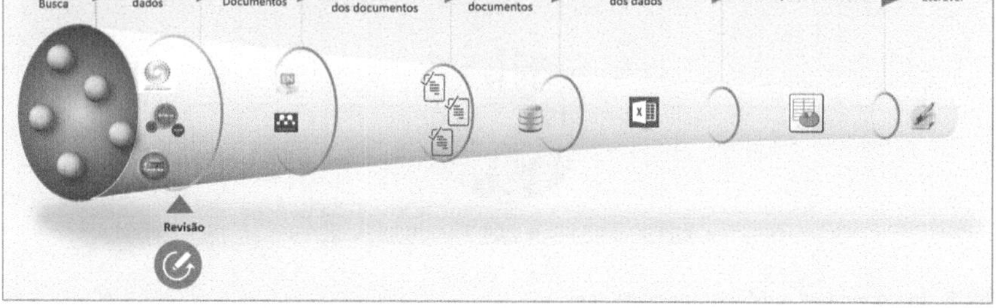

Fig. 1 Systematic search flow [56]

from which 55 papers were selected classified by the Science Direct platform as "research articles". From this selection, the standardization of the articles was initiated through the readings of the titles, abstracts, objectives and keywords. After this filtering, we proceeded to develop the portfolio of articles, by reading them in full, aiming to eliminate those who did not have adherence to the theme under investigation. After performing the aforementioned activities, the composition of the bibliographic portfolio resulted in 21 articles.
Phase 2—Analysis: a spreadsheet was developed containing the following fields as headings for the columns: authors, year, title, publication vehicle, qualification of publications, keywords, countries involved, type of research, type of application (Learning Factories, Serious Games, Simulations, others). The results of this analysis are presented in item "4. Results and discussions".
Thus, Phase 3 (Synthesis) was initiated through the construction of the Knowledge Matrix [56] seeking to extract and organize the data arising from the analysis of the articles. Aiming at consolidating the results obtained, Phase 4 (Write) was prepared, resulting in this research.

4 Results and Discussions

This section describes the analysis of the results of the literature review in order to establish relationships and significance between the search terms, systematically identifying a set of relevant factors to the research and introducing the discussion of these results.

Figure 2 shows the incidence of publications per year starting from 2015 and in Fig. 3 shows the segmentation of articles by type, according to the Science Direct database.

It is noted that in 2017, there were the largest number of publications and the type "research articles" constituted the vast majority of publications.

The categorization and quantification of the 21 identified training practices are shown in Fig. 4, with 33.33% representing "Learning Factories", 47.62% the practice through "Serious Games", 4.76% reflecting "Gamification" and 14.28% representing the application of training through "Simulation".

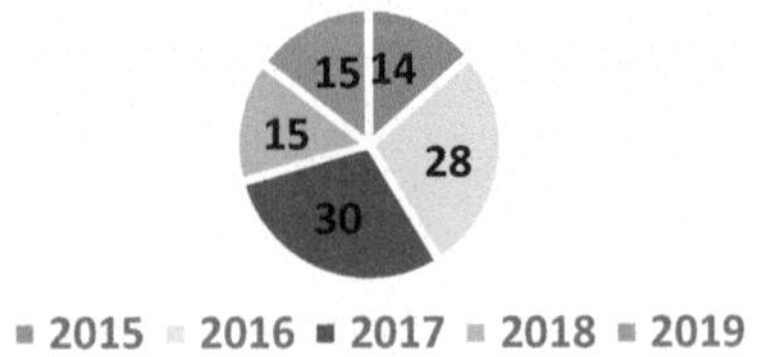

Fig. 2 Incidence of publications per year

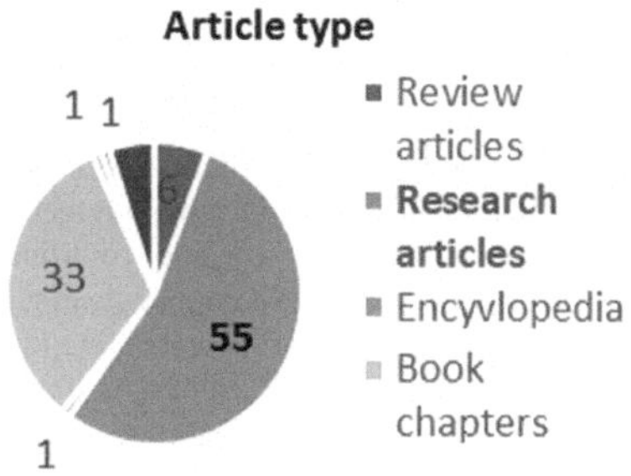

Fig. 3 Article type (the authors)

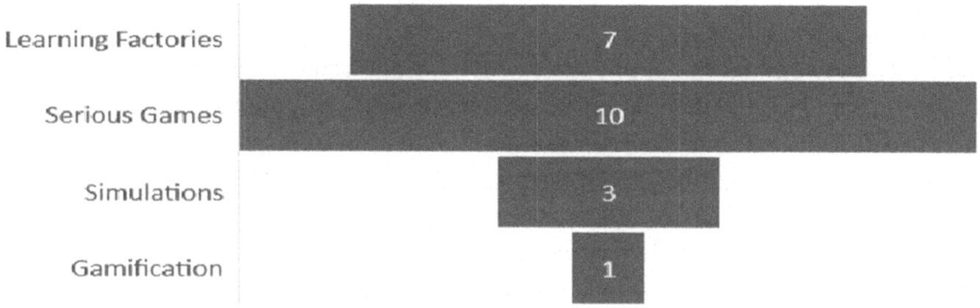

Fig. 4 Training practices (the authors)

It is important to highlight that of the 47.62% of the practices identified through the "Serious Games", 50% of them were applied with the use of digital information and communication technologies.

In following, Table 2 presents examples of practices studied for "Learning Factories" and "Serious Games".

It should also be added that the "Serious Games", in the studied context, have their positive aspects, but also their limitations, as shown in Table 3.

However, while several studies indicate that games can provide an improve learning experience (strengths), the academia and industry are still working very distant on this, a fact indicated by the limited amount of empiricals studies available and for the limitations that make it difficult to disseminate games as learning resources (Table 3).

Table 2 Examples of "Learning Factories" and "Serious Games" (the authors)

Digital learning factories	Serious Games
• ESB Logistics Learning Factory • Virtual learning factory of McKinsey & Co. • KTH XPRES Lab • Karlstad Lean Factory	• LEGO®-based games • Virtual simulation game at IFA learning factory • Simulation game "Lean Leadership"

Table 3 Strengths and limitations of "Serious Games" (as authors)

Strengths	Limitations
• Experimental learning; high scalability; enable individual/personalized learning ways; make it easy the get of knowledge; work skills such as leadership, group work, planning, strategic management and self-learning; improve productivity and engagement	• It demands resources in advanced information technology; the need to consider coordinating a variety of components in game design and evaluation; understanding of the concept of games as leisure only (prejudice); difficulty on integrating with the academic curriculum; difficulty of acceptance by teachers

5 Conclusion

The study showed that the application of Lean training through "Serious Games", "Learning Factory" and their compositions, mixers and derivations, as they are based on experimental learning, constitutes an alternative to conventional training for a possibly more effective production, especially if there is an integrated range of digital teaching concepts, suggesting the connection of the learning factory method with the concept of digital simulation games, however, the scientific community still produces few studies on the topics covered, consequently, limits the measurement and evaluation of the results that can enhance an assertive proposition.

It is true, however, that he enormous advance in the development of information and communication technologies has greatly contributed to the further development of digital and virtual learning factories as well as of ganıes, making it possible to produce them increasingly realistic, dynamic and interactive, which may justify the incidence of 50% of training practices using the "Serious Games" in digital format.

In this sense, the greatest contribution of the present work, from the bibliographic portfolio analysis, is to show evidences, such as the corporative and educational environment, that the "Serious Games" provide the teaching and learning of the principles of Lean.

However, the initiatives are still incipients and the efforts should be made to potentialize the use of Lean tools through qualification by games as learning resources, where it proposes the promotion of scientific research in the sense of develop sufficient empirical study that can promote a statistical analysis of the results, which may contribute to the reduction of the limitations of the application of "Serious Games".

References

1. Lean Institute Brasil. https://www.lean.org.br. Visited on 06/19/2019
2. Gallardo CAS, Granja AD, Picchi FA (2014) Productivity gains in a line flow precast concrete process after a basic stability effort. J Constr Eng Manage 140(4)
3. Tiagy S, Cai X, Yang K, Chambers T (2015) Lean tools and methods to support efficient knowledge creation. Int J Inf Manage 35:204–214

4. Montes J (2017) Millennial workforce: cracking the code to generation Y in your company [e-book]. Lulu Publishing Services, North Carolina. Available at: Google Books. books.google.com
5. Commission of the European Communities (2005) Mobilising the brainpower of Europe: enabling universities to make their full contribution to the Lisbon strategy. Communication from the Commission Office for Official Publications of the European Communities, Luxembourg
6. Oníria. https://www.oniria.com.br. Visited on 05/08/2019
7. Luttik K (2017) Lean learning/learning lean at Scania. In: European lean educators conference. ELEC, Nijmegen
8. Cachay J, Wennemer J, Abele E, Tenberg R (2012) Study on action-oriented learning with a learning factory approach procedia. Soc Behav Sci 55:1144–1153
9. Jovane F, Westkämper E, Williams DJ (2009) The ManuFuture road: towards competitive and sustainable high-adding-value manufacturing. Springer, Berlin
10. Abele E, Reinhart G (2011) Zukunft der Produktion: Herausforderungen, Forschungsfelder, Chancen. Hanser, Munich
11. Chryssolouris G, Mavrikios D, Mourtzis D (2013) Manufacturing systems—skills & competencies for the future. Procedia CIRP 7:17–24 (46th CIRP conference on manufacturing systems)
12. Jäger A, Mayrhofer W, Kuhlang P, Matyas K, Sihn W (2012) The "Learning Factory": an immersive learning environment for comprehensive and lasting education in industrial engineering. In: 16th World multi-conference on systemics, cybernetics and informatics, vol 16 no 2, pp 237–242
13. Rentzos L, Doukas M, Mavrikios D, Mourtzis D, Chryssolouris G (2014) Integrating manufacturing education with industrial practice using teaching factory paradigm: a construction equipment application. Procedia CIRP 17:189–194 (47th CIRP conference on manufacturing systems)
14. Abele E, Metternich J, Tisch M (2008) Learning factories for research, education, and training. J Eng Educ 97(1):5–11
15. Schunk DH (1996) Learning theories: an educational perspective, 2nd edn. Prentice Hall, Englewood Cliffs, New Jersey
16. Schacter DL, Gilbert DT, Wegner DM (2009) Psychology. Worth Publishers, New York
17. Kirkpatrick D (1996) Great ideas revisited: revisiting Kirkpatrick's four level model. Train Dev 54–59
18. European Commission Implementing the Community Lisbon Programme (2006) Proposal for a recommendation of the European Parliament and of the council on the establishment of the European qualifications framework for lifelong learning. European Commission, Brussels
19. Bloom BS, Engelhart MD, Furst EJ, Hill WH, Krathwohl DR (1956) Taxonomy of educational objectives: the classification of educational goals. McKay, New York
20. Erpenbeck J, Rosenstiel LV (2007) Handbuch Kompetenzmessung: Erkennen, verstehen und bewerten von Kompetenzen in der betrieblichen, pädagogischen und psychologischen Praxis, 2nd edn. Schäffer-Poeschel, Stuttgart
21. Dehnbostel P (2007) Lernen Im Prozess der Arbeit. Waxmann, Münster, Nova York, Munique, Berlim
22. Adolph S, Tisch M, Metternich J (2014) Challenges and approaches to competency development for future production. J Int Sci Publ Educ Altern 12:1001–1010
23. Kolb DA (1984) Experiential learning: experience as the source of learning and development. Prentice Hall, Englewood Cliffs, Nova Jersey
24. Crawley E, Malmqvist J, Ostlund S, Brodeur D (2007) Rethinking engineering education: the CDIO approach, 1st edn. Springer, New York
25. Bonwell CC, Eison JA (1991) Active learning: creating excitement in the classroom. School of Education and Human Development, George Washington University, Washington, D.C.

26. Lamancusa JS, Zayas JL, Soyster AL, Morell L, Jorgensen J (2006) The learning factory: industry-partnered active learning—Bernard M. Gordon prize lecture
27. Jorgensen JE, Lamancusa JS, Zayas-Castro JL, Ratner J (1995) The learning factory: curriculum integration of design and manufacturing. In: 4th World conference on engineering education, pp 1–7
28. Abele E, Metternich J, Tisch M, Chryssolouris G, Sihn W, El Maraghy H, Hummel V, Ranz F (2015) Aprendendo Fábricas para Pesquisa, Educação e Treinamento. 5ª Conferência patrocinada pelo CIRP sobre Fábricas de Aprendizagem. Procedia CIRP 32:1–6
29. Micheu H-J, Kleindienst M (2014) Lernfabrik zur praxisorientierten Wissensvermittung: Moderne Ausbildung im Bereich Maschinenbau und Wirtschaftswissenschaften. Zeitschrift für wirtschaftlichen Fabrikbetrieb (ZWF) 109(6):403–407
30. Wagner U, AlGeddawy T, ElMaraghy H, Müller E (2012) The state-of-the-art and prospects of learning factories. Procedia CIRP 3: 109–114 (45th CIRP conference on manufacturing systems)
31. Industrie 4.0 (2016) International benchmark, options for the future and recommendations for manufacturing research. In: Acatech (ed) HNI, Paderborn University; WZL, RWTH Aachen University, Paderborn, Aachen
32. Prinz C, Morlock F, Freith S, Kreggenfeld N, Kreimeier D, Kuhlenkötter B (2016) Learning factory modules for smart factories in Industrie 4.0. Procedia CIRP 54:113–118 (6th CIRP-sponsored conference on learning factories)
33. Thiede S, Juraschek M, Herrmann C (2016) Implementing cyber-physical production systems in learning factories. Procedia CIRP 54:7–12 (6th CIRP-sponsored conference on learning factories)
34. Wank A, Adolph S, Anokhin O, Arndt A, Anderl R, Metternich J (2016) Using a learning factory approach to transfer Industrie 4.0 approaches to small- and medium-sized enterprises. Procedia CIRP 54:89–94 (6th CIRP-sponsored conference on learning factories)
35. Gräßler I, Pöhler A, Pottebaum J (2016) Creation of a learning factory for cyber physical production systems. Procedia CIRP 54:107–112 (6th CIRP-sponsored conference on learning factories)
36. Seitz K-F, Nyhuis P (2015) Cyber-physical production systems combined with logistic models—a learning factory concept for an improved production planning and control. Procedia CIRP 32:92–97 (5th CIRP-sponsored conference on learning factories)
37. Kesavadas T (2013) V-learn-fact: a new approach for teaching manufacturing and design to mechanical engineering students. In: ASME 2013 international mechanical engineering congress and exposition, vol 5, pp 1–6
38. Weidig C, Menck N, Winkes PA, Aurich JC (2014) Virtual learning factory on VR-supported factory planning. Collaborative systems for smart networked environments. In: 15th IFIP WG 5.5 working conference on virtual enterprises, Amsterdam, The Netherlands, pp 455–462
39. Celar S, Turic M, Dragicevic S, Veza I (2016) Digital learning factory at FESB University of Split XXII naučna i biznis konferencija YU INFO, pp 1–6
40. Haghighi A, Shariatzadeh N, Sivard G, Lundholm T, Eriksson Y (2019) Digital learning factories: conceptualization, review and discussion. In: The 6th Swedish production symposium (SPS14). https://conferences.chalmers.se/index.php/SPS/SPS14/paper/viewFile/1729/401. Visited on 06.11.2019
41. Manesh HF, Schaefer D (2010) A virtual factory approach for design and implementation of agile manufacturing systems. Am Soc Eng Educ 15(111):1–12
42. Manesh HF, Schaefer D (2010) Virtual learning environments for manufacturing education and training. Comput Educ J 77–89
43. Abdul-Hadi G, Abulrub AN, Attridge A, Williams MA (2011) Virtual reality in engineering education: the future of creative learning. In: IEEE global engineering education conference (EDUCON)—learning environments and ecosystems in engineering education, pp 751–757
44. Cassandras C, Deng M, Hu J-Q, Panayiotou C, Vakili P, Zhao C (2004) Development of a discrete event dynamic systems curriculum using a web-based “Real-Time” simulated factory. In: Proceeding of the American control conference, Boston, Massachusetts, p 1307

45. Chi X, Spedding TA (2006) A web-based intelligent virtual learning environment for industrial continuous improvement. In: IEEE international conference on industrial informatics 2006, Singapore, pp 1102–1107
46. Dessouky MM (1998) A virtual factory teaching system in support of manufacturing education. J Eng Educ 87(4):459–467
47. Dessouky MM, Verma S (2001) A methodology for developing a web-based factory simulator for manufacturing education. IIE Trans 33(3):167–180
48. Gadre A, Cudney E, Corns S (2011) Model development of a virtual learning environment to enhance lean education. Procedia Comput Sci 6:100–105
49. Goeser PT, Johnson WM, Hamza-Lup FG, Schaefer D (2011) VIEW-a virtual interactive web-based learning environment for engineering advances in engineering. Education 2(3):1–24
50. Ong S, Mannan M (2004) Virtual reality simulations and animations in a web-based interactive manufacturing engineering module. Comput Educ 43(4):361–382
51. Watanuki K, Kojima K (2007) Knowledge acquisition and job training for advanced technical skills using immersive virtual environment. J Adv Mech Design Syst Manuf 1:48–57
52. Gredler ME (2004) Games and simulations and their relationships to learning. In: Jonassen D (ed) Handbook of research on educational communications and technology, 2nd edn. Lawrence Erlbaum Associates Publishers, Mahwah, New Jersey, pp 571–581
53. Deterding S, Dixon D, Khaled R, Nacke L (2011) From game design elements to gamefulness: defining "Gamification". MindTrek' 11, Tampere, Finland, pp 9–15
54. Djaouti D, Alvarez J, Jessel JP, Rampnoux O (2011) Origins of serious games. In: Ma M, Oikonomou A, Jain L (eds) Serious games and edutainment applications, London, pp 25–43
55. Cook DJ, Mulrow CD, Haynes RB (1997) Systematic reviews: synthesis of best evidence for clinical decisions. Ann Intern Med 126(5):376–380
56. Ferenhof HA, Fernandes RF (2019) Passos para construção da Revisão Sistemática e Bibliometria. v. 3.02: https://www.researchgate.net/publication. Visited on 06/10/2019

Lean and TRIZ: From the Problems to Creative and Sustainable Solutions

Anabela Carvalho Alves, Paulo Sousa and Helena Navas

Abstract Lean Thinking philosophy pursuit continuous improvement and continuous innovation towards economic and eco-sustainable growth by new solutions. Such solutions involve eco-efficient products (goods and/or services) and cleaner production to produce them. Nevertheless, sometimes, there are technical contradictions that inhibited companies to achieve good solutions. TRIZ is a methodology capable to solve such contradictions and Lean could use it. Lean and TRIZ are on the road to achieve sustainable development goals, directly, the Goal 12, responsible consumption and production. This paper presents some case studies where this alliance is recognized and credited. These case studies were collected and synthetized by: type of work developed, company/sector, main TRIZ and Lean principles and tools used, problem(s) to solve and products/solutions designed. Building on the results, the authors systematized that Lean and TRIZ impact the design of creative solutions. Many of such solutions were developed in students' master dissertations that reveals the importance of teaching Lean allied with TRIZ.

Keywords Lean Thinking · TRIZ · Case studies

1 Introduction

Several literature reviews [1–7] were published to demonstrate the importance and influence of Lean Production (LP) [8, 9] in the industry and services, over the last 25 years [10]. Its guiding principles, when well implemented, create wealth for the companies and, consequently, for all, by eliminating wastes [11]. Wastes are all

A. Carvalho Alves · P. Sousa
Department of Production and Systems, School of Engineering, ALGORITMI R&D Center, University of Minho, 4800-058 Guimarães, Portugal
e-mail: anabela@dps.uminho.pt

H. Navas (✉)
Department of Mechanical and Industrial Engineering, Faculty of Sciences and Technology, UNIDEMI, Universidade Nova de Lisboa, 2829-516 Caparica, Portugal
e-mail: hvgn@fct.unl.pt

M. Rossi et al. (eds.), *Proceedings of the 6th European Lean Educator Conference*, Lecture Notes in Networks and Systems 122,
https://doi.org/10.1007/978-3-030-41429-0_11

activities that do no adds value in the client point of view, so he/she is not willing to pay for [12]. According to Russel and Taylor [12] (cited by Shirwaiker and Okudan [13]) waste includes anything other than the minimum amount of equipment, materials, space and time that are essential to add value to the product. Based on the Toyota Production System (TPS) of the Japanese company Toyota, designed to face the post-war hard times [14–16], this system intended to "doing more with less". Less means less human effort, less development time for new products, fewer stocks, less of everything.

To do more with less, TPS mentors employed known techniques at that time such as Standardized work and created many others such as Jidoka and JIT that are the TPS pillars. Sometimes these innovations just resulted from what they learned from others contexts, e.g. the supermarket concept when they visited the US which inspired them to create the Kanban tool [17]. TPS has been growing along with generations.

The innovative creation process was always related to some innate genius mind but sometimes it is the need that enables this process, as was the case of TPS. Nevertheless, an engineer named Genrikh Altshuller believed that systematic invention is possible. Through this, his dream was to transform engineering creativity into a logical discipline [18]. This was the origin of TRIZ a Russian acronym that means "Theory of Inventive Problem Solving" (TIPS). Meanwhile, some companies and organizations are introducing TRIZ, but not so many as desired and to a slow pace [19] that could be accelerated if the obstacles are bypassed by combining, for example, TRIZ with other approaches, such as Lean Production [20].

Problem-solving is on the top of the 4P (4P stands for Philosophy, Processes, People and Partners, Problem-solving) model pyramid of the Toyota way [21] and in the center of continuous improvement efforts and learning. This means that Lean and TRIZ are aligned in the same purpose: solving companies' problems by providing what the customers (external and internal) want and, ultimately, social and ecological needs. Nevertheless, this should be done without compromising the future of generations [22]. For this, designing eco-friendly products and processes at economic cost is a challengeable task where a systematic creativity process could help. This paper intends to review the literature for papers, dissertations, thesis and/or books that present case studies employing Lean and TRIZ to design better products and waste-free production systems.

This paper is structured into five sections. After the objectives introduced in the section, a brief literature review is presented. The third section presents the research methodology. The case studies raised in the databases and repositories are presented in section four as well as a discussion. Final remarks were wrapped in section five.

2 Brief Literature Review

Lean Production System was the designation used by Krafcik [8] to describe the Toyota Production System. According to this author, Toyota was a great innovator concerning the Fordism system when they take the minds and hands of the philosophy of the craftsman era allied with work standardization and glued by teamwork performance, instead of an individual. Moreover, their management was beyond seeing the operators as more than a "pair of hands" and truly exploit their "chie", the wisdom of experience [23]. By doing that, each person in the company (operator, manager, executive) become a thinker and become responsible to continuously improve their workstation, product or process [24].

It is in the production system that Toyota also is distinguished from other systems. Their executives soon realized that the Fordism system like many others western systems, at that time, were buffered systems, i.e., had high inventories to face existent problems: broken machines, defects, high absenteeism, line quality problems, among others [8]. They transform their system into a Lean Production System by reducing the inventories to a minimum, training people in diverse operations to cover the absence of colleagues, incentive teamwork, and promote continuous improvement of search and solve problems, any kind of problem.

The reduction of inventories and their impact on Economic Development across the world is recognized by Sanidas and Shin [25]. These authors, by using 31 years of relevant panel data of 88 and up to 152 countries and panel data econometric techniques, found a robust positive relationship between reduction and smaller changes in inventories and economic growth and economic development across the globe. They credited Lean Production and JIT techniques for that achievement. Additionally, Alves et al. [26] advocate Lean Engineering for achieving global development.

Lean Thinking [11] is nowadays accepted as philosophy and a way of thinking [27, 28] that demands a transformational culture. This philosophy helps the management of an organization, ruled by certain principles, being the last one "pursuit perfection", i.e. constant continuous improvement. In this context, everything is questioned to eliminate wastes: existent products, existent processes and existent production system [29, 30]. Lean is an example of whole-system thinking that once applied reduces/eliminates wastes of all kind, e.g. environmental wastes [31], and improve the productivity of natural resources [32]. Lean-Green is the designation given to Lean Thinking applied to sustainability efforts [33–38].

At the same time, the respect-for-human system created by Toyota [39] gives conditions for all to feel comfortable, without the restriction of hierarchies, where all could become trainers and trainees and manage their operations. All team members are considered industrial engineers in a Toyota company or in a joint venture, such as in NUMMI [8]. So, problems are welcome because they could become improvements opportunities [23, 40]. If hidden, they never will be solved. How to solve them is another question but existent tools could help in such endeavors such as TRIZ methodology [41].

TRIZ was defined as a methodology for solving problems based on logic, data and research and a toolkit of methods to support systematic creativity [42–45]. TRIZ was created by Genrich Altshuller [46] when he analyzed more than 3,000,000 patents to discover the patterns which provide innovative solutions to problems. As so, it was based on past knowledge and ingenuity of many thousands of engineers, it brought the ability to accelerate the project team to resolve issues creatively with an approach structured and algorithmic. Based on study, patterns of problems and solutions, creativity does not depend just on spontaneity and intuitive of individuals or groups. By doing this, it brings repeatability, predictability, and reliability to the creative process.

According to [47], TRIZ Body of Knowledge is based on the premise that the evolution of successful technological systems is not random but is governed by certain laws or prevailing trends. This is also highlighted by [48, 49] that considered technical systems evolving towards the increase of ideality by overcoming contradictions mostly with minimal introduction of resources.

TRIZ methodology involves several methods and tools, more well-known or not, simpler or more complex that have been evolving. Oxford Creativity TRIZ toolkit from Institution of Mechanical Engineers [50] provide ten tools: (1) Thinking in Time and Space; (2) Eight Trends of Technical Evolution; (3) Contradictions; (4) Forty Principles; (5) Seventy-six Standard Solutions; (6) Resources; (7) Ideality; (8) Functional Analysis; (9) Smart Little People; (10) Size-Time-Cost. Additionally, are available web sites that provide effects database for users [51, 52]. As mentioned by Bashkite and Karaulova [53], the method of TRIZ is to break the psychologic barrier, to abstract from the initial problem.

3 Research Methodology

The research methodology used in this study was a literature review of case studies in the main online databases such as Scopus and Web of Knowledge [54]. Because many studies about TRIZ and Lean were not based on these databases, it was decided to search also in the Online Knowledge Library, b-on (https://www.b-on.pt/) to capture works such as master dissertations or Ph.D. thesis. Keywords used were "Lean-TRIZ"; "Lean and TRIZ"; "Lean & TRIZ" and "Lean e TRIZ" in the publication title. The authors are aware that by limiting the search to both words linked, a lot of publications were outside of this range. Attending to the conference requested limited pages number, the authors decided not to expand the search to the abstract and/or keywords. Nevertheless, a search by the two isolated words would result in too many papers and the valued added for this paper would not be achieved, because the focus would be lost.

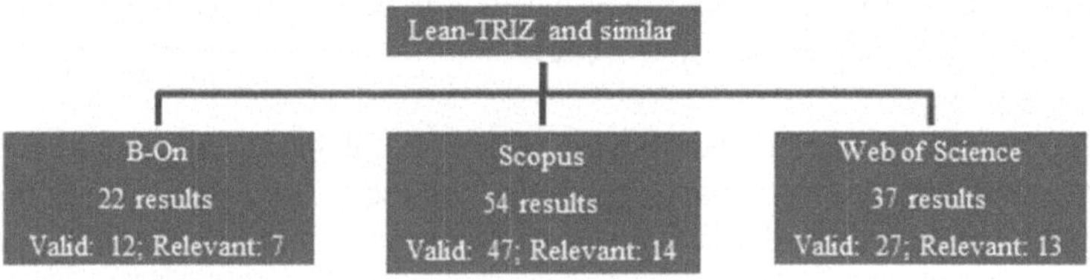

Fig. 1 Search results by Lean-TRIZ and similar words

The search number results are presented in Fig. 1, showing the initial and valid results (after removing the ones that do not address the topic, the repeated ones, and the conferences titles). A paper was considered relevant when a specific case of using Lean and TRIZ to solve a problem was founded. Some papers were too vague and generic, i.e., not addressing a company or specific problem so, they were excluded. This happens because the authors did not consider to include in the search the "case study" word. Although they were also related to the theme researched, it did not serve the purpose of this paper.

Then, each paper was analyzed attending to the main objectives of the paper. Such objectives rely on the research questions collected from each case study:

1. What is the type of work developed (in an academic or industrial environment)?
2. In what company type and sector was developed?
3. What was the problem(s) to solve?
4. What was the product/solution designed?
5. Which were the main Lean and TRIZ tools used?

To collect and analyses the papers, an excel file was used with columns for each piece of information needed to answer the questions, namely, the dimensions type of work developed, company type, company/academic problem, product/solution designed and Lean and TRIZ tools used.

4 Results Analysis and Discussion

This section presents the results obtained from the searches performed in different databases. After this, a discussion is followed.

4.1 Relevant Papers Analysis

This section presents and analyzes the results achieved by the search performed. A total of 113 papers were identified in the search, but only a small percentage 19% (22 papers) were considered relevant for deeper analysis.

From the 22 results of B-On search, just seven publications (32%) were considered relevant for the study: (1) Wang and Chen [55]; (2) da Silva [56]; (3) Hsieh et al. [57]; (4) Pombo [58]; (5) Caro [59]; (6) Madeira [60] and (7) Maximiano [61]. From these, five are master dissertations from Portugal. All present case studies in industrial environments. All were from a different company type and sector, being two from services sector, banking [55] and logistic [60], three from manufacturing (machine tool) [57]; beer and soft drinks [61]; air-conditioning [58] and one from agricultural, specifically, manual vintage (wine company) [59].

These authors, mainly, used simple Lean tools such as VSM, error proofing mechanism; 5S; visual control, standardized work, and SMED. TRIZ tools used were:

Table 1 The collected information from B-On search papers and other documents

References	Problem to solve	Product/Solution designed
[55]	Operation of savings accounts inefficient	Sheet and check mechanism redesigned
[56]	A long time of the changeover process	New screw
[57]	Long cycle time of the machine production	Innovative shape design of machine tools
[58]	Poor organization of the warehouse; defects; limitations of ERP Software	Flow improvement, a defects detection mechanism
[59]	Constant stoppages of teams; long movements of operators	New equipment, new procedures
[60]	Lack of areas for storage of raw material and finished product	Supermarket dimensioned, racks dimensioned
[61]	High setup time, a high number of stoppages, obsolete machines	Standard work instructions and sheets

contradiction matrix, 40 inventive principles, Field-Substance Analysis, ARIZ, ideality. Some also used the Kano Model. Such tools were used to find solutions for the problems identified in Table 1. The authors would like to add these in the table but due to the limited space of the paper dimension, it was not possible to do this. In the original excel file, the authors have additional columns that referred to the type of work, company type and main Lean and TRIZ tools used.

The same analysis was performed for the Scopus papers. From the 54 results, only 14 (26%) were considered relevant: Hammer and Kiesel [62]; Heisler et al. [63]; Alves et al. [64]; Wang et al. [65]; Karaulova and Bashkite [66]; Lee et al. [67]; Jiang and Nguyen [68]; Cabrera and Li [69]; Bashkite et al. [70]; Dewi et al. [71]; Kerga et al. [72]; Moreira [73]; Shirwaiker and Okudan [13]; Stratton and Warburton, [74]. All publications present industrial case studies in sectors so different as construction, electronics or automotive industry.

Additionally, to the already referred Lean and TRIZ tools, in these papers it is possible to find other tools combined with the already discussed such as ergonomics tools, life-cycle management tools, FMEA, SIPOC, OEE, Set-Based Concurrent Engineering, and AHP method. A deeper analysis of these papers showed the problems and proposed solutions to solve them (Table 2).

The search in the WoS produced similar results. From the 37 results, 27 were considered valid but almost all were repeated and the ones relevant were considered in Table 2. From this search, just one paper was relevant that was not considered in Table 2. This paper was from Lanke and Ghodrati [75] that wanted to solve the problems of limited resources in the sintering process but it has the need to produce more new types of magnets. Their solution was to design a new additional product.

Table 2 The collected information from Scopus search papers

References	Problem to solve	Product/Solution designed
[62]	High testing process time	A new testing
[63]	Non-optimization of the manual foaming process for complex foam parts; use of a release agent; high production time	Innovative system design for a laser optical release agent system
[64]	Existing occupational safety issues	Improvement of workplaces' ergonomic quality
[65]	Waste time, occupied space, not reusing old frames	Reduced waste time, free space, new conveyors
[66]	Missing a mechanism to assess the moment condition of the equipment	An integrated method for evaluating the remanufacturability of the used industrial equipment
[67]	Call-center service poor quality	Value engineering based method
[68]	Poor productivity and effectiveness; high waiting time	Standard for quality of raw materials; operation procedure and training plan
[69]	Need to do pavement maintenance and rehabilitation	Ultrasonic sensors to the paver; compaction control system in 2 rollers
[70]	Need to extend the useful life span of lathes and milling machines	Extra options
[71]	High changeover time and non-value added activity in the changeover activities	New tool-rack
[72]	The high cost of adiabatic humidification system (AHS)	New rack for AHS
[73]	Components misplacement	New feature and elongate the track of the tool
[13]	Substitute traditional wood or bricks	Design scrap tires as building materials
[74]	Dependency, fluctuation, sales forecast errors	Design lean/agile supply chain

4.2 Discussion

To start the discussion, the authors provide a big picture of the figures involved, presenting at the Table 3 the resume of the research.

Table 3 Analysis of the papers from the three databases used

Data base	Results	Valid	Relevant	Relevant and not repeated
B-On	22	12 (55%)	7 (32%)	7 (32%)
Scopus	54	47 (87%)	14 (26%)	14 (26%)
WoS	37	27 (73%)	13 (35%)	1 (3%)

As presented, the percentage of the relevant case studies, based in the authors criteria, are similar between WoS and B-on (35 and 32%) respectively and a little below in Scopus (26%) that can lead to the conclusion that the case study's authors do not have a different criteria to choose were to present the papers.

A small percentage (19% of the total) of case studies (that were not repeated) were considered relevant in the search accomplished. An important part was from master dissertations what could show that teaching TRIZ in universities starts to pay off. It is important to notice that TRIZ has been part of the curriculum in many universities [48, 76–79] to train future professionals in such methodology. Research is also been done and publications are being reviewed and published, although, according to Chechurin [44] without the good presentation and articulation as other researches. Additionally, because TRIZ is not an easy topic to learn, the use of games could help on this [80, 81].

At the same time, although TRIZ is being widely used by some major companies, like Samsung, Intel, Ford, Proctor & Gamble, and Mitsubishi to develop better products more quickly [20], more companies need TRIZ to research, develop and innovate in their products and processes [19, 82]. Nevertheless, TRIZ applicability in different areas have been growing, from 23 to 47% in the last years [83].

To potentiate TRIZ use and to this become natural, it is important to explore synergies of TRIZ with others management tools such as Quality Function Deployment [84, 85], Six-Sigma [55], Toyota Kata [86], Lean, Theory of Constraints [87], Axiomatic Design [13, 88], among others [43]. Particularly, relevant and subject of this paper is the synergy between Lean and TRIZ, highlighted by some authors [41, 55, 89–91]. Some case studies founded in the search presented in this paper revealed such synergies and others.

TRIZ is the "hand in glove" to help a Lean organization to growth in a sustainable way by using creative problem solving and a dialectic way of thinking. In a Lean environment, creativity is pulled by the need to solve a problem, inspired by a restless spirit, rooted by simple tools like PDCA, 5Why's, 5W2H, root-cause-analysis, or more complex tools. TRIZ role will be to help to understand the problem as a system, to make an image of the ideal solution first, and to solve the contradictions [49], as the case studies showed.

Contradictions are constant in the engineering field and other's fields. Even in the services field as showed by Wang and Chen [55]. For example, improving functional performance and reducing environmental impacts are not always straightforward in the product's design. To achieve this, concepts as eco-design and eco-efficiency are applied to design new products [13, 92]. TRIZ could help with this achievement [93, 94].

Also, is necessary a change in existing production and consumption patterns of a society or social paradigm change in production and consumption. This need to change was also referred by Womack and Jones [95] when discussed the Lean consumption concept. Responsible consumption and production (Goal 12) are one of the goals of Sustainable Development [96].

As a conclusion, it is possible to remark that Lean, TRIZ, Eco-design among other's approaches are seeking the same and companies need to be aware of such methodologies. How to use them could be a problem and integrated models could help on this [97] as well as monitoring the companies sustainability efforts by suitable indicators. Some of such indicators could be found on [98–102]. Furthermore, Industry 4.0 [103] provides the needed mechanisms to monitor systems, processes, machines and equipment use of materials, energy, water, pollution, helping a company to become Lean [104, 105].

5 Final Remarks

This paper presented a literature review about case studies discussing Lean implementation that simultaneously use TRIZ methodology to create sustainable and eco-efficient solutions for their problems/wastes. Continuous improvement spirit proper of Lean Thinking conduce companies in searching for solutions for their problems. Such problems could be simple or, most of times, imply technical contradictions, which could be solved by TRIZ. Nevertheless, this methodology is far from being known by companies or easy to apply.

The search performed in this paper in three databases, including two of the most reputable, achieved a small percentage, just 19% of case studies published where solutions based on TRIZ were designed for problems/wastes elimination. However, it is important to refer that this does not mean companies are not using, just that there are not many papers published about it, or at least, are not being published in the scientific databases. Probably, more need to be done to promote the publication of works developed in the industrial context (many times by master students) and to share good practices.

Acknowledgements This work has been supported by FCT—Fundação para a Ciência e Tecnologia within the Project Scope: UID/CEC/00319/2019. The author from FCT NOVA acknowledges Fundação para a Ciência e a Tecnologia (FCT—MCTES) for its financial support via the project UIDB/00667/2020 (UNIDEMI).

References

1. Amaro P, Alves AC, Sousa RM (2019) Lean thinking: a transversal and global management philosophy to achieve sustainability benefits. In: Lean engineering for global development. Springer International Publishing, Cham, pp 1–31
2. Bhamu J, Sangwan KS (2014) Lean manufacturing: literature review and research issues. Int J Oper Prod Manage 34(7):876–940
3. Jasti NVK, Kodali R (2015) Lean production: literature review and trends. Int J Prod Res 53(3):867–885
4. Negrão LLL, Godinho Filho M, Marodin G (2016) Lean practices and their effect on performance: a literature review. Prod Plann Control 28(1):1–24
5. Panizzolo R, Garengo P, Sharma MK, Gore A (2012) Lean manufacturing in developing countries: evidence from Indian SMEs. Prod Plann Control Manage Oper 23(10–11):769–788

6. Silva C, Tantardini M, Staudacher AP, Salviano K (2010) Lean production implementation: a survey in Portugal and a comparison of results with Italian, UK and USA companies. In: Sousa R, Portela C, Pinto SS, Correia H (eds) Proceedings of 17th international annual EurOMA conference—managing operations in service economics, Universidade Católica Portuguesa, 6–9 June 2010, Porto, Portugal, pp 1–10
7. Stone KB (2012) Four decades of lean: a systematic literature review. Int J Lean Six Sigma 3(2):112–132
8. Krafcik JF (1988) Triumph of the lean production system. Sloan Manage Rev 30(1):41–52
9. Womack J, Jones DT, Roos D (1990) The machine that changed the world: the story of lean production. Rawson Associates, New York
10. Samuel D, Found P, Williams JS (2015) How did the publication of the book The Machine That Changed The World change management thinking? Exploring 25 years of lean literature. Int J Oper Prod Manage 35(10):1386–1407
11. Womack JP, Jones DT (1996) Lean thinking: banish waste and create wealth in your corporation. Free Press, New York
12. Russell RS, Taylor BW (1999) Operations management, 2nd edn. Prentice Hall, Upper Saddle River, NJ
13. Shirwaiker RA, Okudan GE (2011) Contributions of TRIZ and axiomatic design to leanness in design: an investigation. Procedia Eng 9:730–735
14. Ohno T (1988) Toyota production system: beyond large-scale production, 3rd edn. Productivity Press, New York
15. Shingo S (1989) A study of the Toyota production system: from an industrial engineering viewpoint. Productivity Press, New York
16. Monden Y (1983) Toyota production system: practical approach to production management. Engineering & Management Press
17. Holweg M (2007) The genealogy of lean production. J Oper Manage 25(2):420–437
18. Fey V (2008) Genrikh Altshuller—the creator of TRIZ. [Online]. Available: https://www.aitriz.org/articles/TRIZFeatures/30383038-466579.pdf. Accessed: 30 July 2018
19. Spreafico C, Russo D (2016) TRIZ industrial case studies: a critical survey. Procedia CIRP 39:51–56
20. Rantanen K, Conley DW, Domb ER (2018) Simplified TRIZ: new problem-solving applications for technical and business professionals, 3rd edn. CRC Press, Boca Raton
21. Liker JK (2004) The Toyota way: 14 management principles from the world's greatest manufacturer. McGraw-Hill, New York
22. Brundtland GH (1987) Our common future. Oxford paperbacks, no. A/42/427. World Commission on Environment and Development, p 400
23. Takeuchi H, Osono E, Shimizu N (2008) The contradictious that drive Toyota's success. Harv Bus Rev (June):96–104
24. Alves AC, Dinis-Carvalho J, Sousa RM (2012) Lean production as promoter of thinkers to achieve companies' agility. Learn Organ 19(3):219–237
25. Sanidas E, Shin W (2017) Lean production system and economic development across the world today. Int J Econ Manage Sci 06(06)
26. Alves AC, Kahlen F-J, Flumerfelt S, Siriban-Manalang AB (2019) Lean engineering for global development. Springer International Publishing, Cham
27. Bhasin S, Burcher P (2006) Lean viewed as a philosophy. J Manuf Technol Manage 17(1):56–72
28. Yamamoto Y, Bellgran M (2010) Fundamental mindset that drives improvements towards lean production. Assembly Autom 30(2):124–130
29. Alves AC, Sousa RM, DInis-Carvalho J (2016) Redesign of the production system: a hard decision-making process. In: IEEE International conference on industrial engineering and engineering management, vol 2016-January, pp 1128–1132
30. Alves A, Sousa R, Dinis-Carvalho J, Moreira F (2015) Production systems redesign in a lean context: a matter of sustainability. FME Trans 43(4):344–352
31. US-EPA (2007) The lean and environment toolkit

32. Lovins A, Lovins LH, Hawken P (2007) A road map for natural capitalism. Harv Bus Rev July–August:172–183
33. Abreu MF, Alves AC, Moreira F (2017) Lean-green models for eco-efficient and sustainable production. Energy 137:846–853
34. Alves A, Moreira F, Abreu F, Colombo C (2016) Sustainability, lean and eco-efficiency symbioses. Springer International Publishing, Cham
35. Florida R (1996) Lean and green: the move to environmentally conscious manufacturing. Calif Manage Rev 39(1) (Routledge, New York, NY)
36. Helper S, Clifford PG (1997) Can green be lean? 404
37. Maxwell J, Rothenberg S, Schenck B (1993) Does lean mean green: the implications of lean production for environmental management. MIT, International Motor Vehicle Program, July 1993
38. Moreira F, Alves AC, Sousa RM (2010) Towards eco-efficient lean production systems. In: Balanced automation systems for future manufacturing networks, vol 322. IFIP Advances in information and communication technology, pp 100–108
39. Sugimori Y, Kusunoki K, Cho F, Uchikawa S (1977) Toyota production system and Kanban system materialization of just-in-time and respect-for-human system. Int J Prod Res 15(6):553–564
40. Spear SJ (2004) Learning to lead at Toyota. Harv Bus Rev 82(5):78–86, 151 (May)
41. Maia LC, Alves AC, Leão CP (2015) How could the TRIZ tool help continuous improvement efforts of the companies? Procedia Eng 131:343–351
42. Altshuller GS (1995) Creativity as an exact science: the theory of the solution of inventive problems. Gordon and Breach Publishers, London
43. Chechurin L, Borgianni Y (2016) Understanding TRIZ through the review of top cited publications. Comput Ind 82:119–134
44. Chechurin L (2016) TRIZ in science. Reviewing indexed publications. Procedia CIRP 39:156–165
45. Mazur G (1995) Theory of inventive problem solving (TRIZ)
46. Lerner L (1991) Genrich Altshuller: father of TRIZ. Russian Magazine Ogonek. https://www.aitriz.org
47. Litvin S, Petrov V, Rubin M, Fey V (2007) TRIZ body of knowledge. TRIZ developers summit, vol 18, pp 1–9
48. Nakagawa T (2011) Education and training of creative problem solving thinking with TRIZ/USIT. Procedia Eng 9:582–595
49. Nakagawa T (2012) Creative problem-solving methodologies TRIZ/USIT: overview of my 15 years in research, education, and promotion. Bull Cult Nat Sci Osaka Gakuin Univ 64:1–32
50. Institution of Mechanical Engineers (2012) TRIZ toolkit. [Online]. Available: https://www.imeche.org/knowledge/industries/manufacturing/triz/toolkit. Accessed: 26 June 2012
51. Oxford Creativity (2015) Effects database. [Online]. Available: https://wbam2244.dns-systems.net/EDB_Welcome.php
52. Product Inspiration (2015) Product inspiration. [Online]. Available: https://www.productioninspiration.com/
53. Bashkite V, Karaulova T (2012) Integration of green thinking into lean fundamentals by theory of inventive problems-solving tools. In: 23rd DAAAM international symposium on intelligent manufacturing and automation 2012, vol 1, pp 345–350
54. Mongeon P, Paul-Hus A (2016) The journal coverage of web of science and Scopus: a comparative analysis. Scientometrics 106(1):213–228
55. Wang F-K, Chen K-S (2010) Applying lean six sigma and TRIZ methodology in banking services. Total Qual Manage Bus Excell 21(3):301–315
56. da Silva JCFT (2013) Using TRIZ methodology with lean production techniques applying SMED method in industrial environment. Universidade do Minho
57. Hsieh H-N, Chen J-F, Do Q (2015) Applying TRIZ and fuzzy AHP based on lean production to develop an innovative design of a new shape for machine tools. Information 6(1):89–110

58. Pombo MIC (2015) Metodologias TRIZ e Lean numa Indústria de Unidades de Tratamento de Ar e de Ventilação. Universidade Nova de Lisboa
59. Caro LMG (2016) Inovação no Processo de Vindima Manual – Aplicação da Filosofia Lean e Metodologia TRIZ. Faculty of Sciences and Technology, Universidade NOVA de Lisboa
60. Madeira GFRC (2016) Aplicação da metodologia TRIZ e da filosofia lean para o dimensionamento do layout logístico da Schnellecke para produção de novos componentes automóveis. Faculty of Science and Technology, FCT, Universidade Nova de Lisboa
61. Maximiano DGF (2017) Aplicação da Metodologia TRIZ e da Filosofia LEAN nas Linhas de Enchimento em Indústria Alimentar. Faculty of Sciences and Technology, Universidade NOVA de Lisboa
62. Hammer J, Kiesel M (2019) A TRIZ and lean-based approach for improving development processes. In: Advances in systematic creativity. Springer International Publishing, Cham, pp 101–114
63. Heisler P, Gick SD, Franke J (2018) The way to a simple and environment-friendly production design of complex foam parts—innovative solutions for the polyurethane manufacturing industry. Procedia Manuf 17:579–586
64. Alves JF, Navas HVG, Nunes IL (2016) Application of TRIZ methodology for ergonomic problem solving in a continuous improvement environment. In: Advances in intelligent systems and computing, vol 491, pp 473–485
65. Wang C-N, Huang Y-F, Le T-N, Ta T-T (2016) An innovative approach to enhancing the sustainable development of Japanese automobile suppliers. Sustainability 8(5):1–19
66. Karaulova T, Bashkite V (2016) Decision-making framework for used industrial equipment. Eng Econ 27(1):23–31
67. Lee LJ-H, Leu J-D, Huang Y-W (2015) A value engineering based method of configuring ICT-based customer service centers. In: 2015 2nd international conference on information science and control engineering, pp 86–91
68. Jiang J, Nguyen T (2015) Process improvement by application of lean six sigma and TRIZ methodology case study in coffee company. Int J Appl Innov Eng Manage 4(3):208–219
69. Cabrera BR, Li GJ (2014) A lean-TRIZ approach for improving the performance of construction projects. In: 22nd annual conference of the international group for lean construction: understanding and improving project based production, IGLC 2014, pp 883–894
70. Bashkite V, Karaulova T, Starodubtseva O (2014) Framework for innovation-oriented product end-of-life strategies development. Procedia Eng 69:526–535
71. Dewi SR, Setiawan B, Susatyo Nugroho WP (2013) 5S program to reduce change-over time on forming department (case study on CV Piranti Works temanggung). In: IOP conference series: materials science and engineering, vol 46, no 1
72. Kerga ET, Taisch M, Terzi S (2013) Set based concurrent engineering innovation roadmap, vol 415
73. Moreira D (2012) Component misplacement prevention on the ICOS tape & reel process using TRIZ & lean. In: IPC APEX EXPO 2012, vol 1, pp 716–728
74. Stratton R, Warburton RD (2003) The strategic integration of agile and lean supply. Int J Prod Econ 85(2):183–198
75. Lanke A, Ghodrati B (2014) Reducing defects and achieving business profitability using innovative and lean thinking. In: IEEE international conference on industrial engineering and engineering management, pp 1026–1030
76. Belski I (2011) TRIZ course enhances thinking and problem solving skills of engineering students. Procedia Eng 9:450–460
77. Bušov B, Žídek J, Bartlová M (2016) TRIZ already 35 years in the Czech Republic. Procedia CIRP 39:216–220
78. Navas HVG, Cruz Machado V (2017) Lean teaching experiences in Universidade NOVA de Lisboa and the TRIZ-LEAN model. In: Lean education: an overview of current issues. Springer International Publishing, Cham, pp 67–76
79. Stratton R, Mann D, Otterson P (2010) The theory of inventive problem solving (TRIZ) and systematic innovation-a missing link in engineering education? December 2010

80. Thurnes CM, Zeihsel F (2015) Gamificated linking of lean and TRIZ for training and education. In: Proceedings of the 6th international conference on mechanics and materials in design, pp 1337–1342
81. Giebens G (2009) Using a board game to introduce the use of ideation techniques based on TRIZ. In: Proceedings of the 4th European conference on entrepreneurship and innovation, pp 141–149
82. Alves AC, Leão CP, Maia LC, Navas HVG (2016) Understanding if and how TRIZ is used in the Portuguese reality. In: The 2016 international conference on systematic innovation
83. Goldense B (2016) TRIZ is now practiced in 50 countries. Machine Design
84. Domb E (1995) Using TRIZ to enhance quality functional deployment. [Online]. Available: https://www.realinnovation.com/content/c080922a.asp. Accessed: 18 June 2012
85. Ulwick BAW, Zultner RE, Norman R (2006) Moving the house of quality to systematic innovation. RealInnovation.com. [Online]. Available: https://www.realinnovation.com/content/c071008a.asp. Accessed: 18 June 2012
86. Toivonen T (2015) Continuous innovation—combining Toyota Kata and TRIZ for sustained innovation. Procedia Eng 131:963–974
87. Anosike AI, Lim MK (2013) Integrating lean, theory of constraints and TRIZ for process innovation. In: Ogunleye J, Heger D, Richter UH (eds) Short research papers on knowledge, innovation and enterprise. KIE conference book series, Berlin, pp 64–74
88. Slocum MS (2004) Computer aided comprehensive design for six sigma (DFSS) and axiomatic design (AD): computer aided performance excellence (CAPE), vol 156
89. Bligh A (2006) The overlap between TRIZ and lean. University of Rhode Island, pp 1–10
90. Ikovenko S, Bradley J (2004) TRIZ as a lean thinking tool. In: 2004 ETRIA TRIZ future conference
91. Thurnes CM, Zeihsel F, Hallfell F (2014) TRIZ for waste-elimination in a 'Lean Production'-environment. TRIZ J 1–8
92. Aoe T (2007) Eco-efficiency and ecodesign in electrical and electronic products. J Clean Prod 15(15):1406–1414
93. Pacheco DADJ, Ten Caten CS, Navas HVG, Jung CF, Cruz-Machado V, Lopes GHN (2016) Systematic eco-innovation in lean PSS environment: an integrated model. Procedia CIRP 47:466–471
94. Russo D, Regazzoni D, Montecchi T (2011) Eco-design with TRIZ laws of evolution. Procedia Eng 9:311–322
95. Womack J, Jones DT (2005) Lean solutions: how companies and customers can create value and wealth together. Simon & Schuster, New York, USA
96. United Nations (2018) The sustainable development goals: 17 goals to transform the world. [Online]. Available: https://www.un.org/sustainabledevelopment/sustainable-development-goals/. Accessed: 03 Aug 2018
97. Muiambo C, João I, Navas H (2018) The integration of TRIZ methods with eco-design and lean design: a literature review and future research directions to the development of a new model. In: Proceedings of the MATRIZ TRIZfest 2018 international conference, 13–15 September 2018, Lisbon, Portugal, pp 123–135
98. Abreu MF, Alves AC, Moreira F (2019) The lean-green BOPSE indicator to assess efficiency and sustainability. In: Lean engineering for global development. Springer International Publishing, Cham, pp 259–291
99. Carvalho H, Machado VH, Barroso AP, de Almeida D, Cruz-Machado V (2019) Using lean and green indexes to measure companies' performance. In: Lean engineering for global development. Springer International Publishing, Cham, pp 293–318
100. Harik R, El Hachem W, Medini K, Bernard A (2015) Towards a holistic sustainability index for measuring sustainability of manufacturing companies. Int J Prod Res 53(13):1–23
101. Orsato RJ, Garcia A, Mendes-Da-Silva W, Simonetti R, Monzoni M (2014) Sustainability indexes: why join in? A study of the 'Corporate Sustainability Index (ISE)' in Brazil. J Clean Prod 96:161–170

102. Rashidi K, Farzipoor Saen R (2015) Measuring eco-efficiency based on green indicators and potentials in energy saving and undesirable output abatement. Energy Econ 50:18–26
103. Kagermann H, Wahister W, Helbig J (2013) Recommendations for implementing the strategic initiative INDUSTRIE 4.0: securing the future of german manufacturing industry
104. Bittencourt V, Alves AC, Leão CL (2019) Lean thinking contributions for Industry 4.0: a systematic literature review. In: 9th IFAC conference on manufacturing modelling, management and control (MIM19)
105. Bittencourt V, Saldanha F, Alves AC, Leão CP (2019) Contributions of lean thinking principles to foster Industry 4.0 and sustainable development goals. In: Lean engineering for global development. Springer International Publishing, Cham, pp 129–159

Recapturing the Spirit of Lean: The Role of the Sensei in Developing Lean Leaders

Eivind Reke, Daryl Powell, Sandrine Olivencia, Pascale Coignet, Nicolas Chartier and Michael Ballé

Abstract *Background*: In the late 70s and early 80s when Toyota started to develop its supplier network as well as its overseas plants in the West, European and American executives where taught the Toyota Production System and Total Quality Management in the form of visits from Toyota's own TPS experts, sometimes dubbed Sensei. Their style was highly unusual compared to that of the Western consultant. However, the results were often spectacular with unthinkable levels of quality and productivity improvement, not through data analysis and prescribed actions, nor through implementation of best practices, but through learning exercises that where designed for teaching TPS and TQM to the CEO and/or the company's executives. In most cases today, external Lean support is carried out by consultants, who typically attempt to either implement a fully developed "Operational Excellence" program or carry out a tools-based productivity analysis and suggest countermeasures in the form of a feedback report with recommendations. However, some managers still work with Sensei, where the onus is on challenging and teaching the thinking behind TPS,

E. Reke (✉)
LOS Norge, Trondheim, Norway
e-mail: eivind.reke@losnorge.no

D. Powell
SINTEF Manufacturing AS, Raufoss, Norway
e-mail: daryl.powell@sintef.no

S. Olivencia
Lean Institute France, Saint-Jean-de-la-Ruelle, France
e-mail: sandrineolivencia@gmail.com

P. Coignet · M. Ballé
Telecom Paris, Palaiseau, France
e-mail: coignetpascale@gmail.com

M. Ballé
e-mail: m.balle@orange.fr

N. Chartier
AramisAuto, Arcueil, France
e-mail: Nicolas.chartier@aramisauto.com

M. Rossi et al. (eds.), *Proceedings of the 6th European Lean Educator Conference*, Lecture Notes in Networks and Systems 122,
https://doi.org/10.1007/978-3-030-41429-0_12

constantly looking for the next step for quality improvement, productivity improvement, lead-time reduction and cost-reduction. *Purpose*: Previously, research into Lean transformation has focused solely on the implementation of Lean as a set of best practices. In this paper we explore the workings of Sensei and the role they play in developing Lean leaders, who themselves lead the organization's Lean transformation. *Research*: We carried out several exploratory case studies in the form of interviews with executives who have worked with Sensei rather than following the classic consultancy-led best practice implementation program.

Keywords Leadership · Lean Thinking · Learning · Sensei · Consultant

1 Introduction

"You are no good" [1]. A typical statement from a so-called Lean Sensei. The first examples of the application of the Toyota Production System (TPS) outside of Japan were supported by Japanese TPS experts, many of whom had worked with Taiichi Ohno himself. Their style was completely at odds with the Western-style consultants or executive coaches, which is why they were often referred to as "Sensei". These early Sensei would take executives to the Gemba and challenge them on their willingness to learn, and on their discipline [1]. Each Sensei typically had his own starting point. However, it was not uncommon to start with quality. Not accepting bad parts is after all an essential part of TPS. Some executives saw the strategic implications of what the Sensei was trying to teach them and in the early- to mid-1980s, the West's first Lean leaders began to emerge.

The stories of the pioneering Lean companies such as Danaher, Lantech and Wiremold were not studied by academics during the time at which the success stories unfolded. However, the stories of these companies can be found in the literature in the form of books. Most famously in "Lean Thinking" [2], but also in "The Lean Turnaround" [1], "Better Thinking, Better Results" [3] and "Leading the Lean Enterprise Transformation" [4]. However, when one looks at the subject of implementing Lean in the peer-reviewed scientific literature, there is an abundance of publications in leading international journals covering critical success factors [5, 6], barriers to implementation [7, 8], what to expect from corporate Lean programs [9] the impact of Lean implementation on workers conditions [10], the impact of Lean implementation on performance [11], frameworks for implementation [12], and so forth. Although some of these studies have considered the role which external consultants play in the lean implementation, few have discussed the difference between working directly with a Sensei as opposed to the traditional (internal or external) consultancy-driven program model. Womack and Jones [2] mentioned that Chihiro Nakao, with his "special Sensei treatment", was instrumental in ten of the companies mentioned in "Lean Thinking". Sisson and Elshennawy [13] also found that an external Sensei to guide the development of executives was a success factor at four well-known companies (Danaher, Boeing, UTC and Autoliv). A longitudinal study of lean implementation in a Norwegian public service department [14] show how dominant project thinking is in the consultancy/host organisation symbiose. Even when the emphasis is on the development of people within the organisation implementing lean, it is still done based on a project methodology. Furthermore, [15] discuss how, like the consultant, the Sensei has no hierarchical power. However, unlike the consultant, the Sensei will have authority through the expertise and charisma. In the eyes of executives, the Sensei is a teacher while the consultant assumes a servant role.

In this paper, we present the results of interviews with a set of executives who have worked with Sensei in different eras. From the early days of Lean in the 80s and 90s, to the diffusion era of Lean in the 2000–2010s. Most of the executives had first worked with a consultancy-led implementation program, but later changed the approach to a Sensei-supported transformation.

2 Research Method

Our research design is driven by a lack of research in the extant literature that describes the Sensei-tradition. As such, we adopt an exploratory case-study approach. We draw on findings from a set of semi-structured interviews that have been conducted with executive managers who have worked first-hand with both Sensei and traditional Western-style consultants. Interviews with a Toyota veteran and his student were also carried out. In total, 6 interviews were conducted. The Sensei-executive relationships which form the units of analysis in this study are shown in Fig. 1.

The main aim of the interviews was to uncover the subtle differences between the Sensei-supported approach versus the consultant-driven approach, as well as any similarities in the practices of the Sensei. In addition, we consider the impact that working with a Sensei has had on the thinking and practice of the executives themselves.

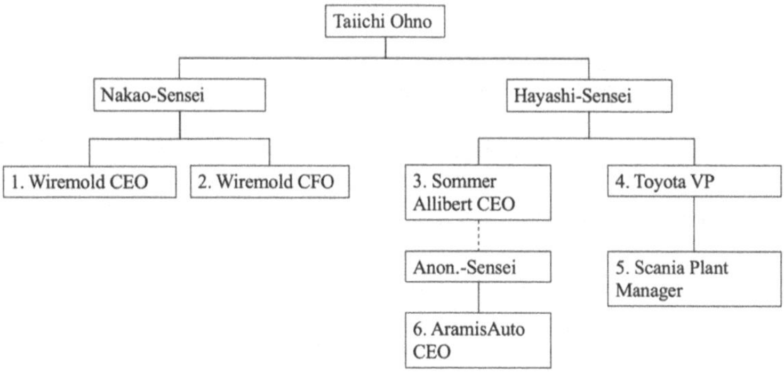

Fig. 1 Sensei lineage and interviewees/units of analysis

3 Findings

Each of the executives interviewed mentioned a common operational trait that separated the Sensei from the consultant. The consultants would often spend a lot of time gathering and analysing company data and come up with improvement suggestions in the form of feedback reports. In contrast, the Sensei would never write or present anything in report form. They all taught TPS through practical exercises. Some of them would also get actively involved in Kaizen activities, but not all of them. When they did it was to prove a point. For example, by moving a machine that had not been moved in many years in order to realize instantaneous productivity gains. The consultants that the executives had used were more concerned with data analysis and reports, or higher-level strategic thinking, completely disconnected from the specific realities on the production floor.

3.1 Sensei Practices

We can summarise our findings as four common practices that the Sensei would use to help the CEO and the organization better understand their own work and teach them how to solve their own problems. We suggest that the purpose of these practices is to stimulate "helicopter thinking", helping the executives see the strategic significance of operational reality: (1) The Sensei would point to a *specific example* of something that the executives would have to explore deeper in order to better understand a core topic. (2) They trained the executives and their organisation to test hypotheses by systematically practicing *Plan-Do-Check-Act* (PDCA), often allowing the executives to fail, even though they knew beforehand that whatever they were doing was destined to fail. (3) They would *teach TPS* through practical exercises. And (4) They would always push the executives towards discovering and taking *the next step*, particularly when the executives could not see this for themselves.

3.1.1 Specific, Practical Examples to Deeply Understand a Core Topic

This could sometimes be quite dramatic. The executives we interviewed told the story of two Japanese men in suits and ties who were on the shop floor with pry bars moving a machine on their own. Another one told the story of when he showed the American Toyota Sensei the company's state-of-the-art automated in-process warehouse, when the Sensei commented that it was so impressive it could easily be mistaken for something that adds value. In both of these cases, the Sensei was showing a detailed, practical example of something he wanted the executive to explore deeper.

3.1.2 Hypothesis Testing Through PDCA

The Sensei would not ask "why don't you do this?" Instead he would ask questions that would allow the executives to clarify their current hypothesis and then test it by using the PDCA cycle. One interviewee reported that every time they were asked about what improvements they wanted to make, the Sensei would force them to identify up to seven different ideas of how they could achieve what they wanted to do.

3.1.3 Teaching TPS Through Practical Exercises

Since TPS is a complete system of production that consists of many different parts that can be both counter-intuitive and difficult to grasp, the Sensei would mostly teach the system through practical exercises. We found that the Sensei would generally start teaching straight away with no classroom presentations first, often setting what seemed like impossible targets to achieve to force people to think outside of the box. Or introducing tools such as takt-time, to make the executives think about balancing to avoid overburden.

3.1.4 Next Step Thinking

Even when the executives we spoke to had made good progress in between Sensei visits, the question was always what do you want to improve next? Who's in the team? Where is the waste and what can we do about it? For example, we found that the Sensei would introduce what they thought their executive students were ready for and would sometimes play down what they thought was possible to achieve. When one of our informants asked what his Sensei thought was possible the Sensei answered 50% of everything. 50% cost-, lead time- and inventory reduction. However, The Sensei could already see the potential was far greater, but he thought that this was a big enough target for the executive to grasp. The executive of course thought his Sensei was crazy to suggest such massive improvement potential.

3.2 *Consultancy Practices*

In contrast to the Sensei approach, the executives we interviewed that had worked with consultants described the consultancy-led programs as costly and ineffective with regards to changing the way the company operated. However, the consultancies were often easily available to the executives when they were looking for external help to solve their problems. The executives described two main traits of the consultancy approach that they had experienced.

3.2.1 Report

The consultant would suggest carrying out an overall analysis of the performance of the company. A large team of consultants would come into the organisation, talk to everybody, gather heaps of data, tear it apart and write up a big report. The report would then be condensed into a PowerPoint presentation and presented in a meeting room with suggestions for improvements. The report would usually be very expensive, and it would not challenge the executives to change their own thinking.

3.2.2 Partner and Teams

The other approach reported was that of a senior partner working with a team of young consultants on one or two lines, machines or areas of the production. At first, the internal teams working on these isolated projects would be quite happy. However, there was often little progress, and any changes would often return back to its original condition after a short time. Again, several of our informants reported that the method did not change their own thinking. Instead they were putting pressure on the isolated projects to get results by setting high objectives for the teams. Regardless of whether the improvement work was carried out by external consultants or if the external consultants were supporting internal teams, any progress was usually slow and expensive.

3.3 Relationship

There also seemed to be a common type of relationship, one of mutual respect, between the Sensei and the executive based on trust of expertise, in contrast to the relationship one finds between executives and consultants, where the consultant often will be servant to the executive [15]. This special-type of relationship allows the Sensei to give direct and often controversial feedback to the executives at their own Gemba. For example, one interviewee was told to "*learn how to do your job properly*," while another was told that "*everything was no good*".

4 Conclusion

The "Lean Sensei" role most likely emerged when Toyota wanted to spread its management thinking and production practices to its suppliers and overseas plants [16]. Like most other practices within the company, the Sensei role was not designed. Rather, it evolved based on the practices of those who had worked with and under Ohno in production and possibly under Hasegawa in product development. Although the Toyota-style Sensei can in some instances be likened to a consultant, the findings of this work make some very clear distinctions between the traditional Western-style consultant-led approach and the Sensei-supported approach to Lean implementation. It also points towards a specific set of practices when working with leaders:

1. The Sensei shows you a specific, practical example of something you will have to explore in depth to understand a core topic.
2. The Sensei will train you to test your hypothesis by systematically practicing the PDCA-cycle.
3. The Sensei will teach you TPS through practical exercises, allowing you to fail when necessary to better support the learning cycle.
4. The Sensei will always push you to take the next step, even if you cannot see it for yourself.

These 4 practices are possibly specific to how Japanese, American and European Sensei work in the West. Such an approach is in direct contrast to the alternative Japanese-style of "follow my orders" and the Western-style of "here are my findings, this is what you should do" consultant.

Another interesting finding which emerged from this research is how the old-time Sensei, those who had worked directly with Taiichi Ohno, would mostly discuss productivity, pushing their C-suite students to look for the causes of non-productivity, usually related to quality issues—*and* consider productivity improvements as human development exercises. In that sense, the Toyota Production System (TPS) enables productivity increases and cost reduction through encouraging people to learn and understand more deeply the causes of non-productivity. In the mind of the Sensei, the number one cost is a product that does not sell, and the number two cost is quality issues in manufacturing. For employees, the rationale is that cost reduction leads to more volume which means guaranteed employment. As Nampashi Hayashi puts it "*Cost Reduction is human resource development*" [17].

If we compare the approach of the Sensei to how our informants experienced their Western consultant counterparts, we found evidence of the program thinking as described by Holmemo et al. [14], before Lean was called Lean. In contrast, the Sensei teaches Lean through practical exercises, sometimes getting involved in the improvement work, sometimes pointing to deeper strategic lessons by highlighting detailed practical examples, challenging the executives to get a deeper understanding of the workings of their own companies. To quote the Toyota veteran we interviewed; "*TPS is a hard sell, it's about getting people to take ownership in their work and then supporting them to successfully complete the work.*" On the other hand, several of our informants felt that with the Western-style consultancy-driven Lean program model it was impossible to create this type of ownership as the improvement work was not led internally, and the focus for improvement efforts was on short-term cost-cutting rather than long-term, sustainable improvements in quality and productivity. The methods of the consultancy firms our informants had worked with did not have an impact on the thinking of the executives. In contrast, the Sensei method challenged the thinking of the executives, sometimes to the point of conflict. As one of our executives put it, "*a good lean consultant has to be prepared to fire his client*".

For practitioners, this has several implications. Firstly, the work reinforces the fact that the Lean transformation needs to be owned and led by the executive team, but it also suggests that an executive can significantly benefit from a Sensei relationship based on mutual respect and trust of expertise. The executives need to develop some practical knowledge and experience of TPS to truly understand its potential impact. And finally, the starting point should not be immediate cost reduction measures, but rather to develop a better understanding of how the company works, stimulated through teamwork and collaboration by creating a space for people to think, learn, and discover the next step.

5 Limitations and Suggestions for Further Research

As a first attempt to investigate the mysterious role of the Lean Sensei, this exploratory research has several limitations. However, given that the sample size is low, we have managed to consider Sensei-executive relationships from diverse geographical locations and industry sectors. For future research, we suggest a more in-depth, action learning research approach would provide a much more rich and robust data-set for further exploration of the Sensei-supported approach to Lean transformation.

References

1. Byrne A, Womack JP (2012) The lean turnaround, 1st edn. McGraw-Hill, New York
2. Womack JP, Jones DT (1996) Lean thinking, 1st edn. Free Press, New York
3. Emiliani B (2007) Better thinking, better results: a case study and analysis of an enterprise-wide lean transformation, 2nd edn. The Center for Lean Business Management, LLC, Wethersfield, CT
4. Koenigsaecker G (2009) Leading the lean enterprise transformation. Productivity Press, Taylor & Francis Group, New York
5. Achanga P, Shehab E, Roy R, Nelder G (2006) Critical success factors for lean implementation within SMEs. J Manuf Technol Manage 17:460–471. https://doi.org/10.1108/17410380610662889
6. Netland TH (2015) Critical success factors for implementing lean production: the effect of contingencies. Int J Prod Res. https://doi.org/10.1080/00207543.2015.1096976
7. Jadhav JR, Mantha SS, Rane SB (2014) Exploring barriers in lean implementation. Int J Lean Six Sigma 5:122–148. https://doi.org/10.1108/IJLSS-12-2012-0014
8. Bhasin S (2012) Prominent obstacles to lean. Int J Product Perform Manage 61:403–425. https://doi.org/10.1108/17410401211212661
9. Netland TH, Ferdows K (2014) What to expect from a corporate lean program. MIT Sloan Manage Rev Summer:83–89
10. Huxley C (2015) Three decades of lean production: practice, ideology, and resistance. Int J Sociol 45:133–151. https://doi.org/10.1080/00207659.2015.1061859
11. Rahman S, Laosirihongthong T, Sohal AS (2010) Impact of lean strategy on operational performance: a study of Thai manufacturing companies. J Manuf Technol Manage 21:839–852. https://doi.org/10.1108/17410381011077946

12. Belhadi A, Touriki FE, El Fezazi S (2016) A framework for effective implementation of lean production in small and medium-sized enterprises. J Ind Eng Manage 9:786–810. https://doi.org/10.3926/jiem.1907
13. Sisson J, Elshennawy A (2015) Achieving success with lean. Int J Lean Six Sigma 6:263–280. https://doi.org/10.1108/IJLSS-07-2014-0024
14. Holmemo MD-Q, Rolfsen M, Ingvaldsen JA (2016) Lean thinking: outside-in, bottom-up? Total Qual Manage Bus Excell 29:148–160
15. Holmemo MD, Powell DJ, Ingvaldsen JA (2018) Making it stick on borrowed time: the role of internal consultants in public sector lean transformations. TQM J 30:217–231
16. Ballé M, Handlinger P (2012) Learning lean: don't implement lean, become lean. Reflections 12:17–32
17. Hayashi N (2018) Toyota production system and the roots of lean

Lean and Industry 4.0—How to Develop a Lean Digitalization Strategy with the Value Stream Method

Markus Schneider, Mathias Michalicki and Sven Rittberger

Abstract This paper presents a methodology that allows manufacturing companies to develop an individual digitalization strategy. The Lean Digitalization Strategy is based on Value Stream Mapping and aims to lead companies on their digital transformation journey. This field-tested approach is put into practice by a workshop series. In six steps, companies achieve a common understanding of Industry 4.0 and Lean, understand the long-term vision and overall challenge, grasping the current state of their value stream, defining its target condition. This is the basis for the workshop participants to select process-oriented technologies towards the defined target condition and digitally connect them.

Keywords Value Stream Method · Digitalization · Strategy · Industry 4.0

1 Introduction

Today we live in a VUCA world. This abbreviation stands for volatility, uncertainty, complexity and ambiguity [1]. Our environment is becoming increasingly volatile. Customer requirements change ever faster and are increasingly unpredictable. Short-cyclical demand for individual products with new or improved performance and increasing division of labor lead to increased complexity in value chains. All this leads to the fact that companies are desperately seeking ways to solve these challenges. In this context, digitalization and Industry 4.0 appear to many as "promises of salvation". However, after trying several ad hoc measures, companies quickly realize that there

M. Schneider · S. Rittberger
Landshut University of Applied Sciences, Am Lurzenhof 1, 84036 Landshut, Germany
e-mail: markus.schneider@haw-landshut.de

S. Rittberger
e-mail: sven.rittberger@haw-landshut.de

M. Michalicki (✉)
PuLL Beratung GmbH, Schweriner Str. 1, 84036 Landshut, Germany
e-mail: mathias.michalicki@pull-beratung.de

M. Rossi et al. (eds.), *Proceedings of the 6th European Lean Educator Conference*, Lecture Notes in Networks and Systems 122,
https://doi.org/10.1007/978-3-030-41429-0_13

is no digitalization strategy behind them. Mostly there exists no target condition for Industry 4.0 that could lead the way into a digital future.

The aim of this paper is to show how to create a Lean-compatible digitalization strategy in a customer- and process-oriented way using the Value Stream Method 4.0.

2 Industry 4.0

2.1 Definition

The term Industry 4.0 is a synonym for the fourth industrial revolution and describes the vision of a future industrial society under the influence of the internet where the physical world of manufacturing will merge with the virtual world. People, machines, objects and systems will operate in networks and exchange information in real time. It will be possible to meet individualized customer requirements (batch size one) with greater flexibility and robustness as well as optimal resource utilization [2, 3].

2.2 Obstacles and Problems of Digitalization in Practice

When trying to implement Industry 4.0 technologies, manufacturing companies encounter certain obstacles and problems. According to [4], Industry 4.0 is a mainly technology-driven approach (technology-push). Although technologies are available in a wide and constantly growing range, their selection and implementation are done in a mostly unstructured manner. There is a great focus on experimentation and gathering experience. But only few companies employ a holistic and target-oriented decision-making process that aligned the many technologies with an overall strategy. *In summary, there exists a methodological gap for developing an individual digitalization strategy which allows to systematically select and implement Industry 4.0 technologies.*

3 Developing a Lean Digitalization Strategy with the Value Stream Method

3.1 The Link Between Lean and Industry 4.0

For almost three decades, Lean Management has been acknowledged in research to help transforming organizations so they can compete under ever-changing market conditions. Especially in a VUCA world, enabling employees to plan and execute lean, waste-free value chains should remain an essential part of a digital strategy.

Lean and Industry 4.0 pursue very similar objectives: Lean aims to meet customer requirements in terms of the highest quality, low costs and short lead times which are a prerequisite for short delivery times. In addition, Industry 4.0 intends to achieve the individualization of products and services mostly by new business models. While the objectives are similar, the approaches to achieving them differ significantly. Lean focuses intensively on the dimensions people and processes. It is therefore suggested to first optimize all processes and structures of a company following the Lean philosophy. Only in a subsequent step, it should be examined if, with the help of Industry 4.0 technologies, the companies' processes can be further optimized towards the targets set by Lean [5].

In this context, technology can only be an enabler, not an end in itself. A company's long-term goals should not be changed for Industry 4.0. It is solely determined by a company's customers and their requirements. Digitalization and Industry 4.0 only offer (new) technical building blocks to realize customer requirements more effectively. *Therefore, the authors' clear credo is Lean before Industry 4.0.*

3.2 Value Stream Mapping—A Method for Developing a Customer-Oriented Target Condition

A digitalization strategy does not mean the rigid implementation of a plan, but the gradual convergence towards a vision. Value Stream Mapping is a methodology which derives strategic steps (customer-oriented target conditions) from a company's long-term goals or vision in four steps [6]:

1. In consideration of *a long-term vision* (*True North*) and an *overall challenge*,
2. with a first-hand, informed grasp of the *current condition* (*Value Stream Analysis*),
3. a next *target condition* on the way to the vision is defined (*Value Stream Design*),
4. Moving towards the target condition uncovers *obstacles* to be worked on.

Value Stream Mapping therefore aims to analyze the current state and to develop a future state for all processes a product passes along its way to the customer. The Value Stream Design describes the target condition mentioned above. It directs all improvement activities towards an overall challenge. The Value Stream Design is used to specify requirements for improving single processes as well as for selecting enabling technologies towards a global optimum. Figure 1 illustrates the relation between the terms described.

Value Stream Mapping is the basis of the Lean Digitalization Strategy, because it is the process mapping method that is best suited for visualizing the interaction of the flow of information and material along a value chain. Their synchronization is one of the core aspects of a Digital Transformation [8]. The result is, that its standardized visualization provides a common language for a better communication between manufacturing and IT departments.

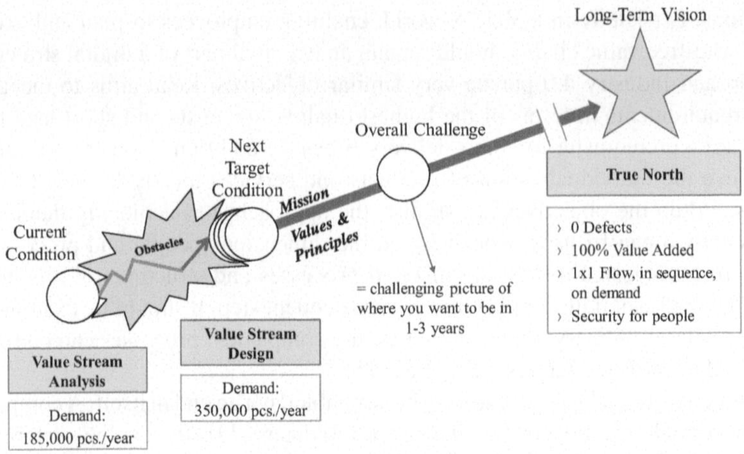

Fig. 1 Developing a customer-oriented target-condition with Value Stream Mapping [7]. Described by means of the fictional medium-sized manufacturing company BBW Inc.

3.3 The Methodology for Creating a Lean Digitalization Strategy

Figure 2 provides a comprehensive overview of the methodology for creating a Lean Digitalization Strategy. The six steps were field-tested in several workshop series with German manufacturing companies.

First Step: Fundamentals of Industry 4.0 and Value Stream Mapping. Before implementing the Lean Digitalization Strategy, it is essential to start with a common understanding in digitalization and Lean. Kick-off begins with the basics of digitalization and Industry 4.0. Furthermore, if a company lacks the necessary fundamentals of Lean, additional time is to be invested in the core concepts of Lean as well as in Values Stream Mapping.

Second Step: Defining the Overall Challenge. It is the responsibility of the executive management to define and share a clear and bright vision—the company's True

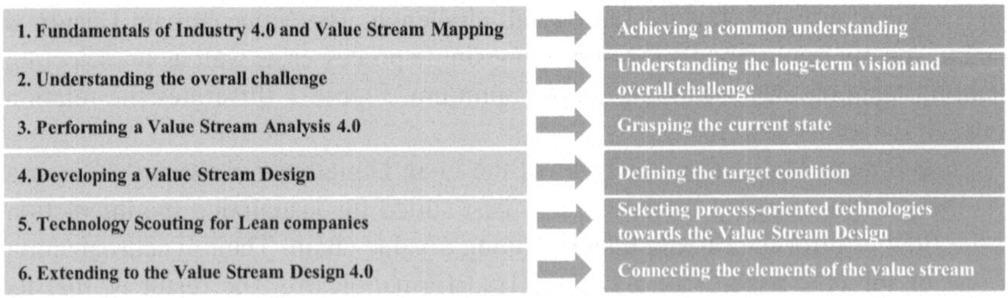

Fig. 2 The six steps of the Lean Digitalization Strategy

North. A vision is an aspirational description of what an organization intends to achieve in the long-term future. How to get there is not defined. If the employees understand, in which direction a company is to develop, they also understand which actions are necessary to get there. As a result, employees in every department and hierarchy are motivated to act independently in the direction of the common vision.

The steps of developing a Lean Digitalization Strategy are subsequently described by means of a simplified example from the workshop series. The fictional medium-sized manufacturing company BBW Inc. produces transport dollies. After being trained in the fundamentals of Industry 4.0 and Lean, the executive management defines the company's True North (see Fig. 1): The long-term vision of BBW Inc. includes zero defects, 100% value-added activities, one-piece flow in sequence and to the demand of the customers, as well as secure processes for all employees.

In a subsequent workshop, the True North is presented to the participants, so that everyone understands the overall challenge of the Lean Digitalization Strategy.

Third Step: Performing a Value Stream Analysis 4.0. Next, the current state needs to be grasped. A well-established tool for this is the Value Stream Analysis.

As described in [9], the purpose of this tool is to illustrate the current state of a company's material and information flows. Since the Lean Digitalization Strategy aims to systematically select and implement Industry 4.0 technologies, a second type of information flow is to be analyzed: The digital connectivity of the elements of the value stream. The Value Stream Analysis 4.0 is an extension of the Value Stream Analysis and helps to additionally map data flows between order or process information and analog or digital storage media (see Fig. 3). Analogous to the Value Stream Analysis, it is used to identify information logistical waste [8].

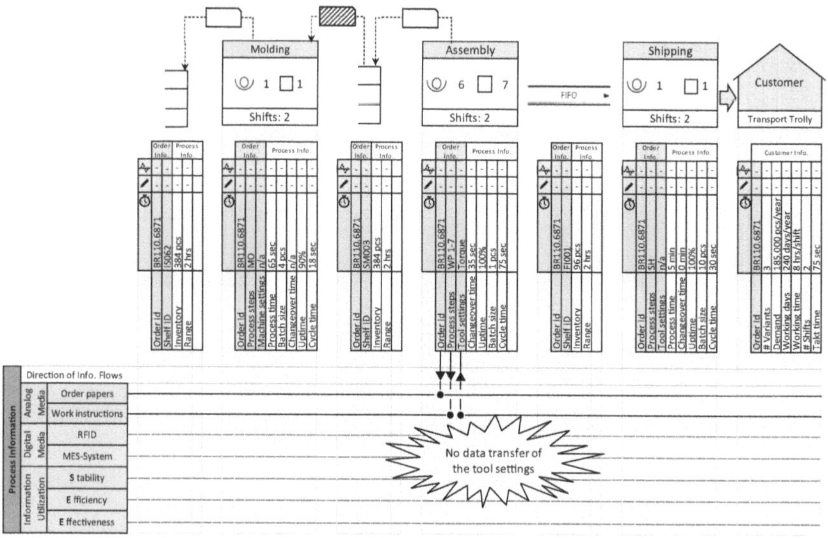

Fig. 3 Value Stream Analysis 4.0 of BBW Inc.

In the example of the BBW Inc., the current condition is a yearly demand of 185,000 transport dollies leading to a takt time of 75 s (see Fig. 1). A detailed analysis of the final assembly reveals, that for the different screwing process, the torque of the screwdrivers is set manually according to paper work instructions (see Fig. 3). This leads to quality issues and cycle time losses due to changeovers. From an information flow point of view, there is no data transfer between the order ID, the process step and the torque of the tools.

Fourth Step: Developing a Value Stream Design. The Value Stream Design is the preferred method to create a customer-oriented target condition, which should be achievable in the next six to twelve months and derived from the True North. A detailed description for creating a Value Stream Design can be found in [9].

In order to create a target condition for the value stream of the transport dolly, the workshop participants of BBW set a target condition of increasing yearly production from 185,000 to 350,000 pieces to meet future customer demand (see Fig. 1). This requires a reduces takt time from 75 to 40 s. To meet this requirement, final assembly needs to reduce its setup time and thus cycle time as illustrated in Fig. 4.

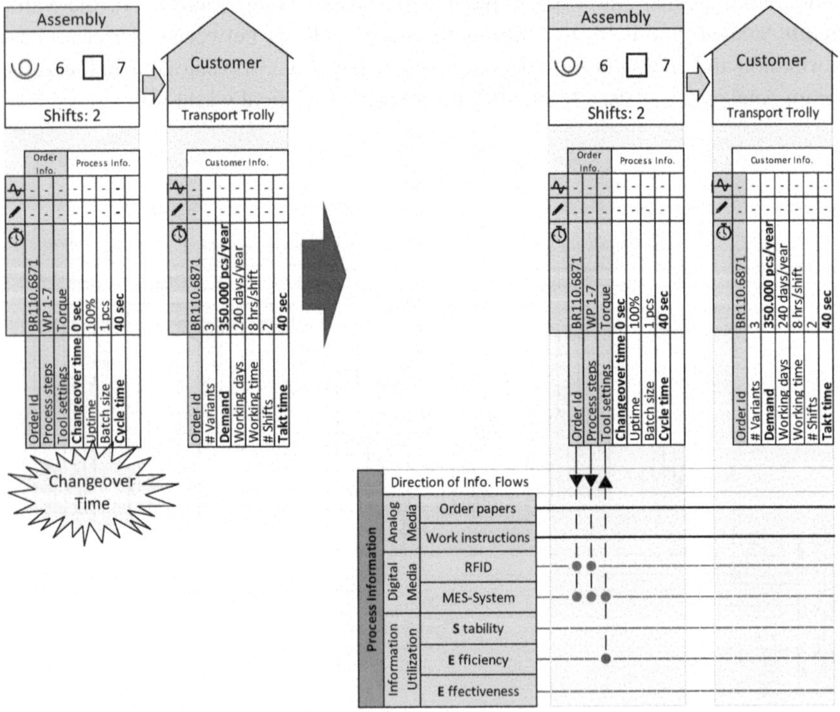

Fig. 4 Simplified Value Stream Design (left) and Value Stream Design 4.0 (right) of BBW Inc.

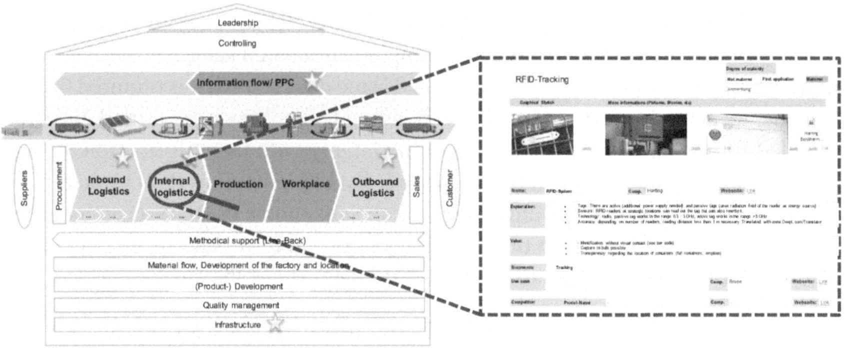

Fig. 5 Extract from the Technology Catalogue as part of the Technology Scouting

Fifth Step: Technology Scouting for Lean Companies. On the way to the Value Stream Design, obstacles occur which are not always predictable and plannable. They show on what to work on next. When recalling the credo "Lean before Industry 4.0" or process before technology, these obstacles are to be first tackled with process solutions. If and only if these potentials are exploited, the use of connected technologies can help to achieve the target condition.

While selecting suitable technologies, several questions are to be answered: What is the current state of the art? Which solutions are available today and which in the near future? If technologies are not known, they cannot be integrated into the problem-solving process. Hence, the first step is to expand the solution space. Technology Scouting serves this purpose.

Technology Scouting comprises the structured observation and early recognition of changes, potentials and relevant expertise in technological developments and processes. Methodically, this involves the use of a catalogue of technologies that is structured along the process of a value stream. For each technology there is a profile with the most important information (see Fig. 5).

Regarding the example of BBW Inc., the workshop begins with a *technology-push* approach. This allows a comprehensive overview of technological possibilities. It is followed by a *market-pull* approach: Based on the target condition of Value Stream Design, possible solutions from the Technology Catalogue are discussed by asking the following questions:

1. *Which aspect of the value stream needs to be improved?* What needs to be done next in order to reduce changeover times in final assembly of BBW Inc.? The workshop team identifies that losses occur during the manual setup of screwdrivers between different screwing processes and product variants. An automated setup allows to reduce changeover times and therefore cycle time losses.
2. *Which industry 4.0 technologies help us to realize this?* Now the question is which technology can contribute to solving the problem. This is where Technology Scouting provides an answer. The profiles help with the process-oriented

selection of certain technologies. In case of BBW, RFID-tracking and upgraded screwdrivers are best-suited for achieving an automated setup.

3. *What is the value added for the customer?* Once it is clear what is required to overcome the next obstacle, and the corresponding technology is identified, the value added for the customer is to be determined. Only with this information the usefulness of the technology can be assessed. By implementing RFID-tracking for automated tool setup, the value added for the customers of BBW is an increased efficiency. The elimination of setup times allows the workers in final assembly to deliver transport trollies to the increasing demand of the customers.

Sixth Step: Extending to the Value Stream Design 4.0. Until now, only technologies for single process steps such as workstations are selected. However, a considerable value added of digitalization is the connectivity of data and objects. This is a key step towards an improved synchronization of material and information flows and therefore faster and more efficient value streams. The visualization of connectivity and of data points is realized with Value Stream Design 4.0.

For the implementation of an automated tool setup in final assembly, the semi-finished transport trolleys are equipped with RFID-tacks. When checking in at a process step in final assembly, the screwdrivers automatically match the product variant (order ID) and the process step with the MES-System. It then retrieves information for the suitable torque setting (see Fig. 4).

4 Summary

This paper presents a methodology that allows manufacturing companies to develop an individual digitalization strategy. The Lean Digitalization Strategy show how to create a Lean-compatible digitalization strategy in a customer- and process-oriented way using the Value Stream Method 4.0.

Starting with the Value Stream Analysis 4.0, digitalization gaps are revealed. The Value Stream Design is used to develop an ideal target condition from the customer's point of view. In order to achieve this target condition, suitable technologies are then selected with the Value Stream Design 4.0 and finally connected via an IIoT platform.

References

1. Mack O, Khare A, Kramer A, Burgartz T (2016) Managing in a VUCA world. Springer, Heidelberg
2. Kagermann H, Wahlster W, Helbig J (2013) Recommendations for implementing the strategic initiative INDUSTRIE 4.0. Final report of the Industrie 4.0 working group. acatech—National Academy of Science and Engineering, Munich
3. Porter ME, Heppelmann JE (2015) How smart, connected products are transforming companies. Harvard Bus Rev 93(10):96–114

4. Bloechl SJ (2017) Radical process improvements in material provision with Industry 4.0 technologies. In: 10th International doctoral students workshop on logistics. Conference proceedings, Magdeburg
5. Enke J, Glass R, Kress A, Hambach J, Tisch M, Metternich J (2018) Industrie 4.0. Competencies for a modern production system. Procedia Manuf 23:267–272
6. Rother M (2010) Toyota Kata. Managing people for improvement, adaptiveness and superior results. McGraw-Hill, New York, NY
7. Rother M, Aulinger G (2017) Toyota Kata culture. Building organizational capability and mindset through kata coaching. McGraw-Hill, New York, NY
8. Meudt T, Metternich J, Abele E (2017) Value stream mapping 4.0. Holistic examination of value stream and information logistics in production. CIRP Ann 66(1):413–416
9. Rother M, Shook J (2003) Learning to see. Value stream mapping to add value and eliminate muda. Lean Enterprise Institute, Cambridge, MA

A Framework for Implementing Lean Through Continuous Improvement and Hoshin Kanri. A Case Study in Guanxi Culture

Silvia Gubinelli, Vittorio Cesarotti and Vito Introna

Abstract Nowadays, the literature recognizes Lean as an integrated system. Despite its widespread adoption, literature highlights a lack of standard and success cases in the implementation of Lean. Hoshin Kanri (HK the literature) is recognized as a component for implementing new management systems and can play an essential role in Lean deployment. Cultural challenges are recurring issues in the literature. Organizational culture and workers' cultural context could affect how HK and Lean are performed. This paper aim to provide a framework for Lean Deployment based on the principles of Hoshin Kanri, which allows the implementation of Lean through Continuous Improvement mechanisms. The innovation consists in incorporating the activities of the organizational culture setting in a Lean Deployment framework. The paper presents the implementation of a framework in the Chinese site of an international company. The case study highlights the importance of understanding organizational implications about lean implementation and possible resistance given by the Guanxi environment.

Keywords Lean deployment · Continuous improvement · Hoshin Kanri · Guanxi · Lean enterprise

1 Introduction

After the success achieved in Japan, starting from the eighties, Lean and HK techniques began to spread among western companies [1]. Thomas L. Jackson and Dan Jones recognized Hoshin Kanri as the heart of Lean and one of the four characteristics

S. Gubinelli (✉) · V. Cesarotti · V. Introna
Department of Enterprise Engineering, University of Rome "Tor Vergata", Via del Politecnico 1, 00133 Rome, Italy
e-mail: silvia.gubinelli@uniroma2.it

V. Cesarotti
e-mail: cesarotti@uniroma2.it

V. Introna
e-mail: vito.introna@uniroma2.it

M. Rossi et al. (eds.), *Proceedings of the 6th European Lean Educator Conference*, Lecture Notes in Networks and Systems 122,
https://doi.org/10.1007/978-3-030-41429-0_14

of highly successful business [2]. HK let companies become learning organization by applying PDCA systematically at all level of the organization and business involving in both directions: Top-down (concerned with Vision and long term plans), Bottom-up (improvement played by workers). In this way, the HK is the frame and the meeting point between the continuous improvement approaches (bottom-up) and the strategic needs (top-down) [3–5]. The main stages of the HK process may be common among a set of companies, but its implementation tends to differ because of sundry management styles and organizational culture. Corporate and workers' culture are universally acknowledged as an enabling factor for the correct implementation of lean and continuous improvement (CI) [6–8]. Despite this, literature about Hoshin Kanri and Lean Deployment lacks successful frameworks [9]. Practitioners highlight the presence of cultural challenges as issues and no studies have been carried out about the importance of the context of belonging [10–12].

This paper aims to provide a framework for Lean Deployment which allows the successful implementation of Lean over time and helps practitioners to transform their company into a lean organization through CI mechanism. The innovation consists in providing guidelines for the cultural setting in international companies, with a focus on environments characterized by guanxi cultures.

2 Literary References of the Framework

The literary analysis underlying this research was carried out by the selection and the complete reading of conference proceedings and research papers from international peer reviewed journals. The source considered was Scopus. The papers were selected according to the topics (title, abstract and keyword): Lean Deployment, HK and implementations of CI strategies. Nicholas [13] and Jolayemi [14] articles were the most representative one literature reviews for our analysis. Jolayemi [14] analyzed the main processes involved in HK applications. Nicholas [13] analyzed main sources about HK and identified the factors related to successful implementation and sustainment of QM/LP initiatives containing within HK methodology and practices. Da Silveira et al. proposed a good literature review of HK tools [15] and roadmap [16]. Literature highlights how HK practices vary by organizations, but studies revealed, all somewhat conform to the 10-step process in the following table [17, 13]. John Nicholas identified seven themes commonly associated with HK [13, 18] that are related to the processes in Table 1: (1) Vision, strategy, long- and medium-term goals, (2) Cascade goals and objectives, (3) Catchball and CFM, (4) Means/ends and targets, (5) Objectives linked to daily work, (6) Review and control, (7) Plan-Do-Check-Act (PDCA) [13].

On the basis based on the existing literature of the existing literature we defined the assumptions at the basis of this work and identified three reference pillars. The hypotheses underlying this framework are:

Table 1 HK strategic planning and implementation process

(1) Analyse organisational and environmental data for strategic planning
(2) Develop mission/purpose relating the company to its customers
(3) Develop a philosophy addressing what the organization cares about
(4) Develop a vision that defines the organization's direction and aspirations
(5) Develop long- and medium-term objectives and strategies to achieve the vision. Senior and division managers use catch ball to develop objectives
(6) Develop annual plans to achieve long- and medium-term objectives. Senior and division managers use catch ball to create plans that include: • A 'vital few' objectives that will bring 'breakthrough' improvements • Annual strategies/means to achieve the objectives • Targets for expected results • Means/actions to be taken to achieve the desired results • Measures to monitor progress and check whether strategies were appropriate
(7) Deploy policies: engage the entire organization to align plans with the organizsation's strategic direction; cascade plans to every level using catch ball
(8) Implement plans and daily management: deploy annual plans to achieve breakthrough objectives while controlling and improving business fundamentals (daily management)
(9) Review progress: identify problems, take corrective action, prepare revisions to plans
(10) Standardize processes and work tasks: retain gains resulting from breakthrough and routine improvements

Source Nicholas [13], (adapted from Jolayemi [17])

1. Align strategy and the operations is necessary to achieve a long-lived and effective lean implementation. Continuous improvement mechanisms are the pillars of lean success;
2. HK is the best tool to achieve this goal with a view to continuous improvement;
3. A consensus-building process to establish the cultural fundamental is necessary to achieve co-ordinated HK and Lean activities.

This research is based on three pillars: HK roadmap given by Jackson [2], Catch ball process proposed Tennant and Roberts [1] and framework for strategic change in organizations given by Pettigrew and reviewed by Da Silveira et al. [19].

Jackson roadmap for HK deployment (2006) presented a reviewed FAIR model and focuses on the implementation of Lean Manufacturing and Six Sigma. Jackson stresses the concept of nested PDCA, given by the systematical application of the cycle at every level of the organization [2]. About this model, Da Silveira et al. [15] stated: *"Jackson's model for Hoshin Kanri places it not only as a framework of strategic management, but also for developing different layers of organizational capabilities, especially the capabilities required to achieve lean production"* [15].

The second pillar is Pettigrew's framework (1987) for strategic change in organizations. It considers the relationships between the "content" of change, the "context" in which change occurs and the "process" through which change is undertaken [19, 20]. Da Silveira et al. [19] proposed a review of the model consisting of two macro phases: Context and Structure. Their analysis focuses on the content of strategic change and considers factors like organizational structure, design of information systems, and organizational capability [19]. Even Da Silveira et al. [15] proposed a collection of guidelines for applying HK for strategy collected in Context and Process dimensions. Their study focus on organizational culture, capabilities, alignment, integration and review. HK is considered as an organization-wide holistic system [15].

This research focuses on cultural factors that characterize the workers mindset and constitute the field of action for managers. Cultural challenges as recurring issues in the implementation of Japanese management models [1]. Da Silveira et al. [19] about this stated: "*For instance, some of the central components of HK are the concepts of Nemawashi and Catchball, which are deeply rooted in the Japanese way of thinking and their leadership style*" [19]. So the third pillar consists of theories related to the catch ball activities. The concepts of Nemawashi and Catch ball are some of the central components of HK. Tennant and Roberts [1] defined the catchball process as *"necessary for successful implementation of Hoshin Kanri in an environment of cross-functional management."* This process ensures feedback in multi-directional horizons, requires a company commitment to employee involvement and continuous improvement [1]. Tennant and Roberts [1] identify the main steps of catch-ball process design: Management style identification, Definition and consensus-building [1]. The first stage aims to examine the prevailing management style within the organization, in order to identify judgments criteria. The second one aims to involve people as team members to achieve common goals [1]. Catchball process derives from the Japanese negotiation and decision making approach known as "Nemawashi" [15]. Sagi [21] defined Nemawashi as *"a consensus building technique prior to the meeting that aims at removing obstacles in decision making or approval of a proposal from the audience."* This process involves informal meetings, premeeting and small one-on-one meetings. The scope is to seek approval so that the process can take less time to get everyone at the same level of understanding. The decision-making process results diffuse, recursive and multi-step [21].

Catch ball activities must be tailored to the local culture. This research focuses on the implementation of lean in the Chinese cultural context, so it's necessary to consider Guanxi Culture.

3 Guanxi Culture

The term Guanxi first appeared in the 1980s, when western societies began to do business in China. Zhang and Seock-Jin defined Guanxi as "*relationship or social connections based on mutual interests and benefits, which is achieved by exchanging favors and giving social status between guanxi parteners*" [22], Guanxi is not only a cultural factor but also an institutional force and governance structure and have a strong influence on the business [22, 23]. Zhang and Seock [22] presented three guanxi related concepts: Guanxi Bases (GB), Guanxi Practices (GP) and Guanxi Exchange (GE). GB is the link that connects two people belonging to the same Guanxi network. This can be from family or social nature. GP are the activities that the individual must implement to stay in the network. The nature of the relationship and guanxi practices influence the quality of the relationship. The GP allows for an active guanxi relationship in which individual people can take advantage of the guanxi base to achieve his or her own goals. The GE constitute the object of the practices carried out in the network. From a lean perspective, guanxi has a strong impact on the relationships between individuals and between the different company functions. Therefore, it is important to align the guanxi network to the value stream flows. To go from stranger to close guanxi partners Zhang and Seock [22] identified two phases to pursue:

- Guanxi initiation phase: find a link point and establish a guanxi base to become an insider of a particular guanxi circle.
- Guanxi cultivation a maintenance phase: exchange of gifts and favors, interactive behaviors and social activities.

The theory of social exchanges explains how to maintain this relationship by identifying what to exchange and which rules follow [24]. Yang and Wang [24] identify three main guanxi exchange mechanisms: favor (renqing in chinese), trustworthy (xinyong) and face (Mianzi). In business language, renqing is attributable to exchanging resources, where resources are intended as tangible, intangible (opportunities, affection etc.). Renqing can be paid only by a specific person in a specific way, through appropriate time. Zhang and Seock delineates the differences between renqing and other exchange resources, and explore renqing rules [22]. The reliability in following and respecting these rules defines the Xinyong credit and also the Mianzi. From the business point of view, xinyong is based on interpersonal trust, business ethical integrity and on past reputation of returning favors. Mianzi consists of the individual reputation, the social status, the ability to influence the other members of the guanxi network. Save Mianzi is a primary objective in Chinese society, in fact, Chinese people rely more on personal trust (xinren in Chinese) than contract or system trust at the organizational level and put personal guanxi first [22, 23]. Guanxi is a personal network, this is not related to the company. Literature about organizational guanxi recognizes the critical role of supervisor–subordinate guanxi (SSG). About this topic please see Ren and Chen [25].

4 Proposed Framework

This research aims to define a structure applicable to international companies operating in different contexts. The proposed framework takes up elements of successful models already present in the literature and recomposes them in a structure aimed at enhancing the central role of the individual and cultural context. It is not prescriptive with respect to the methods to be implemented in the different phases. So, practitioners are free to choose the methods and tools they consider most appropriate for the circumstance. The framework has two macro-phases: **Context** and **Structure**. Each of these has steps and sub-steps. The two distinct phases let to separate the recursive activities from the design of the CI structure. Therefore, the Context phase is related to the design of the CI structure. It is necessary only during the initialization of the CI organizational structure and when the internal context or organizational structure is modified, so it is necessary to modify the structure. In reverse, the STRUCTURE phase is repeated in every improvement cycle, lasting one year or 18 months.

The structure of the framework is based on Da Silveira et al. review of Pettigrew model (2017) [19]. The CONTEXT phase combines the analyses of the external context with the steps of the catch ball process reviewed by Tennant and Roberts [1]. In the STRUCTURE phase, the framework takes up the recursive approach of "nested PDCA" promoted in Jackson's roadmap [2].

The CONTEXT macro-phase is dedicated to analyzing the CONTEXT in a double point of view:

- C1. External Leanness analysis: identification of the LEANNESS level of the company with respect to the socio-economic CONTEXT in which the organization is located (supply chain policy, infrastructures present in the territory, etc.)
- C2. Cultural Leanness analysis: identification of the LEANNESS level present in the organizational culture and the relationships between workers.

For the external Leanness analysis, the literature proposes different tools and methods of supply chain analysis, assessment tools. The framework is not prescriptive. C2 focuses on Catch ball activities through the steps presented by Tennant and Roberts [1]:

- C2.1 Management Style Identification
- C2.2 Consensus building.

C1 and C2.1 activities can be played at in parallel.

In the step C2.1, the HK improvement team identifies managerial styles, maps existing relationships, carries out communicative registers and existent procedures. In this step the company must identify the following team for the implementation of the HK approach:

- The Hoshin Team (HT): composed of executive manager with great both functional and relational power, usually the first line of managers
- Tactical Team (TT): cross-functional team of middle managers dedicated to the care of a specific value stream
- Operations Team (OT): composed by the operational line of the various business functions that deals with the basic daily activities and reports to the tactical team.

For each value stream, are identified: a dedicated Tactical Team (VSTT), one or more Operations team and HT member who lead the groups. For more information about the work of these teams, refer to the Jackson roadmap [2]. Based on the team membership, training is provided on the lean, six sigma and CI. The team are the improvement engine, on these teams it is necessary to build consensus (C2.2).

The Consensus Building Activities are strongly influenced by the cultural belonging of the worker. In China, the role of guanxi is central. For the management of guanxi, the framework takes up the passages illustrated by Zhang and Seock-Jin [22]:

- Guanxi Initialization
- Guanxi Cultivation and Maintenance.

To guarantee the engagement of the members of a Chinese organization, it is necessary to build a Guanxi network that reflects the HK structure. Process mapping, metrics individuation and KPI setting (activities P2, P3, P4) constitute the link point. For these reasons, these activities of stage C2.1 are carried out simultaneously with the activities of the PLAN phase (first stage of STRUCTURE macro-phase) (Fig. 1). The STRUCTURE phase includes the following substep:

- Plan:
  - Process Mapping, KPI identification and Setting (S1).
  - AS IS Situation Analysis, gap analysis and identification of improvement area.

  This phase ends up with the definition of the improvement strategy (S2).

- Implementation of improvement project and daily activities maintain the CI approach:

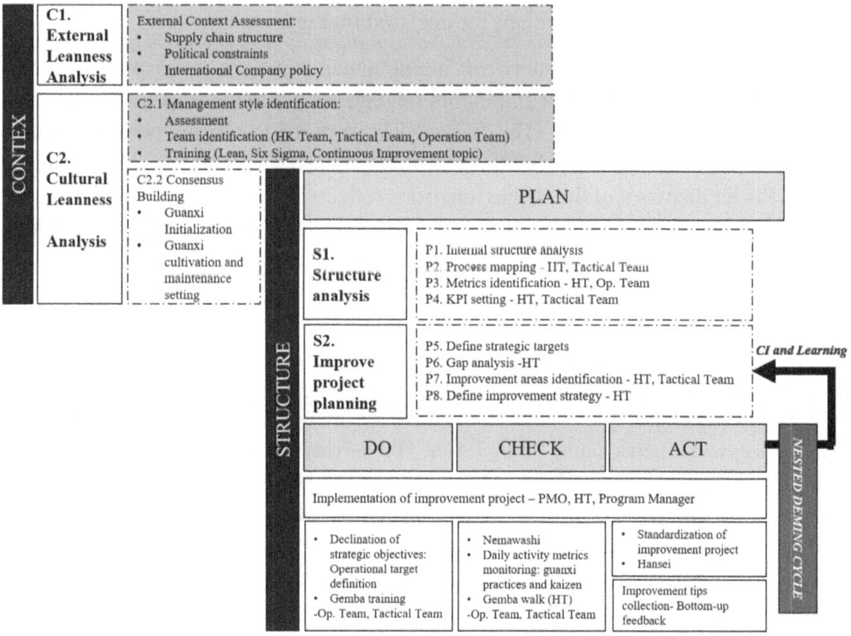

Fig. 1 Structure of the proposed framework

- Do: Strategy declination in the different levels of the organization to provide measurable indicators of the improvement. Project implementation.
- Check: Indicators analysis as enabling factor for the CI implementation and performance control of the improvement projects.
- Act: Improvements standardization and bottom up feedback.

These substeps are implemented at all different levels of the organization to allow the workers to measure their contribution to the strategy performances. For more information about the "nested-PDCA" refer to Jackson's roadmap [2].

To set the consensus, the Mentor role (who connects people from different guanxi circle) is played by a super-parties person with high commitment, with expertise in CI and with high xinyong (trustworthy) and mianzi (face) credits.

Usually, the highly hierarchical structure of Chinese companies means that every function constitutes a different guanxi circle. The value stream analysis activities carried out in multi-functional groups allows the creation of a VS guanxi network. The forcing elements that push guanxi exchange mechanism into the TT and HT are:

- KPI definition activities: The S1 phase, each function defines internal KPIs and Value stream KPIs (trans-functional process KPIs). The trans-functional KPIs oblige resources of different functions, even parigrade, to collaborate with a continuous feedback cycle. This let to initialize a guanxi circle for each value stream.
- Unanimous approval mechanism for decision making.

The setting of the guanxi network maintenance practices takes place through the definition of procedures for indicators management and the improvement of the trans-functional process KPIs. The framework proposes to use the periodic review activities of the trans-functional KPIs to implement the network practices. Transfunctional KPIs let the work of the single functions reflects in the common performances, allowing the creation of Guanxi exchange mechanisms. The implementation of lean visual management techniques and the definition of one point lesson dedicated to daily activities of maintenance and monitoring of flows encourages the collaboration between the subjects of the network by leveraging the concepts of xinyong (trustworthy) and mianzi (face). At the end of the CONTEXT macro phase and step S1, the organizational structure is ready to implement the IC.

In Step S2, the strategic objectives of the year are deployed at different company levels in terms of metrics and KPIs. Later, HT carries out the gap analysis between the value of current metrics/KPIs, the value that the HT hoped to have (target) and the value deriving from the new strategic objectives. Based on this analysis, two lines of activity are defined: Daily activities and Improvement projects.

According to the Jackson model, Daily activities are carried out at all levels of the organization following the Nested PDCA in a Kaizen perspective through lean principles and tools. For major improvements (as process changes, procedures, layouts and existing standards), HT sets dedicated breakthrough projects. The improvement projects refer to a metric/KPI. They are carried out by the TT of the value stream

(VS) and led by the HT member of the VS. The projects must be conducted following the theories of Project Management, the PMO function is necessary. In the DO, CHECK, ACT phases, these two lines of activity are brought in parallel.

The PDCA cycle can be repeated both daily and for a whole year, depending on the task. The lessons learned (hansei) are standardized in each cycle and contextually the analysis of the gap is re-examined when the new strategic objectives for the following year S2 are analyzed.

4.1 Gefran Case

GEFRAN is an Italian multinational company, specialized in designing and manufacturing sensors, systems and components for the automation and control of industrial processes. The company now has headquarter is in Italy and operates in the major markets through six production branches in Brazil, China, Germany, India, Switzerland and the US. The case presented takes place within a research and innovation project that aims to create a scalable model for the implementation of lean at a global level in order to create a learning international company. In this ambitious project, a pilot project was implemented in the Chinese site based in Shanghai: GSDT. The framework has been faithfully implemented.

The C2.1 External Context Assessment was played by a Consultant who also trained resources on CI, Lean and Six Sigma topics. First was assessed the headquarter (Italy) (5 days), and suddenly GSDT, China site, (15 days). In phase C2. (Cultural Leanness Analysis), Consultant conducted individual and group interviews on the first managerial lines of GSDT to identify the existing social network within the organization (15 days). Based on the results of this assessment, the analysis of the organization chart and individual skills, General managers and the Consultant formed the HT, TT and OT. The assessment on workers showed an almost exact correspondence (5 functions out of 7) between the social network and the organization chart. The results were in agreement with the strongly hierarchical culture typical of the Chinese population. All the seven functions were operating as guanxi circles isolated from the other functions as separate silos, but these did not correspond to the value streams. To overcome those problems, it was decided first to form the HT with the first managerial line of the main functions: Sales, Finance, Operations, Quality, Human Resources, Purchasing, and Repair. Then, the HT was trained on the Lean principles and process mapping tools. Consensus building was implemented using the P2, P3, P4 activities that involved HT and the Consultant, who held the role of connection person among the various guanxi circles. The training activity combined with the strong commitment given by the general manager, allowed him to take the role of Mentor and obtain a high level of xinyong (trust-worthy) and mianzi (face). The consultant was responsible for organizing meetings with the people related to the individual value streams and coordinated the mapping and definition activities of the metrics that were carried out personally by the HT members and their functional team members involved in the flow. Value Stream Tactical Teams (VSTT) were created.

VSTT, supervised by the Mentor, mapped the processes. The maps were approved in periodic meetings held by the HT and the general manager, in which unanimous approval was required. The approval processes required four meeting. After the second meeting, Nemawashi trials emerged. After the mapping activities, HT resulted effective in decisions making.

VSTT identified metrics and indicators have been identified for each process. HT, supervised by consultant composed the value stream KPIs. The functions were evaluated and monitored on internal metrics and transitional KPIs. The rules for the interaction and revision of the metrics in VSTT constitute guanxi practices. Communication rules and tools were defined after a second training session on HT and TT on lean management topic. To build on the concepts of trustworthy (xinyong) and face (Mianzi), visual data collection boards have been created in common areas.

After this setting activity of the organizational structure. GM and HT carried out a gap analysis between the values of the metrics measured (as is) and the expected values. The HT identified improvement areas and a program of improvement projects. GSDT started six local improvement projects and two global improvement projects in combination with the Italian site.

5 Conclusions

The implementation of the framework highlights the centrality of cultural aspects. The proposed framework, not being prescriptive, is easily adaptable to pre-existing tools in the company.

In CI structure' design, the training and the right teams' identification are crucial. The social network is confirmed as an enabling factor for the CI. In the implementation of the model, the progress of the daily kaizen activities, in terms of time and effectiveness, were proportional to the quality of the guanxi network.

The process mapping and the definition of the KPIs allowed the initialization of the Guanxi networks. While daily monitoring of the metrics and periodic analysis of the KPIs guaranteed the maintenance of the network.

The lean visual management and communication tools resulted effective support for the creation of favor, trustworthy (xinyong) and face (Mianzi).

Acknowledgements The authors thank Gefran Company for believing in this project research. The authors are grateful for receiving continuous wholehearted support and innovative vision provided by Mr. Fabrizio Merz, General Manager of GSDT. The authors express our heartiest thanks to the GSDT' improvement team that made this project practically feasible with their daily commitment.

References

1. Tennant C, Roberts P (2001) Hoshin Kanri: implementing the catchball process. Long Range Plann 34:287–308
2. Jackson TL (2006) Hoshin Kanri for the lean enterprise, developing competitive capabilities and managing profit. Productivity Press, New York
3. Witcher BJ, Butterworth R (1999) Hoshin Kanri, how xerox manages. Long Range Plann 32(3):323–332
4. Ounda H, Ahmed K (2016) A proposed systematic framework for applying Hoshin Kanri strategic planning methodology in educational institutions. Eur Sci J (12)16:158–194
5. Witcher BJ, Butterworth R (2001) Hoshin Kanri: policy management in Japanese-owned UK subsidiaries. J Manag Stud 38(5):651–674
6. Stamm D (2004) Kinda, sorta lean. Ind Eng 36:2–22
7. Testani M, Ramakrishnan S (2013) Leader-centric Hoshin planning: a systemic approach for sustaining an enterprise-wide lean transformation. In: IIE annual conference and expo 2013
8. Testani M, Ramakrishnan S (2011) Lean transformation leadership model: leadership's role in creating lean culture. In: 61st annual IIE conference and expo proceedings
9. Mirdad W, Eseonu C (2017) A cause-effect strategy map for lean process transformation. Int J Syst Syst Eng 8(2):121–146
10. Negrão L, Filho M, Marodin G (2016) Lean practices and their effect on performance: a literature review. Prod Plann Control, 1–24
11. Cesarotti V, Gubinelli S, Introna V (2019) Lean implementation in different countries: a literature. In: Proceedings of the summer school F. Turco, Brescia
12. Hines P, Holwe M, Rich N (2004) Learning to evolve: a review of contemporary lean thinking. Int J Oper Prod Manag (24)10:994–1011
13. Nicholas J (2014) Hoshin Kanri and critical success factors in quality management and lean production. Total Qual Manag Bus Excell, 1–15
14. Jolayemi JK (2008) Hoshin kanri and Hoshin process: a review and literature survey. Total Qual Manag Bus Excell 19(3):295–320
15. DaSilveira WG, De Lima EP, Deschamps F, Da Costa SEG (2018) Identification of guidelines for Hoshin Kanri initiatives. Int J Prod Perform Manag 67(1):85–110
16. Da Silveira W, De Lima E, Deschamps F, Da Costa S (2015) Guidelines for Hoshin Kanri: successive refinements with experts. In: IIE annual conference and expo 2015
17. Jolayemi J (2009) Policy deployment: a review and comparison of two best practices models. Total Qual Manag Bus Excell 20(8):877–902
18. Witcher B, Chau V (2007) Balanced scorecard and Hoshin Kanri: dynamic capabilities for managing strategic fit. Manag Decis 45(3):518–538
19. Da SilveiraWG, De Lima EP, Da Costa SEG, Deschamps F (2017) Guidelines for Hoshin Kanri implementation: development and discussion. Prod Plann Control 28(10):843–859
20. Pettigrew A, Whipp R (1993) Managing change for competitive success. Wiley-Blackwell
21. Sagi S (2015) "Nemawashi" a technique to gain consensus in Japanese management. Int J Arts Humanit Manag Stud 1(4):23–28
22. Zhang C, Seock-Jin H (2017) Guanxi culture: how it affects the business model of Chinese firms. In: The Chinese business model: originality and limits. Elsevier Asian Studies Series, pp 19–40
23. Chang A, Guo C, Zolin R, Yang X (2014) Guanxi as a complex adaptive system: definition, description and underlying principals. J Asia Bus Stud (8)2:81–103
24. Yang Z, Wang CL (2011) Guanxi as a governance mechanism in business markets: its characteristics, relevant theories, and future research directions. Ind Mark Manag 40(4):492–495
25. Ren H, Chen CW (2018) Why do Chinese employees engage in building supervisor-subordinate guanxi? Chin Manag Stud 12(1):148–163

Advanced Technologies Supporting the Implementation of Lean/Green Supply Chain Management Practices and Its Influence on the Performance

Helena Carvalho, Carina Pimentel, Susana Azevedo and Angel M. Gento

Abstract The lean/green supply chain paradigm supports a better performance in terms of economic, environmental and social dimensions. This paper describes a research about the deployment of advanced technologies such as internet of things (IoT), RFID, big data analytics and cloud computing to support the implementation of lean/green supply chain management practices and its influence on the performance. To this end, four lean/green supply chain management practices were studied: using green purchasing guidelines and sourcing from environmental sources, using reusable packaging to materials delivery, just-in-sequence and single sourcing. This paper main contribution is to provide a managerial perspective of how the advanced technologies could support lean/green supply chain practices helping to achieve a better supply chain performance.

Keywords Lean/green · Advanced technologies · Supply chain · Performance

---

H. Carvalho (✉) · C. Pimentel
Departamento de Engenharia Mecânica e Industrial, Faculdade de Ciências e Tecnologia, UNIDEMI, Universidade Nova de Lisboa, Caparica, Portugal
e-mail: hmlc@fct.unl.pt

C. Pimentel
e-mail: carina.pimentel@ua.pt

C. Pimentel
DEGEIT, GOVCOPP, University of Aveiro, Aveiro, Portugal

S. Azevedo
Departamento de Gestão e Economia, CEFAGE-UBI, University of Beira Interior, Covilhã, Portugal
e-mail: sazevedo@ubi.pt

A. M. Gento
Escuela de Ingenierías Industriales/Organización de Empresas y CIM, University of Valladolid, Valladolid, Spain
e-mail: gento@eii.uva.es

M. Rossi et al. (eds.), *Proceedings of the 6th European Lean Educator Conference*, Lecture Notes in Networks and Systems 122,
https://doi.org/10.1007/978-3-030-41429-0_15

1 Introduction

Companies are facing challenges in what is concerning the adoption of "new" technologies such as internet of things (IoT), radio frequency identification (RFID), big data analytics and cloud computing. Even though these advanced technologies have already been implemented in several companies around the world, there is still lacking knowledge about their effectiveness in supporting supply chain management (SCM) practices [1]. At the same time, the adoption of these technologies should be done considering its effects on the supply chain (SC) triple bottom line (TBL) performance [2]; this is, considering the improvements (or not) in the economic, social and environmental perspective. This paper intends to contribute to providing some guidance on the topic. Thus, the main research objectives are: (1) identify how companies are deploying the advanced technologies within their SCM processes, (2) determine which advanced technologies are enablers of lean/green SCM practices, and (3) identify the advanced technologies implementation effects on TBL performance.

The paper is organised as follows. In Sect. 2, the main theoretical concepts will be presented. An overview of lean/green SCM practices, advanced technologies, and performance measures under study is presented. Section 3 presents the empirical study and its main results. Moreover, finally, in Sect. 4, the main conclusions are drawn, and research limitations and future research topics are presented.

2 Theoretical Background

2.1 *Lean/Green Supply Chain Management Practices*

The lean and green paradigms share the common goal of reducing "waste". From a lean perspective "waste" refers to all activities that do not add value to the product or service. In the environmental perspective, the goal is to reduce materials consumption, emissions and effluents, and improve eco-efficiency. Therefore, the deployment of lean and green paradigms creates synergies. As was previously argued, lean and green paradigms had been considered a source of competitive advantage for companies and SCs. This motivated the development of numerous works about the lean/green topic in the last years, e.g. [3]. Grounded on the results achieved by Azevedo et al. [4] about the influence of lean and green upstream SCM practices on sustainability, this research considers the following set of lean/green SCM practices:

- *Using green purchasing guidelines and sourcing from environmental sources*: Although the purchasing of green materials represents a cost, it can create economic value such as reduced disposal and liability cost, while improving the organisation's resource conservation and public image. Adopting green purchasing practices avoids buying-in waste and reduces environmental costs.

- *Using reusable packaging*: The use of reusable packages requires co-operation with suppliers and helps to reduce storage and recovery delays, which represent operational cost savings and at the same time, are environmentally correct.
- *Just-in-sequence*: It implies that the delivery of materials and components into the assembly line is defined in a pre-determined sequence according to the production order and for a unique identification number. Adopting a just-in-sequence approach in SC represents a high time-dependency.
- *Single-sourcing*: The integration of socially responsible suppliers into companies' upstream SCs promotes the companies' commitment to maintaining relationships with them and creates a high level of trust among partners. On the other hand, the supplier produces a restricted quantity of materials or components that are dedicated to a single customer (the producer); this involves an easier system of control.

2.2 Advanced Technologies

The business digitalisation extends the concept of smart manufacturing to the SC context. A smart SC or a SC4.0 is a new interconnected business system which extends from isolated, local, and single-company applications to SC wide systematic smart implementations [5]. The SC4.0 would possess technologies and capabilities such as interconnectivity, fully enabling data collection and real-time communication across all SC stages, intelligent decision-making, and efficient and responsive processes to serve customers better. In this paper, four main bundles of smart technologies are considered: internet of things (IoT), radio frequency identification (RFID), cloud computing and big data analytics. These technologies were chosen because they were considered by Carvalho et al. [6] as the most used to support SC business processes.

The potential for using the IoT in SC is huge [7]. However, its real business value, when applied to the SC, has not yet been fully recognised or thoroughly addressed. As the IoT shortens the distance between suppliers and manufacturers, it also contributes to optimise total inventory levels and lower the total costs through data analysis and real-time decision making [8]. For example, Wurth USA, an auto-parts supplier, implements IoT to monitor their inventory level, control purchasing quality, and make their inventory replenishment decisions [9]. IoT technologies can also influence the sustainability performance offering potentials for end of life processes [10].

The utilisation of RFID tags and sensors represents an effective way of collecting and processing field data from manufacturing facilities with benefits in productivity, quality, responsive production planning and control [11]. Leading automotive manufacturers have been at the forefront of adopting RFID and ubiquitous computing technologies to alleviate their advanced manufacturing systems, as Volvo Trucks [12].

Big data analytics can also influence the sustainability performance of the automotive SC. Davenport [13] defends that, to generate competitive advantage, the integration of big data and predictive analytics tools, methodologies, and resources must occur throughout the company. The popular and professional press is beginning to mention how big data can and should be used to enhance sustainability measures [14].

Cloud computing is another technology that has contributed to enhancing the sustainability of many companies in the automotive industry. Relational capabilities, including forging proper incentives, sharing information and setting common goals, tend to align behaviours among SC partners, thereby leading to greater social and environmental performance [15]. Scarce research exists on analysing cloud computing's implications in collaboration and economic/environmental performance. This enhanced collaboration will impact not only economic performance dimensions but environmental performance dimensions as well between SC partners [16].

2.3 *Supply Chain Performance Measurement*

To assess the impact of lean/green SCM practices on performance, a TBL approach should be considered. The literature provides some examples of how lean/green SCM practices can influence sustainability. Franchetti et al. [17] define a set of green goals that can be implemented through lean thinking such as select low environmental impact materials; avoid toxic or hazardous materials; choose cleaner production processes; maximise energy and water efficiencies; design for waste minimisation; design for recyclability and reuse of materials. The authors assert that this symbiosis between lean and green helps to address also inventory control, which reduces space needed in a facility, thereby reducing heating, cooling, and energy costs; and logistics, an area in which waste can be cut through leaner transport and packaging of goods. In the literature, there are some examples of synergies reached with lean/green. For example, Markley and Davis [18] described that Ford Company implemented recyclable plastic containers for shipping their car parts, improving process efficiency because the new containers are handled more easily by plant workers and reducing transportation costs by over 25%. In this study, the influence of lean/green SCM practices on sustainability is assessed using the performance measures in Table 1. The performance measures were selected from a previous study developed by Azevedo et al. [4].

3 Research Methodology

A research methodology based on a survey method was selected since it allows collecting the perception of managers about the relations that could exist between the deployment of a set of advanced technologies and the effective implementation of lean/green SCM practices, as well as, numerical comparison of differences among

Table 1 Performance measures used in this research [4]

Performance measures	Characterisation
Distributed economic value	Direct economic value generated and distributed on an actual basis
Environmental cost	Sum of all costs to fulfil environmental protection responsibilities, e.g. implementation of laws, regulations and policies, and operations to prevent adverse impact on the natural environment
Supplier environmental screening	Supplier monitoring of environmental risks and weak points
CO_2 emissions	Directly correlated to fossil fuel consumption
Corruption risk	Conduct codes adoption and employing programs to regulate anti-corruption practices among suppliers
Sourcing from local suppliers	Sourcing from local suppliers contributes to increasing regional incomes and social wellbeing

companies. Once we are approaching a recent topic, the nature of our research is exploratory. Karlsson [19] explains that exploratory survey research takes place during the early stages of research on a phenomenon when the objective is to gain preliminary insight into a topic and provides the basis for more in-depth survey research. A questionnaire instrument was designed, based on previous literature, revised and discussed among the research team. Also, the team members agreed on aims, approach and interpretations of the data regarding the survey creation and application.

The sampling process started with the selection of one professional from the Portuguese' automotive industry and another from the Spanish' automotive industry. Then, using the snowball technique, these two professionals were asked to fill the questionnaire and to invite other colleagues (working in Portuguese or Spanish companies of different sizes and positions in the automotive SC) to participate in the study. The findings resulted from the analysis and synthesis of the data collected until March 2019.

4 Empirical Study

Resulting from the snowball technique, professionals from 9 companies agreed to collaborate with this study and give their perception about the performance impacts of using advanced technologies in supporting the lean/green SCM practices. Seven of them work in companies producing car parts/components, and two at car manufacturers. They have different functions such as Logistic Manager/Coordinator (4), Plant Manager (3), Information Systems Manager (1) and R&D Manager (1). Figure 1 provides an overview of the lean/green practices implementation and Fig. 2 presents the level of technology adoption within the companies under study.

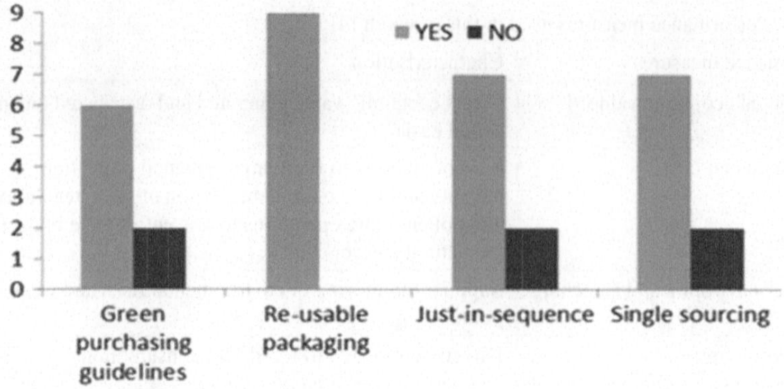

Fig. 1 Implementation of lean/green practices within the sample. *Note* Only eight responses were obtained for the green purchasing guidelines

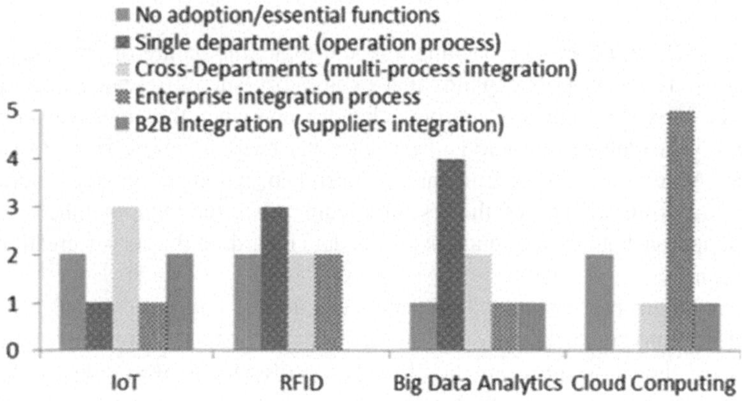

Fig. 2 Advanced technologies adoption within the sample

The results show that all nine companies use reusable packaging. Just-in-sequence and single-sourcing are usual day-to-day practices for seven companies. About the green purchasing guidelines, one manager could not answer to this question, and two managers recognised a low level of implementation for this practice.

The managers were also asked to give their perception about the use of advanced technologies in the effective implementation of lean/green SCM, using a four-point Likert-scale where 1 means "Not at all" and 4 means "To a great extent". Table 2 provides the average values of such perceptions. Although the reduced size of the sample does not allow performing statistical analysis, the results provide some important indications: all managers agree that IoT and RFID are enablers of just-in-sequence, and the RFID is an enabler of reusable packaging. Also, some technologies are not

Table 2 Importance of advanced technologies in the implementation of lean/green practices

	Just-in-sequence	Single sourcing	Green purchasing guidelines	Reusable packaging
IoT	3.9	2.6	2.5	2.8
RFID	3.4	3.1	2.9	3.5
Big data analytics	2.9	3	2.9	2.1
Cloud computing	3	2.4	2.9	2.4

Likert scale: 1—Not at all, 2—To a small extent, 3—To a moderate extent, 4—To a great extent

considered essential for implementing lean/green practices; e.g. big data analytics is not an enabler of reusable packaging.

For the advanced technologies that were identified as enablers of lean/green practices, the managers were asked about if those advanced technologies could help to increase, to decrease or does not impact the performance measures. There were difficulties in achieving a consensus among the managers. This may be justified because there are direct and indirect effects, as well as short and long-term effects on company performance, derived from the use of advanced technologies. So, it is difficult for managers to have a clear perception of the technology impact. Table 3 presents the relationship in which most of the managers agree.

Attending to Table 3 a consensus exists about the positive impact of using the IoT and the RFID in supporting the just-in-sequence and the reusable packaging practices on the performance measures: distributed economic value, environmental cost, and CO_2 emissions.

Table 3 Impact of using advanced technologies on performance

	Dist. eco. value	Env. cost	Supp. env. Screen.	CO_2	Corrupt. risk	Source Loc. Supp.
IoT applied to just-in-sequence	*Increase*	*Decrease*		*Decrease*	*Decrease*	
RFID applied to just-in-sequence	*Increase*	*Decrease*		*Decrease*	*Decrease*	
RFID applied to reusable packaging	*Increase*	*Decrease*		*Decrease*		*Does not change*

Note The performance measures presented in row 1 are the ones described in Table 1

5 Conclusions

This paper aims to study the deployment of advanced technologies to support the implementation of lean/green SC practices and its influence on performance. To attend this objective, a survey was performed using the perception of professionals from the automotive industry from Portugal and Spain. A set of advanced technologies, lean/green practices and performance measures were identified from the literature review.

Attending to the professional's perceptions, it could be concluded that IoT and RFID are considered as an enabler of just-in-sequence and the RFID of reusable packaging. Also, not all technologies are considered essential for implementing some lean/green practices, e.g., the big data analytics when applied to the reusable packaging. Also, there was not a consensus among managers about the impact of using some advanced technologies in lean/green practices on the performance. Only a positive impact is identified by using IoT and the RFID in the just-in-sequence and the reusable packaging practices on the distributed economic value, environmental cost, and CO_2 emissions. Thus, the main practical implication from this study is the identification of the set of advanced technologies that could be deployed by automotive SCs to support their lean/green SC practices and, also, impact positively the company TBL performance. Besides the contribution of this research, some limitations are identified, to be addressed as future research, such as a larger sample of professionals from the Portuguese and Spanish automotive industry should be used to make both the conclusions more robust and to perform benchmarking analyses between the two countries; also, an in-depth multiple case study of specific automotive SCs should be performed to gain deeper insights about the effects and relationships that emerged from the survey results analysis. To generalise the findings of this research it would also be useful to research other sectors since this research is focused only on the automotive SC.

Acknowledgements The authors would like to acknowledge financial support from Fundação para a Ciência e Tecnologia (grants: UIDB/00667/2020, UID/ECO/04007/2019, UID/CPO/04058/2019).

References

1. Farahani P, Meier C, Wilke J (2017) Digital supply chain management agenda for the automotive supplier industry. In: Oswald G, Kleinemeier M (eds) Shaping the digital enterprise. Springer
2. Treiblmaier H (2019) Combining blockchain technology and the physical internet to achieve triple bottom line sustainability: a comprehensive research agenda for modern logistics and supply chain management. Logistics 3(1):10
3. Azevedo SG, Carvalho H (2019) Lean and green supply chains. In: Handbook on the sustainable supply chain. Edward Elgar Publishing, pp 261–290
4. Azevedo SG, Carvalho H, Duarte S, Cruz-Machado V (2012) Influence of green and lean upstream supply chain management practices on business sustainability. IEEE Trans Eng Manag 59(4):753–765

5. Wu L, Yue X, Jin A, Yen DC (2016) Smart supply chain management: a review and implications for future research. Int J Logist Manag 27(2):395–417
6. Carvalho H, Pimentel C, Azevedo S, Velez J (2018) Advanced technologies supporting smart supply chain business processes. In: ILS 2018 information systems, logistics and supply chain, proceedings, pp 492–500
7. Ben-Daya M, Hassini E, Bahroun Z (2017) Internet of things and supply chain management: a literature review. Int J Prod Res, 1–24
8. Reaidy PJ, Gunasekaran A, Spalanzani A (2015) Bottom-up approach based on Internet of Things for order fulfillment in a collaborative warehousing environment. Int J Prod Econ 159:29–40
9. Bughin J, Chui M, Manyika J (2015) An executive's guide to the Internet of Things. Retrieved from https://www.mckinsey.com/Insights/Bus-siness_Technology/An_executives_guide_to_the_Internet_of_Things?cid=digital-eml-altmip-mck-oth-1508
10. Song ML, Fisher R, Wang JL, Cui LB (2018) Environmental performance evaluation with big data: theories and methods. Ann Oper Res 270(1–2):459–472
11. Azevedo SG, Carvalho H (2012) Contribution of RFID technology to better management of fashion supply chains. Int J Retail Distrib Manag 40(2):128–156
12. Johansson M (2005) Identification of the main factors influencing an RFID implementation in the automotive and pharmaceutical industries. Linkopings University
13. Davenport TH (2006) Competing on analytics. Harvard Business Review
14. Hsu J (2014) Why big data will have an impact on sustainability. The guardian. Retrieved from https://www.theguardian.com/sustainable-business/big-data-impactsustainable-business
15. Parmigiani A, Klassen RD, Russo MV (2011) Efficiency meets accountability: performance implications of supply chain configuration, control, and capabilities. J Oper Manag 29(3):212–223
16. Schniederjans G, Hales D (2016) Cloud computing and its impact on economic and environmental performance: a transaction cost economics perspective. Decis Support Syst 86:73–82
17. Franchetti M, Bedal K, Ulloa J, Grodek S (2009) Lean and green: Industrial engineering methods are natural stepping stones to green engineering. Ind Eng 41(9):24–29
18. Markley M, Davis L (2007) Exploring future competitive advantage through sustainable supply chains. Int J Phys Distrib Logist Manag 37(9):763–774
19. Karlsson C (ed) (2016) Research methods for operations management, 2nd edn. Routledge, UK

The Impacts of Additive Manufacturing Technology on Lean/Green Supply Chain Management Practices

Bruna Torres, Bardia Naghshineh, Gonçalo Cardeal, Duarte Filipe, Helena Carvalho, Paulo Peças and Inês Ribeiro

Abstract Additive manufacturing (AM) is an emerging technology that is changing the supply chain's structure in a context where consumers are looking for more diversified, customizable, and yet more environmentally friendly products and technologies. The literature suggests that AM technology improves the efficiency of a supply chain by contributing to waste reduction, elimination of many assembly steps, and less energy consumption, which in turn results in "leaner" and "greener" production processes. To further investigate such implications of AM technology, this study collects anecdotal evidence from research papers regarding the impact of AM on supporting four of the lean/green supply chain practices. As a result, a theoretical framework is developed; and in a final research stage, it is discussed in an interview with a company manager. This paper's main contribution is to provide a theoretical and managerial perspective on how the AM technology can support lean/green supply chain practices helping to achieve a better supply chain performance.

B. Torres · G. Cardeal · P. Peças · I. Ribeiro
Instituto Superior Técnico, IDMEC, Universidade de Lisboa, Avenida Rovisco Pais, 1049-001 Lisboa, Portugal
e-mail: bruna.ferreira@tecnico.ulisboa.pt

G. Cardeal
e-mail: goncalo.cardeal@tecnico.ulisboa.pt

P. Peças
e-mail: ppecass@tecnico.ulisboa.pt

I. Ribeiro
e-mail: ines.ribeiro@tecnico.ulisboa.pt

B. Naghshineh · H. Carvalho (✉)
Departamento de Engenharia Mecânica e Industrial, Faculdade de Ciências e Tecnologia, Universidade Nova de Lisboa, UNIDEMI, 2829-516 Caparica, Portugal
e-mail: hmlc@fct.unl.pt

B. Naghshineh
e-mail: b.naghshineh@fct.unl.pt

D. Filipe
The Navigator Company S.A., Av. Fontes Pereira de Melo, 1050-117 Lisboa, Portugal
e-mail: duarte.filipe@thenavigatorcompany.com

M. Rossi et al. (eds.), *Proceedings of the 6th European Lean Educator Conference*, Lecture Notes in Networks and Systems 122,
https://doi.org/10.1007/978-3-030-41429-0_16

Keywords Additive manufacturing · Supply chain · Lean/green · Sustainability

1 Introduction

Additive manufacturing (AM) describes a set of technologies based on "a process of joining materials to make objects from 3D model data, usually layer upon layer, as opposed to subtractive manufacturing methodologies" [1]. These technologies are different from conventional manufacturing technologies, e.g., machining, which uses subtractive methods on a raw material piece. Generally, AM processes introduce a series of advantages, amongst which are faster time-to-market, higher customisation and flexibility, material savings, toolless manufacturing of complex geometries and fabrication of free-form enclosed structures [2].

The importance of AM has risen significantly between organisations since it can be applied in various phases of a product's lifecycle supporting sustainable manufacturing processes [3]. With the development of new processes and the capability to work with more materials, it is expected that AM will become a disruptive manufacturing method [4]. There are still questions about its impact on production systems, particularly in what concerns supply chain management (SCM) and how companies could leverage it to achieve sustainable practices. Lean/green SCM approaches have been used to exploit synergies and improve business performance while promoting economic, social and environmental benefits. This paper uses content analysis to deduce a framework with the AM impact on lean/green SCM practices. The framework was discussed in an interview with an industrial manager currently involved in AM adoption projects.

2 Background

2.1 Additive Manufacturing Impacts on Supply Chain

AM promises to transform SCM [5], namely in what is related to the network structure, customer's centricity in the manufacturing processes, SC capabilities, e.g., flexibility and agility [6]. AM technology facilitates small-run productions, increasing the SC agility for promptly respond to changing market demands [7]. It enables value co-creation across an SC due to its versatility and flexibility for distributed working and collaboration [8]. In addition, it supports on-location production [9], with lesser transportation costs and warehousing needs [10]. With the rapid prototyping, the forecasting time horizon decreases allowing stock and inventory costs reductions [11]. The SC literature [5, 12] describes the following areas that are likely to be impacted by AM: mass customisation; resource efficiency; decentralisation of manufacturing; complexity reduction; rationalisation of inventory and logistics; product design and prototyping; and legal and security concerns. Also, [13] raise the question of how AM would affect the SC network design; the forecasting (timing/accuracy); the scale and scope economies; the inventory management (quantities/placement); the warehousing, (size/technology/location); and, the transportation (modal/carrier choice).

2.2 *Additive Manufacturing and Its Impact on Supply Chain Sustainability*

Considering all the above impacts of AM, this technology will also play a critical role in what is concerned with SC sustainability [14–16]. Overall, "AM is expected to become a key manufacturing technology in the sustainable society of the future" [16]. The potential of AM as a sustainable alternative to manufacturing processes is undoubtedly promising and has been explored in the literature [17, 18]. It optimizes the use of raw material in the SC, reducing waste in manufacturing processes and enabling a product with flexible design and more efficient production planning [19]. Despite this, [19] shows that environmental sustainability benefits are not mainly relevant to the AM adoption; companies are being motivated mainly by two critical factors: time reduction and reduction of costs. In this sense it is necessary to study which synergies could appear between the adoption of AM technologies and the set of SCM practices that promote not only a reduction in time and overall costs but also assures a proper balance between the social, environmental and economic performance (i.e. triple-bottom-line TBL).

Green and lean initiatives are considered synergistic to focus on reducing and eliminating waste in production systems [20, 21]. Disney et al. extend the lean to the SC context; they state that lean processes create value through the elimination of "waste" in the SC [22]. Srivastava defined green SCM as "integrating environmental thinking into supply chain management, including product design, material sourcing and selection, manufacturing processes, delivery of the final product to the consumers as well as end-of-life management of the product after its useful life [23]." Several studies, e.g., [24, 25], advocate the use of green and lean SCM practices to achieve better TBL performance. This work will be built upon the four lean/green SCM practices proposed by [25]:

- *Green purchasing guidelines*: although could represent a cost, it reduces disposal and liability costs, while improving the organisation's resource conservation and public image. It avoids buying-in waste and reduces environmental costs.
- *Using re-usable packaging*: this practice is referred to the packages that could be returned to the suppliers and re-used in new deliveries. It requires co-operation with suppliers reducing storage and recovery delays, which leads to operational cost savings and environmental gains.
- *Just-in-sequence (JIT)*: it implies a predetermined sequence to materials and components delivery; this sequence is defined according to the production order. It

results in non-value added activities/resources elimination, namely in what is related to materials and components storage and management.

- *Single sourcing*: it is based on mutual trust between producer and supplier. The integration of socially responsible suppliers into companies' SCs demands a high level of trust among partner, resulting in a simpler and more effective system.

3 Empirical Study

To identify what could be the impacts of AM on lean/green SCM practices, a research process was conducted in two steps: 1st—a dedicated literature review to collect the anecdotal evidence; 2nd—an interview with a company manager. The search process was initiated by searching in SCOPUS database the following set of strings in the fields Article title, Abstract, and Keywords (until March 2019): "Additive manufacturing" AND {"Lean Supply Chain" OR "Green Supply chain"}. The search returned only 2 documents. To overcome this, purposive sampling was used to select documents that provide some insights on AM and the four lean/green SCM. A sample of 13 documents was deeply studied (Table 1). A content analysis was done manually for making inference by systematically and objectively identifying individual information in the texts under study [26].

In order to conduct the content analysis, each paper in the sample was read and analyzed in detail in order to find which AM characteristics, SCM practices and SC performance measures are pointed out in its text. It is important to notice that the objective was not to find the common set of pieces of evidence in the individual papers but to acquire an overview of the anecdotal evidence contained in the sample. For example, what is related to the SC performance, paper [32] refers that AM promotes material waste and energy consumption reduction, and more, paper [34] relates those with JIT: "AM may play a significant role in diminishing waste resources and reducing energy consumption by employing just-in-time production". From these two paper analysis, it is proposed that when using AM the JIT practices are expected to reduce overall waste (including energy). Another example is reduction of the number of entities in AM-based SC referred by [36], which is related to the reduction of the network complexity (which will drive SC performance). The limited number of AM machine and material suppliers point out by [37] as responsible for increasing the supplier's negation power, which will be a significant factor driving SC performance not just from the economic perspective, but also for social and environmental aspects.

The sample analysis does not provide evidence about the impact of AM on green SCM practices "green purchasing guidelines" and "reusable packaging". This can be because AM technology is still immature. The green purchasing is seen as a step after the adoption of AM processes. The re-utilisation of packaging in AM has not been verified. However, two studies considering the impact of AM on the packaging were found stressing that AM products could be relatively fragile and require special

Table 1 Anecdotal pieces of evidence

	Quote/inference	References
Just-in-sequence (JIS)/Just-in-time (JIT)	"Rapid manufacturing" as the just-in-time manufacturing of parts/components, when needed using available technologies such as AM	[24]
	"Individual players can produce parts locally, allowing for the true just-in-time production of parts needed suddenly and more robust supply chain"	[27]
	"AM (generic) operations opportunities have been outlined and are based on 'Manufacturing Systems Engineering', 'Lean Product and Process Development' and 'Lean Manufacturing' principles. These are AM as an enabler of: 'Continuous Improvement' (CI) in product development and the workplace, Just-in-Time (JIT) with related reduction of inventory, mass customisation to tailor products"	[28]
	"(…) consumer products are printed to order, using the consumers own design data practice upon a globally distributed just-in-time supply chain or a manufacture within the consumers own home (…)"	[29]
	"A value can be placed on the speed, versatility, and adaptability of AM, because it allows for just-in-time manufacturing"	[30]
	"The geometric complexity of a part, the choice of small-scale or tailored fabrication, and the just-in-time or short time-to-market requirements are the criteria for the application of AM (…)"	[31]
	"(…) AM offers multiple advantages over conventional manufacturing technologies, including reduced material waste and energy consumption, shortened time-to-market, just-in-time production (…)"	[32]

(continued)

Table 1 (continued)

	Quote/inference	References
	"The benefits, which included design flexibility, no tooling and just-in-time delivery, all contributed to the end customer accepting (…)"	[33]
	"InKjet Printing can achieve a faster response and just-in-time customization"	[16]
	"AM may play a significant role in diminishing waste resources and reducing energy consumption by employing just-in-time production"	[34]
	"For all other industries, too, 3-D printing can become the new *kanban*, a true just-in-time inventory management solution"	[35]
Single sourcing	"(…) AM tends to alleviate this issue by reducing the number of SC layers and suppliers"	[36]
	"There are limited number of AM machine suppliers and material suppliers which gives them high negotiating power (…) the firms often think twice before investing in a new technology where it depends on a specific vendor"	[37]

packaging for safe shipping [38] and suggesting that AM technology could be cost-effective to produce customized packages for products with challenging designs [39].

From the evidence collected in the literature, a theoretical framework representing the impact of AM on SCM practices was elaborated (Fig. 1).

To obtain more insights on the relationships among AM and lean/green SCM practices, a structured-interview was carried out with a Project Manager. He supervises the digital transformation of maintenance process in Navigator (a Portuguese company that produces pulp and paper). He is exploring the use of AM technology to produce spare parts for machines and production equipment. The interview was done by three researchers to assure data triangulation and to avoid data analysis bias.

Firstly, the researchers presented the study and clarified the meaning of each of the four lean/green SCM practices. Then, the main reasons why there is a lack of research on AM and its implication for "green purchasing guidelines" and "reusable packaging" were discussed. It was concluded that as AM technology is still immature, there are not many standards available for AM materials and products which makes the

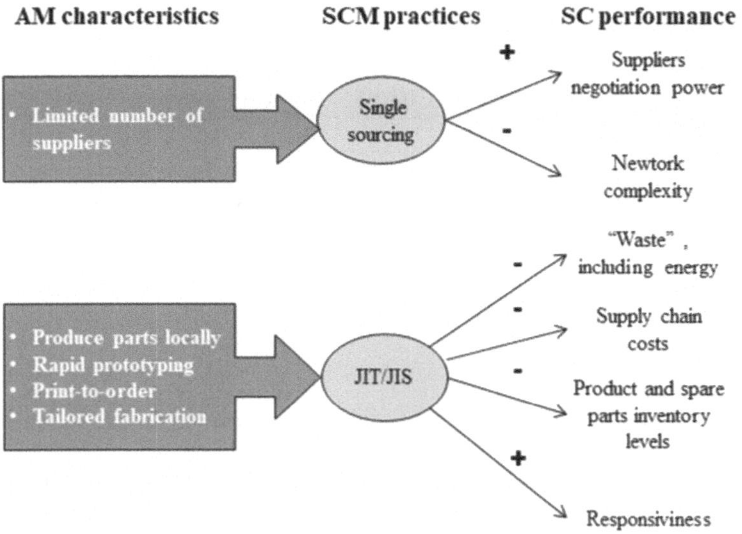

Fig. 1 Theoretical framework

application of green purchasing guidelines unfeasible. Regarding "reusable packaging", it was the manager's perception this is not a viable choice for single part AM production. In the case of AM batch production, it could be a possibility.

In the second part of the interview, the manager's perceptions about the AM characteristics and its implications to the lean/green SCM practices were collected. The AM characteristic "limited number of suppliers" is in accordance with single-sourcing practice. There are few AM suppliers available; e.g., in Portugal, he only identified one supplier for AM metal parts. This reinforces the just-in-sequence practice by encouraging building relationships with a limited number of suppliers.

The AM feature "produce parts locally" is relevant to single-sourcing, as it helps to establish partnerships with suppliers located nearby the factory, creating liability policies between both, and making it possible to use reusable packaging. Supplier and producer can develop specific packaging that meets the materials and parts' requirements. It also improves JIT practice by reducing the transport needs over long distance, helping to reduce the negative environmental impacts along the product life cycle. AM also enables "Rapid manufacturing", creating synergies with the single-sourcing, because it needs in-depth knowledge of product specifications. This also improves the JIT by accelerating production processes and shortening lead times.

According to the manager, AM only enables "print-to-order" if the company relies on one exclusive supplier that already knows design requirements for a specific set of parts/products ordered by the manufacturer. This characteristic is also relevant to JIT. Also, it has implications for re-usable packaging, especially when the supplier knows the product design, and thus a specific packaging can be produced for it, in case of future order placements by the client. This characteristic has no implications

Table 2 Relations derived from the interview

AM characteristics	Lean/Green			
	Single sourcing	JIT/JIS	Reusable packaging	Green purchasing
Limited number of suppliers	√	√	n/a	n/a
Produce parts locally	√	√	√	n/a
Rapid manufacturing	√	√	n/a	n/a
Print-to-order	√	√	√	√
Tailored fabrication	√	√	n/a	√

n/a Not applicable

for green purchasing unless the AM product has specific requirements for environmental hazards. "Tailored fabrication" is very similar to print-to-order in terms of its implications, but it does not have a specific implication with re-usable packaging (Table 2).

4 Conclusions

This study pretends to bring an understanding of how AM characteristics will impact the lean/green SCM practices. To this end, a theoretical model was developed from the literature review. Since the theoretical model was derived from a sample of 13 papers, it is biased by the available information extracted from the sample. This means that the model does not cover all the possibilities of AM relations with lean/green SC practices. To overcome this, an interview with a manager was made. The main findings are that AM will impact the adoption of single-sourcing and JIT/JIS (lean practices) because of the limited number of suppliers for AM component/equipment, materials and products, its ability to produce parts locally, rapid manufacturing capabilities, print-to-order and tailored fabrication. Regarding the green SCM practices, little evidence was found. This is mainly explained because of the undeveloped maturity stage of AM technology. Therefore, it is necessary to develop tools, prove efficiency and sustainability to acquire deep and broad levels of AM adoption, gaining legitimacy in new markets.

Acknowledgements Project FIBR3D (POCI-01-0145-FEDER-016414), co-financed by Fundo Europeu de Desenvolvimento Regional (FEDER) and by national funds through Fundação para a Ciência e Tecnologia, Portugal. Project UNIDEMI (UID/EMS/00667/2019) and project KM3D (PTDC/EME-SIS/32232/2017).

References

1. ASTM (2012) Standard terminology for additive manufacturing—general principles—terminology. ASTM Int., West Conshohocken, PA
2. Prakash KS, Nancharaih T, Rao VVS (2018) Additive manufacturing techniques in manufacturing—an overview. Mater Today Proc 5(2):3873–3882
3. Despeisse M, Yang M et al (2017) Sustainable value roadmapping framework for additive manufacturing. Proc CIRP 61:594–599
4. Weller C, Kleer R, Piller FT (2015) Economic implications of 3D printing: market structure models in light of additive manufacturing revisited. Int J Prod Econ 164:43–56
5. Waller MA, Fawcett SE (2014) Click here to print a maker movement supply chain: How invention and entrepreneurship will disrupt supply chain design. J Bus Logist 35(2):99–102
6. Durach CF, Kurpjuweit S, Wagner SM (2017) The impact of additive manufacturing on supply chains. Int J Phys Distrib Logist Manag 47(10):954–971
7. Garrett B (2014) 3D printing: new economic paradigms and strategic shifts. Glob Policy 5(1):70–75
8. Martinelli EM, Christopher M (2019) 3D printing: enabling customer-centricity in the supply chain. Int J Value Chain Manag 10(2):87–106
9. Khajavi SH, Partanen J, Holmström J (2014) Additive manufacturing in the spare parts supply chain. Comp Ind 65(1):50–63
10. Mavri M (2015) Redesigning a production chain based on 3D printing technology. Knowl Process Manag 22(3):141–147
11. Janssen R, Blankers I (2014) TNO: the impact of 3-D printing on supply chain management. TNO, The Hague, Netherlands
12. Mohr S, Khan O (2015) 3D printing and its disruptive impacts on supply chains of the future. Technol Innov Manag Rev 5(11):20
13. Mashhadi AR, Esmaeilian B, Behdad S (2015) Impact of additive manufacturing adoption on future of supply chains. In: ASME 2015 international manufacturing science and engineering conference
14. Ford S, Despeisse M (2016) Additive manufacturing and sustainability: an exploratory study of the advantages and challenges. J Clean Prod 137:1573–1587
15. Niaki MK, Torabi SA, Nonino F (2019) Why manufacturers adopt additive manufacturing technologies: the role of sustainability. J Clean Prod 222:381–392
16. Huang SH, Liu P et al (2013) Additive manufacturing and its societal impact: a literature review. Int J Adv Manuf Technol 67(5–8):1191–1203
17. Holmström J, Liotta G, Chaudhuri A (2017) Sustainability outcomes through direct digital manufacturing-based operational practices: a design theory approach. J Clean Prod 167:951–961
18. Schniederjans DG (2017) Adoption of 3D-printing technologies in manufacturing: a survey analysis. Int J Prod Econ 183:287–298
19. Peng T, Kellens K et al (2018) Sustainability of additive manufacturing: an overview on its energy demand and environmental impact. Add Manuf 21:694–704
20. Dües CM, Tan KH, Lim M (2013) Green as the new lean: how to use lean practices as a catalyst to greening your supply chain. J Clean Prod 40:93–100
21. Ng R, Low JSC, Song B (2015) Integrating and implementing lean and green practices based on proposition of carbon-value efficiency Metric. J Clean Prod 95:242–255
22. Disney SM, Naim MM, Towill DR (1997) Dynamic simulation modeling for lean logistics. Int J Phys Distrib Logist Manag 27(3/4):174–196
23. Srivastava SK (2007) Green supply—chain management: a state-of-the-art literature review. Int J Manag Rev (Wiley Online Library)
24. Azevedo SG, Helena C (2019) Lean and green supply chains. In: Sarkis J (ed) Handbook on the sustainable supply chain. Elgaronline
25. Azevedo SG, Carvalho H et al (2012) Influence of green and lean upstream supply chain management practices on business sustainability. IEEE Trans Eng Manag 59(4):753–765

26. Spens K, Kovács G (2006) A content analysis of research approaches in logistics research. Int J Phys Distrib Logist Manag 36(5):374–390
27. Singamneni S, Yifan LV et al (2019) Additive manufacturing for the aircraft industry: a review. J Aeronaut Aerospace Eng 8(214):2
28. Busachi A, Erkoyuncu J et al (2016) Defining next-generation additive manufacturing applications for the ministry of defence (MoD). Proc CIRP 55:302–307
29. Cozmei C, Caloian F (2012) Additive manufacturing flickering at the beginning of existence. Proc Econ Financ 3:457–462
30. Frazier WE (2014) Metal additive manufacturing: a review. J Mater Eng Perform 23(6):1917–1928
31. Berger U (2015) Aspects of accuracy and precision in the additive manufacturing of plastic gears. Virtual Phys Prototyping 10(2):49–57
32. Huang Y, Leu M (2014) Frontiers of additive manufacturing research and education. University of Florida, Gainesville, FL
33. Vashishtha VK, Makade R, Mehla N (2011) Advancement of rapid prototyping in aerospace industry—a review. Int J Eng Sci Technol 3(3):2486–2493
34. Dilberoglu UM, Gharehpapagh B, Yaman U, Dolen M (2017) The role of additive manufacturing in the era of industry4.0. Proc Manuf 11:545–554
35. Kietzmann J, Pitt L, Berthon P (2015) Disruptions, decisions, and destinations: enter the age of 3-D printing and additive manufacturing. Bus Horiz 58(2):209–215
36. Ivanov D, Dolgui A, Sokolov B (2019) The impact of digital technology and Industry4.0 on the ripple effect and supply chain risk analytics. Int J Prod Res 57(3):829–846
37. Dwivedi G, Srivastava SK, Srivastava RK (2017) Analysis of barriers to implement additive manufacturing technology in the Indian automotive sector. Int J Phys Distrib Logist Manag 47(10):972–991
38. Reeves P, Tuck C, Hague R (2011) Additive manufacturing for mass customization. In: Fogliatto F, da Silveira G (eds) Mass customization. Springer Series in Advanced Manufacturing, Springer
39. Attaran M (2017) Additive manufacturing: the most promising technology to alter the supply chain and logistics. J Serv Sci Manag 10(03):189

Towards Continuous Improvement by Using a Lean-TRIZ Approach

Radu Godina, Helena Carvalho, Pedro Rodrigo and Helena Navas

Abstract The search for excellence and operational benefits forces organizations to adopt new types of techniques and approaches for the systematic creation of innovative solutions. In view of this context, this paper describes the implementation of the Lean-TRIZ approach in a production line in a Portuguese food organization. In a first step, a continuous improvement model was designed for a technical/organizational system, by combining the TRIZ methodology with Lean techniques and with the Kano Model. The initial analysis was carried out through techniques of direct observation, brainstorming and questionnaires to employees. This analysis allowed identifying the problems by identifying the root causes. Then, analytical instruments of the TRIZ methodology were applied. Finally, the level of employee satisfaction with the improvements introduced in the production line was verified through the Kano Model, which showed us positive results both from the point of view of the employee and the company. The main contribution of this paper is to determine, through a real case study, the potential of the Lean-TRIZ approach in helping organizations to implement a continuous improvement process.

Keywords Kano Model · TRIZ methodology · Lean-Triz approach · Lean manufacturing · Autonomous maintenance

R. Godina · H. Carvalho · P. Rodrigo · H. Navas (✉)
Department of Mechanical and Industrial Engineering, Faculty of Science and Technology(FCT), UNIDEMI, Universidade NOVA de Lisboa, 2829-516 Caparica, Portugal
e-mail: hvgn@fct.unl.pt

R. Godina
e-mail: r.godina@fct.unl.pt

H. Carvalho
e-mail: hmlc@fct.unl.pt

M. Rossi et al. (eds.), *Proceedings of the 6th European Lean Educator Conference*, Lecture Notes in Networks and Systems 122,
https://doi.org/10.1007/978-3-030-41429-0_17

1 Introduction

As the world's business environment changes, it becomes increasingly difficult to maintain the competitiveness of the food industry [1]. Customers demand change, technology evolves, and competitive forces change [2]. The competitive challenge is present in the current business areas due to the competitive characteristics of the markets and the survival principle of the stronger, those who offer obsolete or noncompetitive products, services, concepts or systems will be extinct. There are many concepts of high current demand, but competitiveness will be in charge of excluding those with little added value [3]. Thus, the organizations' flow of activities becomes more efficient, including only the activities that create value, aligned in the best possible sequence and with the minimum of interruptions [4].

Systematic innovation and Lean philosophy can complement each other. Organizations feel the need to make their processes more creative and innovative. The process of generating solutions is complex, thus creative, sometimes disruptive or radical solutions are needed. Therefore, with the aid of the Theory of Creative Problem Solving (TRIZ), there are techniques and analytical tools capable of finding creative solutions to solve existing problems [5].

This paper intends to illustrate how combining the Lean philosophy, the TRIZ methodology and the Kano Model, to find inventive solutions allows solving real case problems. The case under study is an industrial unit of transformation and commercialization of meats and respective processed products. This industrial unit follows the highest standards of quality, innovation, hygiene and safety, and competing in a demanding market, at national and international level.

This paper is organized as follows. In Sect. 2 the state of the art is shown, where Lean Manufacturing, TRIZ and Kano Model are briefly addressed. In Sect. 3 the case study is briefly analyzed. In Sect. 4 the result analysis can be found. Finally, conclusions are drawn in Sect. 4.

2 State of the Art

2.1 Lean Manufacturing

Lean manufacturing is defined by turning "leaner" all the characteristics in mass production, meaning less human effort, less defects, less manufacturing space, fewer stocks and less time spent in developing a new product. The challenge is to maximize the quantity of the delivered product and, at the same time, minimize resources, namely raw materials, labor, space, among others [6]. Thus, the Lean concept evolved into a philosophy of thought, called Lean Thinking, which seeks to eliminate waste and pursues continuous improvement within the organization. This thinking applies to all links in a supply chain, services and others [7].

The commitment to Lean Thinking should begin at the top management level and should develop a cascading improvement at all levels of the organization to improve the flow of materials and information as well as process efficiency [8].

2.2 TRIZ Methodology

This methodology is known by its Russian origin acronym, TRIZ, known as the "theory of the resolution of invention-related tasks" or "Theory of Creative Problem Solving" is a specially adjusted methodology for the resolution of new problems in the areas of Science and Engineering [5]. It is considered as a set of tools, used to solve problems and helps to choose the right decision, in an inventive and innovative way, replacing the conventional method of unsystematic trial and error [9]. TRIZ was developed by Altshuller who analyzed a large number of patents from different areas. After this exhaustive analysis, made with the aid of his collaborators, Altshuller concluded that most patents were nothing more than improvements in systems, only a few contained inventions [10].

For Altshuller, level one solutions, by not being innovative, are largely ignored, as are level five solutions, since they require a higher degree of knowledge and understanding, in addition to their minimum percentage of occurrence [11]. One of the key aspects of TRIZ is to recognize and mobilize the necessary resources. Proper resource identification can include any aspects of the system and the environment that are involved. TRIZ considers it important to follow a systematic approach in the search for resources. Resource research focuses on understanding the functional requirements of the solution the engineer is looking for [9].

One of TRIZ's most popular and powerful tools, Substance-Field Analysis, or Su-Field Analysis, is a method to visualize and model the problem in a more simplified manner [12]. Among the advantages associated with this tool is the increase in the ideality of the processes through the analysis of several resources and the use of different "substances" and "fields" [9]. Often, a system is represented by a triangle composed of two substances (S1 and S2) and by a field (F), as shown in Fig. 1.

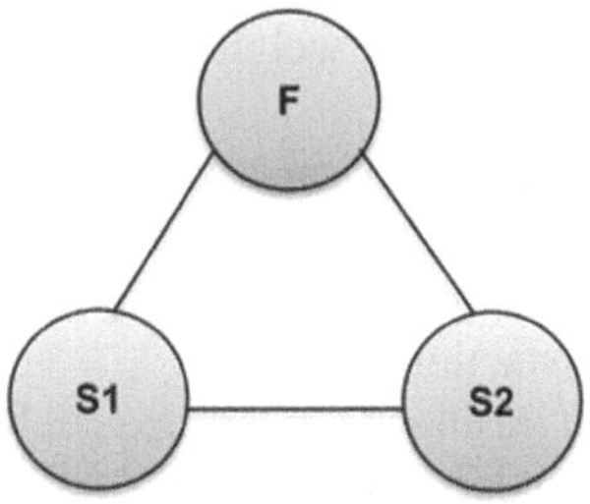

Fig. 1 Basic model of Su-field analysis—complete system

2.3 Level of the Ideality of the Solution

At this stage, the formula of the ideality level is applied, and a solution with a high level of ideality does not mean that it is the ideal solution, since there are no ideal solutions. However, in TRIZ's view, there are certain characteristics of a solution close to the ideal. On the other hand, a low level of ideality means that there are still problems (wastes or contradictions) to be solved. In cases in which the solution obtained has a high level of ideality and if it is concluded that the system has become closer to the ideal, then its implementation begins. For this purpose, a Matrix of Ideality helps identify interactions between technical requirements and also shows the positive and negative effects of these interactions.

2.4 Kano Model

To define specific attributes of a product or service, it is important to define and understand customer needs through market analysis. In [14] it is referred that a customer is someone who is affected by a product or process and can be defined as internal or external, which can be considered as belonging to the organization or not belonging to the organization, respectively.

A closer approximation of the true customer's needs to be matched is through the use of the Kano model. They need to be objective; otherwise, they should be discarded [15]. The use of this model allows to determine the customer's requirements and expectations, distinguishing six types of attributes of a product or service that influence the satisfaction of the product [16], which are as follows:

- Mandatory attribute—In case this attribute is not filled, will cause the client a huge dissatisfaction. These attributes are considered as decisive ones concerning the factor of competitiveness.
- Attractive attribute—A requirement that presents great influence with the client, creating enormous satisfaction when present. However, if this attribute is not achieved, there will be no dissatisfaction on the part of the attribute.
- Linear attribute—These attributes are responsible for incremental customer satisfaction; that is, they depend on the weight it presents in a particular product or service.
- Indifferent attribute—Presence or absence of this type of attribute does not impact the client.
- Contradictory attribute—Attribute that may be present and cause dissatisfaction or be absent and cause satisfaction.
- Questionable attribute—It results from inadequately worded questions or an insufficient level of client understanding.

3 Case Study

The study was developed in a food industrial unit. The operations under study belong to the ham production line, composed of several machines. At the time of the elaboration of this work, this production line had a throughput level higher than usual, due to an increase of customers' orders. Thus, to improve product quality and reduce waste, considering possible failures and some shortcomings in methods currently in use, it was necessary to apply innovative methodologies.

This production line is strongly affected by non-scheduled stops, which required the intervention of maintenance technicians. To evaluate if these types of stoppages could be minimized or even eliminated, as well as if some maintenance tasks could be transferred to the workers, the TRIZ methodology was used, specifically, Su-Field Analysis. The objective is to characterize the production processes present in the ham production line, to identify opportunities for improvement, and to present improvement proposals, as well as implement the improvements and discuss the results.

3.1 The Results of the Ideality Matrix

After presenting the improvement proposals, the Ideality Matrix was used, in order to analyze the difference of the increase/decrease of the level of the ideality of the system. After presenting improvement proposals, it is important to realize if they are improvements that provide an increase in the level of the ideality of the system; this is if they meet the company needs. Therefore, the most important parameters for the production line under study are:

1. Increase productivity;
2. Reduction of the number of faults or unscheduled stops;
3. Cost reduction (costs associated with overtime to meet production orders, maintenance costs, consumable or raw material costs);
4. Reduction of errors made by machine operators due to lack of training;
5. Increase the final product quality;
6. Increase equipment maintainability;
7. Reduction of the time related to the assembly and disassembly of equipment;
8. Increase workers' safety.

Based on the defined parameters, it was possible to create an Ideality Matrix to identify the negative interactions (represented by the sign "−"), the positive interactions (represented by the sign "+") and nonexistent interactions. Table 1 presents the matrix of ideality and the respective interactions. The ideality level of the described scenario was calculated by applying the following equation:

$$Ideality = \frac{\sum number\ of\ positive\ interactions}{\sum number\ of\ negative\ interactions} \quad (1)$$

Table 1 Matrix of ideality applied to the defined parameters

Parameters	1	2	3	4	5	6	7	8
1. Productivity		–	–		–	–		
2. Reliability	+		–		+		+	
3. Costs	–	–		–	–	–	–	–
4. Lack of training	–	–	–		–	–	–	–
5. Quality			–					
6. Maintenance	+	+	–		+		+	+
7. Mounting/dismounting time	–		–					–
8. Safety	–		–				–	

The level of ideality, in this case, was 8/27 ≈ 0.296, which is thus less than 1, and this result is so low because the amount of harmful or negative interactions is greater than the number of positive interactions. Through the analysis of Table 1, it can be observed that the parameters that most negatively affect the system are "Productivity", "Costs" and "Lack of Training". Therefore, to increase the level of ideality, three types of solutions were implemented:

1. *Autonomous Maintenance Checklists*: it will reduce the lost time of production, since it will reduce the small stops for replacement of parts and will allow a better prediction of future faults or maintenance needs. Thus, making all the changes necessary to the operation of the machines to be made during the production lines scheduled stops. Autonomous Maintenance will also allow the 1st and 2nd level maintenance interventions to be performed by the machine operators, thus reducing the loss of time of the maintenance technicians in these tasks, freeing up the time for them to plan and direct the maintenance intervention for other types of maintenance activities.
2. *5S Methodology*: by analyzing the identified problems by Ideality Matrix, it was perceived that there is a deficit in the standardization of work procedures, and it would be important for them to become more organized and systematized. This situation was defined as a non-complex problem, and it was opted for an application of the 5S methodology and standardized procedures, in order to reduce waste. Figure 2 provides an example of the re-organization.
3. *One Point Lessons tool*: the lack of training for unexperienced new employees, employees transferred from other lines, may lead to information failure, material misuse, breach of specifications, production orders or even cause machine malfunctions that can profoundly affect production. After a meeting with management, it was decided to create a manual describing the relevant procedures of all the equipment operating on the production line, which will be used by any employee that needs urgent training.

Applying the above mentioned three solutions caused a significant increase in the ideality of the system, as evidenced by the following result: 22/12 ≈ 1.83.

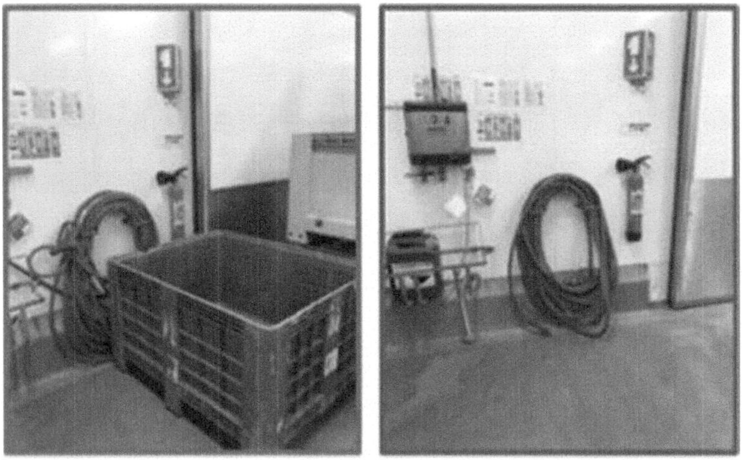

Fig. 2 An example of before (left) and the current practice (right) after 5S implementation

3.2 The Results of the Kano Model

When all the improvements were completed, it became important to realize the level of satisfaction that these would have caused in the company's internal customers, i.e. the ham production line workers. In this way, the Kano Model was used to identify the factors with the greatest influence on the change in workers satisfaction. A questionnaire was developed to gather the worker opinion on several factors, such as follows:

- If the level of motivation to work has increased;
- If the level of motivation to work has not increased;
- Can accomplish the tasks in less time;
- Cannot perform tasks in lesser time;
- The tools and materials needed for your work are more organized;
- The tools and materials needed for your work are no longer organized;
- Use point-to-point lessons or *One Point Lessons* (OPLs);
- Does not use point-to-point lessons or OPLs;
- Can move faster from one location to another on the line;
- Cannot move faster from one location to another on the line;
- The machines would malfunction less;
- The machines would not malfunction less;
- Feel that it can be useful in machine maintenance;
- The employee feels that it cannot be useful in maintaining the machines;
- You feel that you are safer working at your workplace;
- Feel that you are no longer safe when working at your workplace.

Through several iterations it was possible to carry out a more in-depth analysis, allowing to obtain the results that are verified in Table 2, where Q1–Q8 are questions.

Table 2 Results of the application of Kano Model surveys

		Attributes	L (%)	A (%)	R (%)	I (%)	C (%)	Q (%)	Total (%)
Q1	L	Level of motivation	40.0	23.8	8.8	6.3	11.3	10.0	100
Q2	A	Tasks performed in less time	11.3	45.0	13.8	13.8	16.3	0.0	100
Q3	R	Organized tools and materials	11.3	6.3	75.0	2.5	1.3	3.8	100
Q4	A	Point-to-point lessons	2.5	55.0	26.3	5.0	5.0	6.3	100
Q5	I	Faster moves in the workplace	20.0	18.8	17.5	37.5	6.3	0.0	100
Q6	R	Machines have lesser failures	17.5	16.3	31.3	16.3	6.3	12.5	100
Q7	I	Help with machine maintenance	13.8	16.3	16.3	40.0	6.3	7.5	100
Q8	A	Improved job security	2.5	45.0	28.8	6.3	8.8	8.8	100

For this purpose, it is necessary to take into account the meaning of the following indicators: L—Linear attribute; A—Attractive attribute; R—Required Attribute; I—Indifferent attribute; C—Contradictory attribute; Q—Questionable attribute.

By analyzing Table 2, it has been deduced that the satisfaction increases when the level of motivation is higher. The improvements made have led to the worker's feeling better and more confident in the performance of their duties. As well as increasing the feeling of ownership in relation to the machines, since they increasingly interact with them, not only at the level of production, but also at in: (i) maintenance, (ii) lubrication; (iii) small adjustments, (iv) filling work orders until then carried out exclusively by employees assigned to the maintenance department of the company.

Finally, it should be noted that the Kano Model could also have been used before the implementation of the improvements to confirm that they would meet the requirements of the company, in which case it was decided to use it only when the improvements were completed.

4 Conclusions

In this study, a continuous improvement model was designed for a technical/organizational system, by combining the TRIZ methodology with Lean techniques and with the Kano Model for an industrial food unit. By using TRIZ, the Ideality Matrix showed a significant increase in the ideality of the system, while with the Kano Model, it was observed that the employee satisfaction increased when the level of motivation was higher—as indicated by the 40% of the Linear Attribute of the Kano Model. In this way, it can be concluded that the work developed had a positive impact on the line of ham production, allowing not only an increase in productivity but also the reduction of waste and an increase in discipline in working methods, while not overlooking the satisfaction of the employees in the performance of their functions.

Acknowledgements The authors acknowledge Fundação para a Ciência e a Tecnologia (FCT—MCTES) for its financial support via the project UIDB/00667/2020 (UNIDEMI).

References

1. Santeramo FG, Carlucci D, De Devitiis B, Seccia A, Stasi A, Viscecchia R, Nardone G (2018) Emerging trends in European food, diets and food industry. Food Res Int 104:39–47
2. Ehgartner U (2018) Discourses of the food retail industry: changing understandings of 'the consumer' and strategies for sustainability. Sust Prod Consum 16:154–161
3. Leonidou CN, Hultman M (2019) Global marketing in business-to-business contexts: challenges, developments, and opportunities. Ind Mark Manag 78:102–107
4. Ball P, Lunt P (2018) Lean eco-efficient innovation in operations through the maintenance organisation. Int J Prod Econ

5. de Jesus Pacheco DA, ten Caten CS, Jung CF, Navas HVG, Cruz-Machado VA, Tonetto LM (2019) State of the art on the role of the theory of inventive problem solving in sustainable product-service systems: past, present, and future. J Clean Prod 212:489–504
6. Henao R, Sarache W, Gómez I (2019) Lean manufacturing and sustainable performance: trends and future challenges. J Clean Prod 208:99–116
7. Stone KB (2012) Four decades of lean: a systematic literature review. Lean Six Sigma J 3:112–132
8. Antony J (2011) Six Sigma vs lean: some perspectives from leading academics and practitioners. Int J Prod Perform Manag 60:185–190
9. Navas HVG (2017) Problem solving and increase of ideality of complex systems. In: Kahlen F-J, Flumerfelt S, Alves A (eds) Transdisciplinary perspectives on complex systems: new findings and approaches. Springer International Publishing, Cham, pp 305–327
10. Chechurin L, Borgianni Y (2016) Understanding TRIZ through the review of top cited publications. Comput Ind 82:119–134
11. Maia LC, Alves AC, Leão CP (2015) How could the TRIZ tool help continuous improvement efforts of the companies? Proc Eng 131:343–351
12. Alves JF, Navas HVG, Nunes IL (2016) Application of TRIZ methodology for ergonomic problem solving in a continuous improvement environment. In: Arezes P (ed) Advances in safety management and human factors. Springer International Publishing, pp 473–485
13. Navas HVG (2014) Radical and systematic eco-innovation with TRIZ methodology. In: Azevedo SG, Brandenburg M, Carvalho H, Cruz-Machado V (eds) Eco-innovation and the development of business models: lessons from experience and new frontiers in theory and practice. Springer International Publishing, Cham, pp 81–95
14. Juran JM (2003) Juran on leadership for quality. Simon and Schuster
15. Defeo JA, Juran JM (2010) Juran's quality handbook: the complete guide to performance excellence 6/e. McGraw-Hill Education, New York
16. Violante MG, Vezzetti E (2017) Kano qualitative vs quantitative approaches: an assessment framework for products attributes analysis. Comput Ind 86:15–25

An Exploration of the Interplay Between National Culture and the Successful Implementation of Lean Six Sigma in International Companies

Alinda Kokkinou and Ton van Kollenburg

Abstract Multinational firms often initiate continuous improvement programmes such as Lean Six Sigma and roll them out to their foreign subsidiaries. So far, little attention has been given to the role national culture plays in the successful implementation of Lean Six Sigma. Through a review of the Critical Success Factors of Lean Six Sigma, coupled with in-depth case studies of multinational companies currently implementing Lean and Lean Six Sigma, the authors propose a conceptual model that will be refined in subsequent studies using action research. Action research will contribute to improved managerial actions grounded in scientific research for the companies participating in these studies.

Keywords Continuous improvement · Lean Six Sigma · Critical success factors · Culture · Action research

1 Introduction

Operational excellence is a key requirement for firms to become and remain competitive in the global business arena [1]. One way for companies to be efficient with their use of resources and consistently deliver quality to their customers is by investing in the implementation of continuous improvement initiatives such as Lean Manufacturing, Six Sigma, Total Productive Maintenance (TPM), Total Quality Management (TQM) or a variation thereof. Famous case studies at companies such as GE, Motorola, GlaxoSmithKline and Lockheed Martin have demonstrated the potential

A. Kokkinou (✉)
AVANS School of International Studies, AVANS University of Applied Sciences, Hogeschoollaan 1, 4818 CR Breda, The Netherlands
e-mail: a.kokkinou@avans.nl

T. van Kollenburg
AVANS University of Applied Sciences, Centre of Expertise for Sustainable Business, Hogeschoollaan 1, 4818 CR Breda, The Netherlands
e-mail: ajc.vankollenburg@avans.nl

M. Rossi et al. (eds.), *Proceedings of the 6th European Lean Educator Conference*, Lecture Notes in Networks and Systems 122,
https://doi.org/10.1007/978-3-030-41429-0_18

of these continuous improvement initiatives to deliver superior quality and thus satisfy customers while simultaneously significantly reducing waste and costs. These companies have all rolled out improvement initiatives globally, often starting with a single location and rolling out the initiative to their subsidiaries located in other countries. Despite well documented success stories [2] many authors have reported significant failure rates for continuous improvement initiatives [3].

Amongst the possible reasons for failure, McLean and Dahlgaard [3] cite a lack of regard for the organizational culture and environment in which the implementation must take place. Specifically, the authors provide the example of failing to consider cultural and language barriers when implementing such a programme. The following quote is illustrative: "the implementation of certain initiatives can be viewed by some as an attempt to 'import a management technique developed halfway around the world', where it is viewed to be successful in an extremely different culture" [3, 4]. Anecdotal evidence gathered in the course of this study from multinational companies currently rolling out continuous improvement initiatives further supports the notion that continuous improvement initiatives are often rolled out relatively indiscriminately to subsidiaries across the globe. Frequently, this is done by corporate headquarters imposing a uniform "corporate" continuous improvement programme on the subsidiaries.

Even when designers of such initiatives want to take national culture of the subsidiaries into account, little guidance is provided by the existing literature. Despite numerous studies investigating critical success factors (CSFs), little is known about whether and how CSFs differ across national cultures. Furthermore, despite extensive research on the role that organizational culture plays on the success of quality management, researchers have called for further research that examines the effect of national culture on continuous improvement initiatives and the resulting effect on organizational performance [5].

The purpose of the present study is to explore the interplay between national culture and the successful implementation of continuous improvement initiatives. The present study focuses on three related methodologies, namely Lean, Six Sigma and Lean Six Sigma as CSFs for their implementations have received the most attention.

The following section briefly defines Lean, Six Sigma and Lean Six Sigma and the CSFs identified in literature. The section thereafter highlights the main differences in these findings. The fourth section briefly reviews the literature on national culture and highlights several factors that could explain the contradictory findings. The fifth section describes interviews that were conducted with two companies recently or currently undergoing implementation of Lean Six Sigma to contextualize the previous findings, culminating in a conceptual model. This final section also describes the action methodology approach that will be used to conduct further research on the topic.

2 Critical Success Factors of Continuous Improvement

Deming described continuous improvement as consisting of "Improvement initiatives that increase successes and reduce failures" [6]. Companies implementing a continuous improvement initiative typically will adapt one of the most commonly used methods to their organizational needs or create a hybrid. Of these methods, the CSFs for the implementation of Lean, Six Sigma, and Lean Six Sigma have received extensive attention. While originating in manufacturing industries such as automotive and consumer products, Lean, Six Sigma, and Lean Six Sigma have now been adopted by a variety of industries including finance, utilities and healthcare.

2.1 Lean, Six Sigma, and Lean Six Sigma

Six Sigma is a quality improvement methodology introduced by Motorola in 1986 and further popularized by the publicized benefits achieved by GE in the late 1990s [7]. Six Sigma is defined as "an organized and systematic method for strategic process improvement and new product and service development that relies on statistical methods and the scientific method to make dramatic reductions in the customer defined defect rates" [8]. Improvement in the context of Six Sigma is customer driven and requires the company to redirect its attention to the "voice of the customer." As such, when properly implemented, Six Sigma can have a profound effect on an organization's structure and culture.

Lean Manufacturing is often perceived to have originated in Japan in the 1950s and gradually increased in popularity due, amongst other factors, to the simplicity of its approach and tools. Lean as it is now called, seeks to eliminate waste and non-value-added activities. Tools employed by Lean to solve common problems include line balancing, 5S method, Single Minute Exchange of Dies (SMED), and visual management [9].

Lean and Six Sigma are increasingly being combined to benefit from those synergies and complement each other's' shortcomings. Typically, the combined approach adopts the organizational structure and stepwise methodology prescribed by Six Sigma and incorporates the tools and techniques of both methods [9]. Despite their differences, research on CSFs has not discriminated between Lean, Six Sigma, and Lean Six Sigma. Therefore, for the purpose of this paper Lean, Six Sigma, and Lean Six Sigma will be reviewed together and for simplicity, will be referred to as Lean Six Sigma.

2.2 Critical Success Factors

Critical success factors are, as the name suggests, factors that are critical to the success of a system or programme implementation within a company [10]. In the context of Lean Six Sigma implementation, CSFs have been investigated extensively. A recognizable literature stream originates with Banuelas and Antony [11] who applied the definition of critical success factors to the context of continuous improvement initiatives and formulated the following definition: "CSFs represent the essential ingredients without which a project stands little chance of success" (p. 93). In other words, CSFs are needed to ensure the durable success of the implementation of a continuous improvement initiative.

Antony and Banuelas [12] used an extensive review of published case studies such as Motorola, GE, and AlliedSignal to describe twelve CSFs of Six Sigma. These included management involvement and commitment, cultural change, communication, organization infrastructure, training, project management skills, project prioritization and selection, amongst others. In a separate article, Banuelas and Antony [11] attempted to establish how large multinational UK companies prioritized these "ingredients' as the authors called them and found through the use of a survey the highest rated CSFs to be (a) management commitment and involvement, (b) understanding of six sigma methodology, (c) linking six sigma to business strategy, and (d) linking six sigma to customers. The three lowest rating CSFs were (i) linking Six Sigma to suppliers, (ii) training, and (iii) linking Six Sigma to employees.

Their list was refined, augmented and tested by a series of studies replicating their methodology for different size firms and/or in different countries. These included a survey of Dutch manufacturing SMEs [13], a cross-sectional survey of Indian companies [1], a survey of Singaporean service organizations [14], a cross-sectional survey of firms in Brazil [15] and an international sample [16]. The surveys used typically asked respondents to rate the importance of the CSFs to the success of Lean Six Sigma implementation. Furthermore, several studies further built on this line of research using case studies and mixed methods by investigating CSFs of Six Sigma and Lean Six Sigma in Italy [17] and the UK [18].

Across studies, respondents rated management involvement and commitment as the most critical ingredient for the successful implementation of Lean Six Sigma [1, 11–13, 16] Two other CSFs were supported by most studies, namely linking Lean Six Sigma to business strategy and linking Lean Six Sigma to customers [11]. Surprisingly, and despite the attention given to it, cultural change was only flagged as an important CSF in two of the studies mentioned above [14, 17]. However, both these studies investigated CSFs in a non-Anglo Saxon context, namely Italy and Singapore.

Instead of relying on managerial self-reports of perceived importance, Jayaraman et al. [19] extended this line of research by testing the effect of CSFs on operational and organizational performance. Consistent with previous research, Jayaraman et al.

found management engagement and commitment to be related to operational performance, and organizational performance. This was also found for frequent communication and assessment on Lean Six Sigma results, an effective Lean Six Sigma training program and an established LSS dashboard. Jayaraman et al. introduced organizational beliefs and culture as moderating variables by arguing that a "supportive organizational culture is an essential platform for the implementation of lean manufacturing" (p. 438). They found that organizations with higher Lean Six Sigma belief and culture rewarded participants in successful Lean Six Sigma projects. This supports the point made by Monteiro de Carvalho et al. [15] who distinguished between employee intrinsic and extrinsic motivation to participate in Six Sigma projects.

3 The Role of National Culture

Culture is defined as "the collective programming of the mind which distinguishes the members of one human group for another" [20]. National culture assumes people clustered in a geographical location share beliefs, attitudes and customs that differ from people's beliefs and attitudes in a different geographical location. Hofstede's and GLOBE are two frameworks explaining cultural differences.

3.1 Culture Frameworks

In his seminal work on national culture, Hofstede initially identified four cultural dimensions: power distance, uncertainty avoidance, individualism/collectivism and masculinity/femininity by surveying 116,000 IBM employees in 50 countries [20]. He later added two more cultural dimensions: indulgence and long-term orientation.

The "Global Leadership and Organizational Behavior Effectiveness" (GLOBE) Research Program was started in 1991 and built on Hofstede's and others' research [21]. In this project, survey results from 17,300 middle managers from 951 organizations in a variety of industries were used to empirically establish nine cultural dimensions, namely power distance, uncertainty avoidance, humane orientation, institutional collectivism, in-group collectivism, assertiveness, gender egalitarianism, future orientation, and performance orientation. The findings of the survey were used to place countries in country clusters, where cultural similarity was largest among countries within a cluster. The clusters are positioned in a circle where clusters nearer to each other are considered more similar and clusters at the opposite end of the circle other most dissimilar. These clusters differ in how they value the dimensions above. For example, power distance is defined as: "the degree to which members of a collective expect power to be distributed equally" and was measured, amongst others, with "followers are (should be) expected to obey their leaders without question" [22].

Countries such as Brazil (Latin American cluster) and Thailand (Southern Asia cluster) prefer hierarchical decision-making as opposed to the Netherlands (Germanic cluster) where decision-making style is more participative.

3.2 CSFs and National Culture

Management participation and commitment, linking Lean Six Sigma to strategy, and linking Lean Six Sigma to customers were found to be the most important CSFs in the implementation of Lean Six Sigma. Management participation and commitment incorporates aspects of power distance and uncertainty avoidance. In countries with high power distance, employees prefer to be told what to do [20]. Similarly, employees in high uncertainty avoidance cultures are more accepting of leader rules and policies as they reduce uncertainty and clarify expectations.

Management participation and commitment is thus closely linked to leadership. Leadership has been found to be influenced by national culture. As part of the GLOBE project, six leader styles were defined [21]. Each country cluster exhibited different preferences for the leadership styles. For example, the Anglo cluster showed high preference for the performance oriented, participative, and human leader styles, while the South East Asian and Confucian clusters showed preference for the team oriented, humane, self-protective and group-protective leader styles.

The authors therefore propose that the way management participation and commitment is operationalized during the implementation of Lean Six Sigma in a particular national culture will influence the success of the implementation (Proposition 1).

3.3 Inconsistencies Between CSFs as Explained by National Culture

The CSF cultural change, referring to organizational culture, was only flagged as an important CSF in two studies, both conducted in a non-Anglo Saxon context, namely Italy and Singapore [14, 17]. Studies conducted in the UK [12] and the Netherland [13] did not find this CSF to be important. A possible explanation for the contradictory findings is that Six Sigma was developed in the United States. The United States and the UK's culture closely resemble, as they both belong to the "Anglo" cluster according to the GLOBE classification [21]. The Netherlands belong to the adjacent "Germanic" cluster while Italy and Singapore belong to clusters positioned further away (Latin European and Confucian). Cultural change refers to resistance to change [12] which is fueled by uncertainty avoidance, a lack of future orientation and enabled by institutional collectivism and in-group collectivism.

Therefore, the authors propose that the degree of cultural change required for a company to successfully implement Lean Six Sigma will be influenced by the national culture in which the implementation takes place (Proposition 2).

In the context of Lean Six Sigma implementation, rewards and recognitions can be used to motivate employees to be involved in the implementation of Lean Six Sigma. However, whether rewards and recognitions are appropriate may be influenced by the performance orientation of a culture, defined as "the degree to which a collective encourages and rewards group members for performance improvement and excellence" [22]. Furthermore, how rewards and recognitions should be distributed may depend on the degree of in-group collectivism. In-group collectivism is defined as "the degree to which individuals express pride, loyalty, and cohesiveness in their organizations or families" [22].

Therefore, the authors propose that for rewards and recognitions to have a positive effect on the implementation of Lean Six Sigma they will need to be tailored to the organizational and national culture (Proposition 3).

4 Interviews

The purpose of our study being exploratory, the authors were not interested in a test of the above proposition but in gathering evidence to determine whether this line of inquiry was worth pursuing. To that effect, the authors approached two companies operating in multiple countries and whose subsidiaries are in various stages of Lean Six Sigma implementation. After a review of implementation materials including policy documents and training materials, in-depth interviews were conducted with key personnel. For both companies, these were the initiators and principle actors of the Lean Six Sigma initiative. This resulted in two in-depth case studies of companies hereafter referred as A and B.

Company A is a large multinational headquartered in the Netherlands with subsidiaries in over 100 countries. Company A has successfully implemented continuous improvement in its production facilities and is currently rolling out Lean Six Sigma to the non-production processes (marketing, finance, etc.). The implementation is managed from headquarters who provide guidance and training materials and who review the progress. The researchers interact with company A's headquarters.

Company B is part of a large multinational headquartered in Sweden, with production facilities in the Netherlands, Malaysia and Brazil. The company has successfully implemented Lean in two of its production facilities and is current starting up implementation of TPM in its third production facility. The company used a decentralized implementation approach by allowing each production facility to decide on the best way. However, the production facilities assist each other through the sharing of knowledge and key personnel. The researchers are in direct contact with the Dutch subsidiary. For brevity and clarity, a few findings are summarized per proposition:

Proposition 1 *the way management participation and commitment is operationalized during the implementation of Lean Six Sigma in a particular national culture will influence the success of the implementation.*

Company A has experienced Lean Six Sigma implementation in several countries and confirmed this to be the case. Examples discussed included the implementation of Lean Six Sigma in Nigeria, Singapore, Vietnam and the Netherlands. An anecdote related to implementation in Nigeria was particularly informative for this proposition. Employees in the Nigerian subsidiary of Company A needed additional resources for the continuation of their projects but were reluctant to approach management without having obtained results. The high power distance present in Nigeria was impeding progress, but Dutch management did not recognize the problem as they were viewing the situation from a low power distance lens as is customary in the Netherlands.

Proposition 2 *the degree of cultural change required for a company to successfully implement Lean Six Sigma will be influenced by the national culture in which the implementation takes place.*

This proposition was also confirmed by Company A who has classified their subsidiaries in terms of anticipated resistance to change and uses this as a criterion for selecting the order in which the subsidiaries will implement Lean Six Sigma.

Company B indicated that implementation in the Netherlands did initially encounter resistance, but the resistance was overcome by letting the results speak for themselves once the implementation was initiated. The company therefore did not take measures to overcome the resistance to change. Currently, the company is encountering extensive resistance to change in the implementation of Lean at their Brazilian subsidiary and is thinking of measures to overcome it.

Proposition 3 *for rewards and recognitions to have a positive effect on the implementation of Lean Six Sigma they will need to be tailored to the organizational and national culture.*

This proposition was supported by both companies who used rewards and recognitions to extrinsically motivate employees to be involved in Lean Six Sigma projects. Company A indicated that as part of implementation, managers at subsidiaries are advised to establish rewards and recognitions tailored to the national culture in which the subsidiary operate based on advice provided by headquarters. Company A's experience was that not all subsidiaries were eager to use rewards and recognitions. A possible explanation that was given was that in some of the subsidiary countries, individuals were not willing to be recognized individually and perceived the achievement to be a group achievement.

The above findings are not meant to be exhaustive, merely illustrative of the need to examine the interplay between culture and CSFs of Lean Six Sigma implementation.

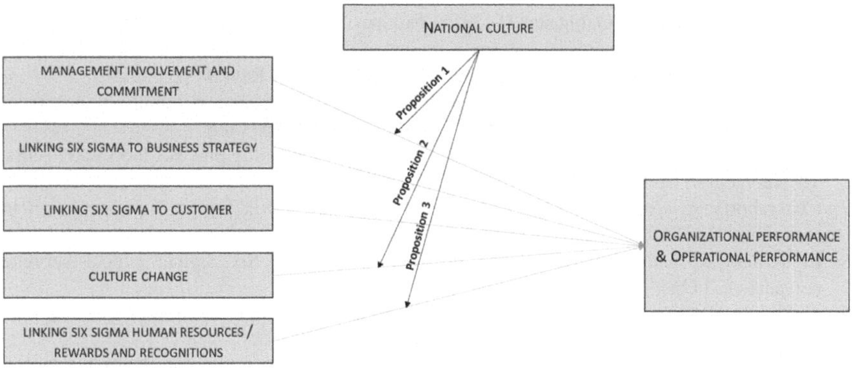

Fig. 1 Conceptual model

5 Conclusions and Further Research

The above findings along with the literature review support the formulation of the following conceptual model (Fig. 1).

The authors propose to continue the line of inquiry of Antony and Banuelas [12] and Jayamaran et al. [19] and investigate the role of CSFs in the successful implementation of Lean Six Sigma as defined by organizational and operational performance improvements. The authors hypothesize that the effect of management involvement and commitment on organization performance and operational performance is influenced by national culture. Similarly, the authors hypothesize that the effects of the CSFs culture change and rewards and recognitions on organizational performance and operational performance will be influenced by national culture.

The study findings show that attempts to quantify CSFs and cultural factors are still premature as these scales do not capture the subtleties of intercultural differences. Even within a single national culture there is extensive variation as the anecdotes above show. Therefore, a qualitative research approach that examines Lean Six Sigma in its context is needed. The authors are currently initiating, in cooperation with Company A and Company B, a multi-year research study that will use action research to further investigate this issue. Action research has been selected due to its unique quality to "integrate social sciences with organizational knowledge to generate actionable scientific knowledge" [23].

Acknowledgements The authors acknowledge funding received by SiA.

References

1. Desai DA, Jiju A, Patel MB (2012) An assessment of the critical success factors for Six Sigma implementation in Indian industries. Int J Prod Perform Manag 61(4):426–444
2. O'Reilly C (1998) New united motors manufacturing Inc. (NUMMI). Stanford Business Case

3. McLean RS, Antony J, Dahlgaard JJ (2015) Failure of continuous improvement initiatives in manufacturing environments: a systematic review of the evidence. Total Qual Manag, 1–19
4. Abrahamson E (2000) Change without pain. Harvard Business Review, pp 75–79
5. Patyal VS, Koilakuntla M (2018) Impact of organizational culture on quality management practices: an empirical investigation. Benchmarking 25(5):1406–1428
6. Juergensen T (2000) Continuous improvement: mindsets capability process, tools and results. The Juergensen Consulting Group Inc., Indianapolis
7. Chakraborty A, Chuan TK (2012) Case study analysis of Six Sigma implementation in Singapore. Bus Process Manag J 18(6):992–1019
8. Linderman K, Schroeder RG, Zaheer S, Choo AS (2003) Six Sigma: a goal-theoretic perspective. J Oper Manag 21:193–203
9. de Koning H, Does RJ (2008) Lean Six Sigma in financial services. Int J Six Sigma Compet Adv 4(1):1–17
10. Rockart J (1979) Chief executives define their own needs. Harvard Bus Rev 57(2):238–241
11. Banuelas R, Antony J (2002) Critical success factors for the successful implementation of Six Sigma. TQM Mag 14(2):92–99
12. Antony J, Banuelas R (2002) Key ingredients for the effective implementation of Six Sigma program. Meas Bus Excell 6(4):20–27
13. Timans W, Antony J, Ahaus K, van Solingen R (2012) Implementation of Lean Six Sigma in small- and medium-sized manufacturing enterprises in the Netherlands. J Oper Res Soc 63:339–353
14. Chakraborty A, Chuan Tan K (2007) A survey on Six Sigma implementation in Singapore service industries. In: Proceedings of the IEEE international conference on industrial engineering, Furama River Front, Singapore
15. Monteiro de Carvalho M, Lee Ho L, Boarin Pinto SH (2014) The Six Sigma program: an empirical study of Brazilian companies. J Manuf Technol Manag 25(5):602–630
16. Laureani A, Antony J (2012) Critical success factors for the effective implementation of Lean Six Sigma. Int J Lean Six Sigma 3(4):274–283
17. Brun A (2011) Critical success factors of Six Sigma implementations in Italian companies. Int J Prod Econ 131(1):158–164
18. Manville G, Greatbanks R, Krishnasamy R, Parker DW (2012) Critical success factors for Lean Six Sigma programmes: a view from middle management. Int J Qual Reliab Manag 29(1):7–20
19. Jayaraman K, Kee TL, Lin Soh K (2012) The perceptions and perspectives of Lean Six Sigma (LSS) practitioners: an empirical study in Malaysia. TQM J 24(5):433–446
20. Hofstede G (1980) Culture's consequences: international differences in work-related values. Sage, Beverly Hills, CA
21. House RJ, Hanges PJ, Javidan M, Dorfman PW, Gupta V (2004) Leadership, culture and organizations: the GLOBE study of 62 societies. Sage Publications, Thousand Oaks, CA
22. Flynn B, Saladin B (2006) Relevance of Baldrige constructs in an international context: a study of national culture. J Oper Manag 24:583–603
23. Cronemyr P, Eriksson M, Jakolini S (2014) Six Sigma diplomacy—the impact of Six Sigma on national patterns of corporate culture. Total Qual Manag Bus Excell 25(7–8):827–841

Application of Lean Product and Process Development in FIRST Robotics Competition

Bas Koomen and Ron Visser

Abstract In FIRST (For Inspiration and Recognition of Science and Technology) Robotics Competition, students have to design, build, and test a competition robot during a building season of six weeks. Lean Product and Process Development promises to shorten product development times and increase knowledge reuse. There is a knowledge gap for the application of Lean Product and Process Development in the context of student competitions. In this paper, we outline an approach to apply Lean Product and Process Development during the preparation and the building season. We hypothesize that the students can frontload knowledge in problem-based learning cycles before the game is published. Once the game is published, students can apply the front-loaded knowledge for the specific requirements of the game. The proposed approach includes an organizational structure, processes, and the use of Product Lifecycle Management software. We are going to test the approach with a larger FIRST Robotics Competition team in The Netherlands. The expected results of this case study are an increased insight in the effectiveness of Lean Product and Process Development and a measurable difference with the traditional design approach. Future research needs to be done on the results of this case study. Also, more similar case studies can be performed to obtain more general knowledge about the effectivity of the methodology.

Keywords Lean product and process development · Problem based learning · FIRST robotics competition · Set based concurrent engineering

---

B. Koomen (✉)
University of Twente, Enschede, The Netherlands
e-mail: s.p.koomen@utwente.nl

Cadmes b.v., 's-Hertogenbosch, The Netherlands

R. Visser
MTA b.v., Helmond, The Netherlands
e-mail: r.visser@m-t-a.nl

Team Rembrandts STEAM Foundation, Eindhoven, The Netherlands

M. Rossi et al. (eds.), *Proceedings of the 6th European Lean Educator Conference*, Lecture Notes in Networks and Systems 122,
https://doi.org/10.1007/978-3-030-41429-0_19

1 Introduction

Educators are looking for ways to present modern industrial approaches to students. Problem-Based Learning is a commonly accepted approach, in which content and practice are holistically integrated [1].

For Lean Manufacturing, Tortorella and Cauchick-Miquel have proposed an initiative with industrial engineering graduates, based on Problem-Based Learning [2]. Their research paper references other research on the benefits of Problem-Based Learning and what the optimal circumstance should be. They conclude that a Problem-Based Learning approach enhances the ability of students to acquire and apply knowledge in real-world situations. Moreover, they can better meet the current demands of organizations and academia.

The challenge in Problem-Based Learning is to find complex, real-life projects that allow students to apprehend and experience the content and enables them to reflect on it within a limited timeframe. If real projects from companies or organizations are not available or feasible within the available time, educators have to use fictive problems.

With Lean Product and Process Development (LPPD), as defined by Ward and Sobek II [3], it is a challenge to find real-life problems in companies. Not many companies are looking into this theory yet, and LPPD projects involve a more profound organizational change before successes are achieved [4]. The consequence is that educators have to divert to fictive problems or serious games [5], and therefore compromise on the real-life aspect of Problem-Based Learning.

We identified the combination of the aforementioned issues (the lack of suitable cases for Problem-Based Learning with LPPD and the complexity of LPPD introduction in real organizations) as a limiting factor to let students experience and investigate LPPD. Therefore, we are using a student robotics competition program as an a context instead of industrial problems.

In this paper, we propose to apply LPPD in *FIRST* Robotics Competition (FRC) [6] (*FIRST* is an acronym for "For Inspiration and Recognition of Science and Technology"). In FRC, student teams collaboratively design, test, and build a robot (see Fig. 1 for an example) within six weeks, based on requirements that change every year. FRC has a high level of complexity and includes mechanical, electrical, and software engineering. The benefit of LPPD is that the students can prepare reusable knowledge in the months preceding the six weeks of the challenge.

We will verify the proposed methodology in a case study, with the Dutch FRC team #4481 (Team Rembrandts), that will take place between August 2019 and May 2020, during which we will measure the effectivity of LPPD. The results of this case study contribute to a better understanding of LPPD in an educational context.

The remainder of the paper has the following structure: in Sect. 2, we elaborate more on the research question and the context. Section 3 explains the LPPD methodology that we will use in the case study. The implementation plan for the case study is described in Sect. 4 and the measurement of the expected results in Sect. 5. Section 6 contains the conclusion and a brief outlook to future research.

Fig. 1 FRC robot (FRC team 4481, Season 2019)

2 Research Question

FRC teams work intensely on the design of their robot during the competition's building season, which lasts only six weeks. Before the start of this period, the exact requirements for the game are unknown. Therefore, it is a challenge to let students work on relevant engineering problems before the announcement of the game requirements.

From experience, we know that the six week period allows for only one design iteration, with little time for prototyping. When the team participates in a regional tournament, the first real test of the design is done. Only if the team participates in a second regional tournament, or qualifies for the World Championships, a next design iteration might take place.

Teams have to make many design decisions in different areas. For each of these areas, knowledge gaps emerge during the building season. Below, we give two examples:

(1) *A shooter that has to shoot (or throw) an object in or on a specific target*. This feature was needed in most of the previous seasons in various forms. The objects have been: balls in various weights and sizes, cubes, frisbees, and crates. The shooter problem is very suitable for knowledge creation in an extensive range of requirements since it is unknown upfront what the exact rules of next year's game will be (Fig. 2).

(2) A drive train that supports the robot and enables it to move around. Here the team can decide to use tracks, tank-drive wheels, omni-directional (Mecanum) wheels, swerve drive, or other solutions. On the one hand, the robot always needs a drive train, and the rules are mostly known. On the other hand, the best fit of the characteristics with the game is unknown upfront. In some years, an agile and fast drive train is optimal. In other years, a sturdy drive train with much grip is better (Fig. 3).

Fig. 2 Ball shooter module (FRC team 4481, Season 2016)

Fig. 3 Drive train (FRC team 4481, Season 2016)

Following the LPPD principles, it should be possible to front-load knowledge of technical solutions for typical challenges in the FRC game structure. Set-Based Concurrent Engineering on sub-problems—with potential ranges of requirements—could enable more design iterations. Reusable knowledge should emerge from these iterations.

Hence, our research question is: *"How can FRC teams use LPPD in order to improve the design outcome during the six weeks of the building season and overcome current knowledge creation and knowledge management challenges?"*

3 Lean Product and Process Development Methodology

It is essential to give a clear definition of LPPD to understand the proposed approach. We based our approach on the theories of Ward and Sobek II [3]. Subsequently, Kennedy [4], Kennedy et al. [7], and Cloft et al. [8] have elaborated on these theories and came up with practical ways to apply LPPD in design processes. In the next paragraphs, we highlight the four pillars of LPPD (Fig. 4), according to Ward and Sobek II [3], and describe how we will apply them in the FRC context.

3.1 *Set-Based Concurrent Engineering*

We can divide an FRC robot into subsystems. Some subsystems will be relevant in each season, like the drive train, steering control, or vision systems. Other subsystems are dependent on the specific game challenges of a season. Here, the team can learn from the past seasons which elements are likely to return in the next season. Systems we have seen are: lifting mechanisms for crates, shooters for balls or frisbees, intake mechanisms for balls or crates, handling devices for circular discs, and some other exotic features. Over the years, the games have used several variations on similar principles. For example, many games had the challenge to shoot balls, but there was variation in the size and material properties of the balls.

In a "traditional" design approach, as schools teach it to most engineering students, alternative solutions to a problem are evaluated in the shortest possible time. One candidate solution is selected, using various selection mechanisms. From that point on, the design process is focused on making the selected option work, with regular unexpected setbacks and rework. Most rework is the result of knowledge gaps, earlier in the process.

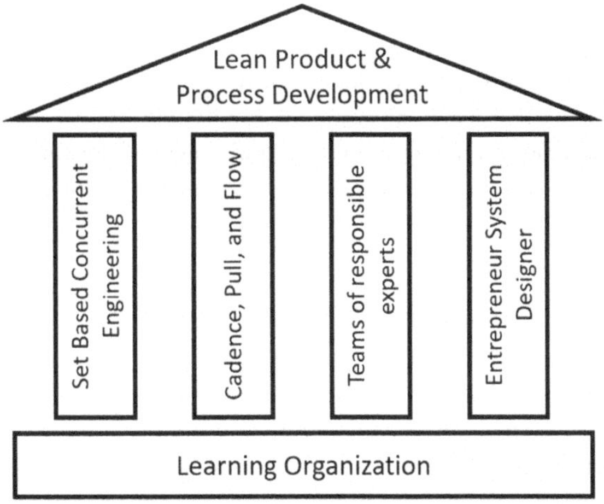

Fig. 4 Ward's "four-pillar model" of LPPD (modified from [9])

In Set-Based Concurrent Engineering, it is the aim to delay the decision, until enough knowledge is gathered to make it safely. Cloft et al. [8] explain how causal maps can help to visualize the decision-making process for a (sub)system. In the example of a ball shooter, the relation between all kind of attributes and functions can be visualized, like target accuracy, ball speed, exit angle, ball spin, ball diameter, ball weight, and ball stiffness.

3.2 Cadence, Pull and Flow

Knowledge is required to make the decisions in the causal map. If not enough knowledge is available to make a decision, a knowledge gap is identified. Once a knowledge gap exists, actions have to be taken to fill the knowledge gap. Consequently, the team will execute the actions in the form of research or tests, and make the resulting knowledge available. In the example of the ball shooter, experiments can be done using different types of balls, different speeds, different angles, different shooter models, different ball spin, and other physical behavior.

The need to fill knowledge gaps ensures the pulling behavior of the process: knowledge is pulled from a need, arising from the causal map. To create cadence, we propose to use regular team meetings to discuss and update the causal map. The frequency outside the six weeks building season can be in terms of once per two or three weeks, inside the building season twice per week. These meetings are integration events, where the acquired knowledge is put together in context.

Furthermore, the team needs to create a constant flow of knowledge. The team can record knowledge in trade-off curves [3]. Again, in the ball shooter example, the effect of backspin of the ball on the target accuracy can be measured for different diameters of balls. Multiple lines in a graph can visualize this knowledge.

Sobek II and Smalley [10] describe how knowledge can be captured in A3 knowledge briefs (K-briefs). The K-brief describes the problem, explains the physics of the problem, describes solution proposals, and describes the decision-making process. These K-briefs may include one or more trade-off curves for generic robot solutions.

The role of the K-briefs is different in the six week building season. In this time frame, the decisions are made specifically for the game challenges. In the K-brief, the team members record the rationale for the final decisions for the game robot. Later, this rationale is reusable as knowledge for the next generation.

3.3 Team of Responsible Experts

Each team member, in his or her specialty, needs to contribute to the overall success. It is not enough to concentrate only on the content of a specific task [3]. Members are expected to collaborate within and across the sub-team and know the context of their work. In the regular integration events, team members will share their findings and listen to others. Asking (why) questions is vital to get to the core of problems.

3.4 Entrepreneur System Designer

Ward and Sobek II [3] introduce the role of the *Entrepreneur System Designer* in the process of LPPD. The Entrepreneur System Designer has a central position in the design team. His responsibility is to keep the focus on the (real) customer interest and the causal map with all design decisions. For FRC, each subsystem design team could have a "subsystem Entrepreneur System Designer" who will focus on potential requirements for the subsystem. Furthermore, the team appoints an Entrepreneur System Designer for the entire robot. This "system Entrepreneur System Designer" will be responsible for the integration of subsystem knowledge for the specific game requirements of a particular season.

4 Implementation Plan

For the experiment with an FRC team in a real game season, we have defined the following implementation steps.

4.1 Create Awareness Among Mentors

The FRC consists of students and mentors. Mentors generally are more experienced engineers. Most of them have participated in FRC in previous years, and gathered experience with the FRC game structure.

To create awareness among the mentors, we will start with some Set-Based Concurrent Engineering and causal map exercises, to experience the contribution effect of this method. Potentially, the Set-Based Concurrent Engineering Serious Game by Kerga et al. [5] could be suitable for this phase.

4.2 Define Subsystems

Next, the team will decide which subsystems they will identify as part of a generic robot. These subsystems must be relatively independent from other subsystems to enable focused prototyping and learning. Also, the number of subsystems should be appropriate for the size of the team. A larger team can work on more subsystems simultaneously than a smaller team.

4.3 Establish Sub-teams (Roles)

For each subsystem, the team will establish a sub-team. Each sub-team needs one leader: the subsystem Entrepreneur System Designer. The size of the team can be depending on the type of subsystem. Some subsystems require a single discipline, like mechanical experts. Other subsystems may also need software and hardware experts.

It may be smart to initiate a dedicated game strategy team for scoring trade-off knowledge. This sub-team might even include experts on game theory [11].

4.4 Train Team Members in the Methodology

When the teams are established, the members need to be trained. The training will focus on the specific elements in the methodology:

- Causal maps. The members will learn how they can build a causal map for a specific design problem. The process is important, how to discuss and generate required decisions.
- Knowledge gaps. From the causal map, knowledge gaps emerge. The members are trained on how to approach the knowledge gap and learn to come up with the right questions.
- A3 K-briefs. The team members need to learn how to write a useable A3 K-brief. We will provide a predefined structure.
- Integration events. We will instruct the sub-teams, how to organize an integration event for the subsystem and the entire robot system.

4.5 Establish Knowledge Platform

It is crucial to establish a common platform to make the acquired knowledge available. The platform could be relatively low-tech, where cloud storage is used to host digital A3 K-briefs, causal maps, and trade-off curves, or a more advanced approach with industry-standard collaboration platforms.

For this case study, Dassault Systèmes will make the 3DEXPERIENCE Platform [12] available in the cloud for the FRC team. The team has already experience with SOLIDWORKS 3D CAD and virtual prototyping, which will be used more extensively during the case study.

4.6 *Coach Teams to Work LPPD-Style*

After having been trained, the team will start their work of knowledge creation, ahead of the six week season start. The team will need coaching of LPPD practitioners to keep the required methodological quality.

5 Measuring the Results

We need to measure two results from the experiment to answer the research question: the quality of the design outcome, and the performance of knowledge creation.

5.1 *Design Outcome*

We can measure the design outcome by the relative performance of the robot in the competition. The organization tracks many scoring data on each element of the game. This performance data is known for previous seasons, so after the next season, we can measure if there is an improvement. In Fig. 5, there is an example of such data from the 2018 game.

Year Specific Statistics

Year Aggregate | Week 1 | Week 2 | Week 3 | Week 4 | Week 5 | Week 6 | Week 7 | FIRST Championship - Houston | FIRST Championship - Detroit

Qualification

Bonus Objective Statistics

Objective	Count	# Opportunities	% Success
Auto Run	3952	4068	97.15%
Auto Switch Owned	1221	1356	90.04%
Auto Quest *	1158	1356	85.40%
Force Played	850	1356	62.68%
Levitate Played	1279	1356	94.32%
Boost Played	1039	1356	76.62%
Climbs (does not include Levitate)	1583	4068	38.91%
Face the Boss *	470	1356	34.66%
"Unicorn Matches" (Win + Auto Quest + Face the Boss)	265	1356	19.54%

Playoff

Bonus Objective Statistics

Objective	Count	# Opportunities	% Success
Auto Run	697	702	99.29%
Auto Switch Owned	227	234	97.01%
Auto Quest *	222	234	94.87%
Force Played	113	234	48.29%
Levitate Played	194	234	82.91%
Boost Played	159	234	67.95%
Climbs (does not include Levitate)	366	702	52.14%
Face the Boss *	130	234	55.56%
"Unicorn Matches" (Win + Auto Quest + Face the Boss)	86	234	36.75%

Fig. 5 Relative scoring data versus potential scoring from 2018 [14]

With the statistical data, the team can measure their relative performance compared to other teams and the performance compared to the potential maximum performance. If this method is valid, a change should be noticeable on longer-term (multiple years from now).

5.2 Knowledge Creation and Management

Thomke and Fujimoto [13] have investigated the effect of "Front-Loading Problem-Solving" at Toyota. They identify two mechanisms of front-loading:

- Project-to-project knowledge transfer, which can be measured by the number of k-briefs that is created to transfer knowledge, or the number of k-briefs that can be used during the six week building season.
- Rapid problem-solving, which is a mechanism of rapid learning, where team members perform many tests (physical or virtual) on a large number of alternatives. Rapid learning would optimally result in a large number of trade-off curves, which can be measured.

Effect of this case study should be measurable in the amount of documented knowledge (A3 k-briefs, causal maps), and the interaction during integration events.

A specific aspect of FIRST competitions is that the organization encourages teams to proliferate their knowledge. LPPD offers a suitable context to improve this, so the knowledge transfer between teams could also be measured if more teams would adopt this methodology.

6 Conclusion and Future Research

We conclude from this conceptual design of the experiment that FRC is a proper context to let students experience the effect of LPPD. FRC offers several aspects that enable a very realistic environment:

- The design is sophisticated and can be divided into subsystems.
- There is a fixed time frame between the publication of the requirements and product delivery.
- Teams can use the months before the six week building season to frontload knowledge.
- There is an opportunity for learning from the past by investigation of previous seasons.

As a next step, we need to evaluate this case study during and after the experiment. We will need to measure short and long term effects for more definitive conclusions.

References

1. Jones BD et al (2013) The effects of a collaborative problem-based learning experience on students' motivation in engineering capstone courses. Interdisc J Probl Based Learn 7(2):2
2. Tortorella G, Cauchick-Miguel PA (2017) An initiative for integrating problem-based learning into a lean manufacturing course of an industrial engineering graduate program. Producao 27(Special issue)
3. Ward AC, Sobek II DK (2014) Lean product and process development, 2nd edn. The Lean Enterprise Institute, Cambridge, Mass
4. Kennedy MN (2003) Product development for the lean enterprise: why Toyota's system is four times more productive and how you can implement it. Oaklea Press
5. Kerga E et al (2014) A serious game for introducing set-based concurrent engineering in industrial practices. Concurr Eng 22(4):333–346
6. FIRST (2019) FIRST Robotics Competition [cited 2019 16–07–2019]. Available from https://www.firstinspires.org/robotics/frc
7. Kennedy BM, Sobek DK II, Kennedy MN (2014) Reducing rework by applying set-based practices early in the systems engineering process. Syst Eng 17(3):278–296
8. Cloft PW, Kennedy MN, Kennedy BM (2018) Success is assured: satisfy your customers on time and on budget by optimizing decisions collaboratively using reusable visual models. Productivity Press, Cambridge, MA
9. Flores M et al (2017) Lean product development best practices, ten industrial success stories. Lean Analytics Association, Carona, Switzerland
10. Sobek II DK, Smalley A (2008) Understanding A3 thinking: a critical component of Toyota's PDCA management system. Productivity Press, New York
11. Webb JN (2007) Game theory: decision, interaction, and evolution. Springer, London
12. Dassault_Systèmes (2019) 3D Experience Platform [cited 2019 1–8–2019]. Available from https://www.3ds.com/about-3ds/3dexperience-platform/
13. Thomke S, Fujimoto T (2000) The effect of "front-loading" problem-solving on product development performance. J Prod Innov Manag 17(2):128–142
14. The_Blue_Alliance (2019) 2018 Insights [cited 2019 1–8–2019]. Available from https://www.thebluealliance.com/insights/2018#year-specific-week-9

Continuous Improvement System: Team Members' Perceptions

José Dinis-Carvalho, Mónica Monteiro and Helena Macedo

Abstract The complexity, competition and fast changing of the global market requires the companies to seek excellence, if they crave for good performance and success. To reach this purpose, it is mandatory the existence of a Continuous Improvement (CI) culture within the companies. This concept, also known as Kaizen, mean "change for the better" and it defends that a very large of small improvements applied and sustained over the time, brings more value than a few big improvements. The objective of this paper is to evaluate the effect of lean teamwork after a Continuous Improvement system had been implemented in Lipor, a Public Organization based in Oporto that treats the municipal waste. The CI system implemented a few years ago, was possible to maintain and adapt over the years through the creation of a Kaizen Team, Natural Teams, Project Teams and the cooperation between them, cultivating this Continuous Improvement culture in the company's core. The results of the survey done to the workers regarding the CI system demonstrated that the majority consider that their work process has improved significantly with Kaizen and highlighted the importance for them to be involved in improving team's processes.

Keywords Continuous improvement · Teamwork · Kaizen · Lean production

J. Dinis-Carvalho (✉) · H. Macedo
Department of Production and Systems, University of Minho, Campus Azurém, 4800-058 Guimarães, Portugal
e-mail: dinis@dps.uminho.pt

H. Macedo
e-mail: helena_macedo@outlook.com

M. Monteiro
Instituto Universitário Lisboa, ISCTE, Av.ª das Forças Armadas, 1649-026 Lisboa, Portugal
e-mail: monicamonteiro@lipor.pt

M. Rossi et al. (eds.), *Proceedings of the 6th European Lean Educator Conference*, Lecture Notes in Networks and Systems 122,
https://doi.org/10.1007/978-3-030-41429-0_20

1 Introduction

Continuous improvement or pursuing of excellence is one of the five principles of the lean thinking [1], being the most important one in insuring future competitiveness of a company. If even successful companies do not work on improving continuously their processes and products, they would gradually loose market to their competitors and eventually close down. Continuous improvement is not a choice, it is an obligation to survive in the future. Continuous improvement (CI), to be effective should be part of daily work as any other operational task. It should be clear, to all employees in a company, that making small improvements everyday should be part of the daily routine tasks. This idea is very well presented by Rother [2] where he presents his interpretation how Toyota achieves CI. Most managers may agree that CI is very important to companies' future but few are willing to spend their human resources on such tasks. As Abraham Lincoln (president of the USA from 1861 to 1865) once said: "If I had six hours to chop down a tree, I'd spend the first four hours sharpening the axe". Continuous improvement can be viewed in a metaphoric way as "sharpening the axe". "Sharpening the axe" takes precious time to the cutting time so most managers are not willing to "waste" precious cutting time in "sharpening the axe" (Continuous Improvement). But "sharpening the axe" improves the future productivity in a way that the time spent in production can be more effective afterwards. In reality every small improvement brings permanent gains and that is precious for organizations. Every small improvement brings the organization to a new current state, with a better standard and better performance. A new small improvement will take advantage of small previous improvements. A big sequence of small improvements become a big improvement. Many improvements wouldn't be possible if a previous improvement was not implemented.

Continuous Improvement can be materialized in many different ways. Some companies assign a person or a team of people to continuously search and identify, through "Gemba Walks" for instance, improvement opportunities and implement improvement actions. A very common approach is by using Value Stream Mapping (VSM) to represent current state observed through Gemba Walk [3], then draw a VSM of the future desired state and take the necessary actions to go from current to future state [4]. More effective CI systems are the ones where everyone in the organization are involved in CI activities as for instance in the Toyota Kata approach [2]. There is not a standard amount of human resources that should be assigned to the Continuous Improvement effort but Koenigsaecker [5] believes that it is critical to have full-time resources dedicated to CI, recommending that 3% of an organization should dedicate full-time to it. This may not be necessary true, since many companies tend to have everyone spending a bit of their working time dedicated to CI instead of having some people dedicated full-time to it. Teamwork seems to be crucial issue since a common feature in every CI system [6] and most CI systems known in literature use teams of workers as one of the layers of responsibility in CI. In each organization, value adding activities are performed by employees, and the human factor has an enormous importance in achieve a better performance. Their creativity and involvement are the fuel to improve the organization's business [7]. In fact, reports show that Teamwork, besides Employees Involvement and Training are the main critical success factors of Continuous Improvement Projects [8]. The main guideline to Lipor's strategy is "Towards Sustainability"—depicts a sustainable management, that combines the three main principles of sustainable development—Economic, Environmental and Social. Addressing the Economic principle, a Continuous Improvement model was implemented in the company a few years ago in the back office department [9]. Based on the good results obtained, the board directors decided to extend it through all organization. This CI system suffered some small changes along the time and nowadays the CI is assigned to teams and everyone is within one team.

This paper describes the dynamic of this CI system, including team development and organization, what kaizen events are being held and which artefacts are used by the teams. Its main objective is to identify the perception of workers regarding the new practices created by the Continuous Improvement system.

2 Lipor Continuous Improvement System

Lipor was founded in 1982 as a Municipalities Association for eight municipalities around the city of Porto in Portugal and it has implemented an integrated waste management, recovered, developed and built infrastructures and organized awareness campaigns for the population, estimated in almost 1 million inhabitants. The Continuous Improvement system adopted by this company is now in a quite mature phase since it is in place for the last five years. The systems has been changing and continuously adapting according to the learnings obtained over many improvement actions and also because of the natural nature of CI concept. This CI system is based on teamwork, events and processes. Regarding the teamwork, the company is organized in Natural Teams supported by a Kaizen Team leading the process. In this company, a Natural Team is composed by a group of workers that normally work together in the same workspace, same production cell, same department, or same section, performing related or similar work, sharing a common team board, pursuing the same team objectives and making part of the company Continuous Improvement effort. On the other hand, the Kaizen Team is responsible to coordinate and support all the CI effort. Kaizen Team is supported by top management and, it is its responsibility to define the CI strategy, decide upon improvement projects and guide every Natural Team. Finally, every time a problem occurs, a multi-disciplinary team is created, with the denomination of Project Teams. These teams are created to develop some specific improvement projects and they cease to exist at the end of that project.

2.1 The Kaizen Team

Kaizen Team plays a key role in the organization being the engine of the Continuous Improvement effort of the whole company. This team needs to be continuously putting energy on the CI systems otherwise it tends to gradually reduce the improvement speed and ultimately stagnate. The Kaizen team defines CI strategies, provides guidance, implements CI audits and monitor the CI activities in the whole company. Elements of the Kaizen team are also responsible to give training and coaching to the natural teams from many different areas, from office to production work.

This team is composed by elements representing some departments. Not every Natural Team is represented in the Kaizen Team but every Kaizen Team element is member of a Natural Team. All 8 Kaizen team members are CI leaders from different departments of the company and were trained by external Kaizen experts on Lean thinking, lean leadership and kaizen tools. The kaizen team leader was selected according to her motivation and knowledge in continuous improvement and the other kaizen team members were also selected by their natural skills in CI as well as their role as CI leaders in their own departments.

2.2 Natural Teams

Natural Teams are characterized by a group of people that work together, performing similar or related tasks and normally in the same room or working area. At Lipor, Natural Teams are represented by one of their elements called "pivot". The company chose to substitute the usual term "team leader" for "team pivot". Natural team's pivot participates in the audit preparation meetings, as he or she is part of the audit team, and also participates in the kaizen teams training. Other team members' role in CI is mainly contributing to team dynamics in identifying problems and improvement opportunities and contribute to problem solving and improvement implementations. The team's pivot is the most important communication channel between the Kaizen Team and all team members. The designation and number of member of the existing natural teams (one natural team per area) are listed in Table 1.

2.3 Project Teams

Project Teams, on the other hand, are created with the purpose to achieve or develop a specific improvement project. They are constituted by the process owner and elements from the teams involved in the process. At the end of each disruptive project the team responsible for the project ceases to exist. Usually, improvement projects or Kobetsu, result from the use of the tool Quality Matrix. In this matrix, the errors/occurrences detected in the various tasks of a given process are identified according to its origin

Table 1 Designation and number of members of all natural teams

Natural team designation	No. of members
International Business Unit	3
Legal and Audit Department	7
Operations and Logistics Department	5
Operations Division	5
Logistics and Infrastructures Division	97
Support Division for Operational Project Implementation	8
Environmental Education and Training Unit	16
Communication, Sustainability and Marketing Unit	10
Supply and Accounting Division	14
Information Systems and Management Division	11
Human Resources Division	6

or intervenient. Consequently, it is possible to determine objectively if it is an occasional or frequent error/occurrence, facilitating the prioritization of the improvement projects. Any new project team that is created frequently requires training on kobetsu methodology. This training on Kobetsu is always provided by Kaizen Team, which in some cases may be represented in the Project Team. In order to monitor the performance of the Project Teams, as well as the state of the teams, a board is created with information concerning the important details of the project.

2.4 Team Artifacts and Events

Team boards play a very important role in the natural teams' management, providing transparency and creating a platform for team development, motivation and performance improvement. They afford a better daily work management, allowing to monitor in a very visual way, task assignment and priorities management as well as monitor the progress of the main team performance indicators. The board structure was created by each team, once the company believes that is important for them to decide what is more relevant for their organization, monitoring and management and by participating in the design process, team members become more engaged in the process.

Moreover, this artefact enables the track of all the ideas for improvements gathered across the team and then helps to analyses the status of ongoing improvement initiatives. Every kind of improvement idea is written down on structured cards and fixed on the board and depending on the progress of implementation, the cards are transferred to the appropriate section of the PDCA circle.

Every day, each team takes some minutes around the board to discuss the plan for the day and to analyze the performance indicators. Time to time it is necessary to introduce changes in the board and start the standardization process again. Changing needs are identified by the team during the daily meetings reflecting the dynamics of Continuous Improvement. In this process it is important for managers to ensure that the boards are used and maintained. With this goal it was conducted regular audits by other teams.

Daily Kaizen is one of the events of Lipor Continuous Improvement System. It aims to develop teams and to increase teamwork on Natural Teams. Team are encourage to have a daily meeting where all team members become aware of their team performance development, monitor standards and goals, solve problems that arise on the work floor and give new improvement suggestions. These meetings also allow Natural Teams to become more autonomous and at the same time turn these practices into routine. Daily Kaizen is therefore vital to maintain the levels of efficiency and to, steadily, improve performance indicators.

The implementation of Daily Kaizen was a challenge especially because habits and routines are difficult to change in people. As a start, it was asked to each work team to prepare a set of performance indicators that could clearly reflect the performance of their daily activities. The next step was to decide which of these were more suitable to be discussed at Daily Kaizen Meetings, and which would give a better picture of the working day. This was done by the team leaders together with their Natural Teams. Finally, everything was compiled on a board where tasks and indicators can be easily checked in a daily basis, allowing an increased level of control and efficiency. After the team boards were assembled, the concept of daily meetings was introduced to easily monitor the daily progress of work in different areas and of different team's members. This tool has proven to be much more useful than initially thought and allowed to manage both the implementation of improvement projects, as well as the daily problem solving requirements.

An important Kaizen event is the Kaizen Team meeting. This event occurs every week in a one hour meeting base, and one day meeting once in a month. In these meetings Kaizen Team analyses the action plan and make adjustments, updates the natural team's performance and issues and take action in order to solve existing problems.

Kaizen Team Audits is another type of event in the Continuous Improvement system. Every month, Kaizen Team performs audits in very Natural Team. The audits are oriented to a specific topic that changes every month. They could be focused on physic 5S in one month, 5S informatics in the other month and on standards in another. The annual external audit is also another important Kaizen Event. The external audit is performed by a lean consultant and the objectives are to get an external evaluation of the existing practices, routines and improvements. From the results of the external audits the company defines new challenges and objectives for next year.

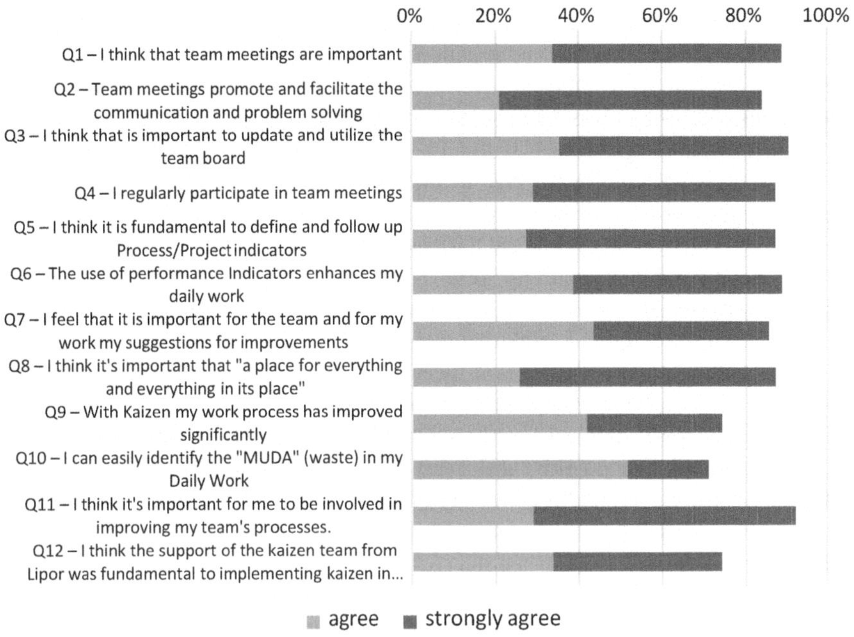

Fig. 1 Percentage of "agree" and "strong agree" answers obtained from the questionnaires

3 Methodology Applied in This Study

In order to understand workers perception and degree of acceptance of several features of the Lipors' Continuous Improvement System, a questionnaire developed with the 12 questions (see Fig. 1). For each question the five possible answers followed a Likert scale were: strongly disagree; disagree; neither agree nor disagree; agree; strongly agree. The questions were devolved to be as neutral as possible to avoid possible bias. The questionnaire was then delivered to 135 out of 182 workers obtained a total of 62 valid replies. We did not manage to deliver (not even by email) the questionnaires to 47 remaining workers. In order to guaranty valid replies from all natural team we had to deliver in person 10 printed questionnaires.

4 Results

The analysis of the survey intends to collect and understand the opinion of the workers about the different activities developed in the context of the CI implemented system, in 4 areas: Team Meeting; Team Board and Performance Indicators Recording; Kaizen Efficiency and Worker's Suggestions. Figure 1 depicts the general positive feedback given by employees regarding the CI movement. They perceived the continuous improvement routines and practices in a very positive way since in average 84% of the answers were either "agree" or "strongly agree". The question with more positive answers (92%) was "Q11—I think it's important for me to be involved in improving my team's processes" with 63% of "strongly agrees". This clearly shows the importance that employees give to their involvement in improving their processes. This is strongly related to the main stream Lean idea related to "respect for people" mentioned by Sugimori et al. [10] as one of the two basic concepts of TPS as well as "use of employee creativity" referred by Liker [11]. The second best positive feedback was given to questions "Q3—I think that is important to update and utilize the team board" and "Q1—I think that team meetings are important". That reflects the importance they perceive in team meeting, the visual management importance of team boards and the involvement in team's processes.

The question with worse performance was "Q10—I can easily identify the "MUDA" (waste) in my Daily Work" with 70% "agree" or "strongly agree" answers. In reality this question is not about the way they perceive the CI system but instead it is about a certain skill that is important for CI.

Regarding Team Meeting, when questioned about the importance of daily/week meetings, 89% of the inquires consider that is important to have these sessions, 84% consider that team meeting promote and facilitate the communication and problem solving, and 87% assumes to regularly attend meetings. On the other hand, relating to performance indicators recording, 90% of the respondents think that is important to update and utilize the team board, 87% consider that is fundamental to define and follow up Process/Project indicators and also 89% have the opinion that the utilization of performance indicators enhances their daily work. The importance of visual management provided by team boards as well as the constant feedback provided by performance monitoring is very well accepted by employees.

Thirdly, when questioned about kaizen efficiency, 87% of the inquires answered that is important "a place for everything and everything in its place", 74% consider that their work process has improved significantly with Kaizen, 71% are able to easily identify the "MUDA" (waste) in their daily work, 92% think that is important for them to be involved in improving team's processes and also 74% mentioned that the support of the kaizen team from Lipor was fundamental to implementing kaizen in their working area. Finally, the analysis of the survey allows us to conclude that 85% of the respondents feel that their improvement suggestions are important for their team and individual work. Written comments were also provided by some members in order to clarify aspects that were not covered in the questionnaire or to provide specific feedback. Some examples are:

"I think we could still get more knowledge and support from the Kaizen team."

"More meetings."

"The Kaizen at Lipor is extremely focused on quantity, getting numbers and does not help in the evaluation of work's quality."

"Keep up the great work!"

"Speed is not a sign of quality. I would have to change / adapt Kaizen to the office areas because many of these areas are conceptual work therefore, having the same method for everything does not work."

"It is necessary to focus on what is fundamental and to minimize the investment of time in bureaucratic processes."

"Very well! You're doing a good job!"

Some less positive aspect were mentioned specially related to the perceived emphasis given by the Kaizen Team to quantity and speed instead of quality of the performed work. Another mentioned issue is related to the implementation of solutions that being conceived in industrial context may not be adequate to office work.

5 Conclusions

In this paper we have described the dynamic of the Continuous Improvement system, team's composition and team's events, as well as tools and artifacts used by the teams to sustain the CI movement. Achieving a successful implementation and maintenance of the CI system was quite challenging as it involved organizational changes on many levels and still requires dedication and proactivity of the workers every day. The overall workers perception regarding the CI approach is very positive since more than 80% of the workers are happy with it. When inquired about the different activities developed in the context of the CI system, the workers highlighted the involvement in improving team's processes as being the most important aspect of the CI movement with 92% of positive answers. The act of updating and utilizing the team board was considered the second most important aspect of the CI with 90% of positive answers. The employees also emphasized the effect of team's events as daily meetings (with 89%) promoting and facilitating the communication and problem solving within the teams. With same level of importance was given to the monitoring of the teams' performance indicators, also with 89% of positive answers. Some employees pointed out some negative aspects of the CI movement related to the importance given by the Kaizen Team to quantity and speed instead of the quality of the performed work and also the use of solutions that may be adequate to industrial work but may not fit office work contexts.

Although the CI routines require energy and dedication from everyone it seems that the gains obtained are worth it by the results in performance, work satisfaction and motivation. Once the implementation phase was finalized, it was very important to focus on monitoring the process and the conduction of regular audits by other teams were an important procedure to ensure that the necessary practices to maintain the CI system were being held.

Acknowledgements This work has been supported by national funds through FCT—Fundação para a Ciência e Tecnologia within the Project Scope: UID/CEC/00319/2019.

References

1. Womack J, Jones D (1996) Lean thinking: Banish waste and create wealth in your corporation. Productivity Press
2. Rother M (2010) Toyota KATA. McGraw-Hill
3. Tyagi S, Choudhary A, Cai X, Yang K (2015) Value stream mapping to reduce the lead-time of a product development process. Int J Prod Econ
4. Rother M, Shook J (1999) Learning to see: value stream mapping to add value and eliminate muda
5. Koenigsaecker G (2009) Leading the lean enterprise transformation. Productivity Press, New York
6. Sanchez-Ruiz L, Blanco B (2016) How do companies implement process management ? The case of Cantabrian companies. Bus Manag Econ Res 2(1):1–9
7. Moica S, Veres Harea C, Marian L (2019) Effects of suggestion system on continuous improvement: a case study. In: IEEE international conference on industrial engineering and engineering management
8. Morales S, Valles A, García-Alcaraz J, Martinez E (2008) Determination of the critical success factors of continuous improvement projects. Int J Ind Eng (special issue 2008):471–480
9. Monteiro MFJR, Pacheco CCL, Dinis-Carvalho J, Paiva FC (2015) Implementing lean office: a successful case in public sector. FME Trans 43(4):303–310
10. Sugimori Y, Kusunoki K, Cho F, Uchikawa S (1977) Toyota production system and Kanban system materialization of just-in-time and respect-for human system. Int J Prod Res 15(6):553–564
11. Liker J (2005) The Toyota way : 14 management principles from the world's greatest manufacturer. McGraw Hill

Lean Green—The Importance of Integrating Environment into Lean Philosophy—A Case Study

Samuel Silva, J. C. Sá, F. J. G. Silva, Luís Pinto Ferreira and Gilberto Santos

Abstract This work is supported by a case study based on the action research methodology with the identification and quantification of the economic and environmental impact in the integration of the two concepts (Lean and Green) in a Portuguese company, where the main manufacturing process is the robotized welding process. In the case study, four Lean tools are implemented: *Kaizen*, Value Stream Map (VSM), *Jidoka* and Total Productive Maintenance (TPM). The Lean tools application allowed to increase the efficiency of the production system, reducing on average 33% of cycle times, reducing energy consumption by 38% and reducing scrap by 66% which demonstrates that Lean tools application leads companies to environmental and economic gains and contributes to better environmental management, enabling companies to differentiate themselves and increase their profits.

Keywords Lean green · Waste reduction · Manufacturing · Increasing productivity · Energy consumption reduction · Sustainable production · Sustainability

---

S. Silva · J. C. Sá (✉) · F. J. G. Silva · L. P. Ferreira
ISEP—School of Engineering, Polytechnic of Porto, Rua Dr. António Bernardino de Almeida, 431, 4200-072 Porto, Portugal
e-mail: cvs@isep.ipp.pt

S. Silva
e-mail: samueljmcsilva@hotmail.com

F. J. G. Silva
e-mail: fgs@isep.ipp.pt

L. P. Ferreira
e-mail: lpf@isep.ipp.pt

J. C. Sá
IPVC—School of Business Sciences, Polytechnic Institute of Viana Do Castelo, Av. Pinto da Mota, 4930-600 Valença, Portugal

G. Santos
IPCA—School of Design, Polytechnique Institute of Cavado Ave, Vila Frescaínha S. Martinho, 4750-810 Barcelos, Portugal
e-mail: gsantos@ipca.pt

M. Rossi et al. (eds.), *Proceedings of the 6th European Lean Educator Conference*, Lecture Notes in Networks and Systems 122,
https://doi.org/10.1007/978-3-030-41429-0_21

1 Introduction

Current concerns about environmental impacts and environment preservation increasingly force organizations to adopt green techniques that allow them to reduce environmental impacts through emission and consumption reductions [1, 2]. In the literature, these Green techniques are also associated with Lean techniques because they share some principles and the literature also demonstrates that applying Lean improves organizations and drives them to success [3]. However, there are still few investigations that evaluate the existence of Green gains with the application of Lean tools [4]. This work was then developed to assess whether the integration of these two concepts leads companies to improve their environmental performance without ever neglecting their efficiency and competitiveness. In order to do so, in this research, four Lean tools are applied: *Kaizen*, Value Stream Map (VSM), *Jidoka* and Total Productive Maintenance (TPM). The objective of the application of these Lean tools is to be able to demonstrate that it is possible to reduce the environmental impacts of companies while reducing their manufacturing costs, thus increasing their efficiency and competitiveness in the market.

This study is divided into six sections. The first of these consists of the introduction; the second presents a brief literature overview; the third describes the used methodology, the fourth section describes results obtained; the fifth section discusses the evolution of the environmental and economic impact of integrating Lean and Green concepts and the results of this implementation; the sixth section presents the final conclusions and suggestions for future studies.

2 Literature Review

2.1 Lean and Green Integration

Lean emerges as a set of techniques that allow the manufacturing sector to reach the desired goals for the final product using less resources, energy and raw materials, and that focuses on the waste reduction or elimination [1]. The diverse literature on the application of Lean techniques and a consequent performance improvement of the organizations is consensual, demonstrating that organizations adopting the Lean philosophy and applying its techniques have high efficiency improvements, thus reducing its costs and increasing its products quality [5–7]. Looking at the Lean manufacturing objectives, it is possible to observe that they lead to the reduction or elimination of seven wastes which, according to Taiichi Ohno, consumes resources but does not add value to the product: defects, overproduction, waiting, transport, handling, over processing and stocks [8]. On the other hand, a green production is defined as a continuous implementation of environmental prevention strategies applied to products, processes and services [9]. Thus, producers can simultaneously adopt Lean and Green techniques to create an environmental position that is the way to reduce costs and risks, thus increasing their profits and improving their brand image to their customers. Firms' efforts to improve their manufacturing processes and increase productivity can create substantial opportunities for improving environmental performance through Lean [10]. Some investigations demonstrate that this is possible in several areas of analysis and in the analysis of these investigations it is possible to find cases where Lean and Green models were developed to reduce the wastes and the environmental impacts in the manufacture [1]. For instance, the implementation of this model allowed reducing the use of resources by 30–40% and the total cost of mass and energy flow by 5–10%, as referred in [11]. In other cases, the impact of the integration of Lean and Green concepts on waste reduction has also been evaluated through a quantitative experimental study that has showed that the integration of these two concepts brings positive impacts on the waste reduction in companies [12] and the impact of the implementation of Lean tools in manufacturing. Within the Lean tools, the Value Stream Map, 5S's, cellular manufacturing, Single Minute Exchange of Die (SMED) and Total Productive Maintenance tools were applied in multiple case studies to reduce environmental impacts in the production process [7, 13, 14]. In the case of the research [13], the SMED tool did not reveal environmental reductions in the company, while all four other tools revealed. However, in another case study where only the SMED tool was applied in order to demonstrate that it is possible to create value and, at the same time, reduce the environmental impact with the adoption of Lean and Green tools, a reduction of 70% of setup time was achieved and an 81% reduction in carbon footprint after application of SMED tool in a metal machining centre was achieved [15]. In another area of analysis, such as the supply chain, Piercy and Rich [16] explored the benefits of Lean practices for sustainability and concluded that Lean operations address a broad range of sustainability outcomes, as well as environmental benefits, and proposes that Lean implementation and environmental performance are interlinked. On the other hand, in the same area of analysis, Fahimnia, Sarkis, and Eshragh [17] developed a tactical supply chain planning model to investigate trade-offs between cost and environmental degradation, and concluded that while some Lean interventions may result in Green benefits, not all practices are in line with Green strategies.

2.2 Lean Tools Towards Green Production

Waste reduction using Lean tools and sustainability techniques are similar and, when combined, both can bring advantages to organizations: cost savings and better environmental performance [18, 19]. *Kaizen*, for example, represents the continuous improvement of the process that creates a sustainable focus on the elimination of any form of waste identified in the manufacturing process [20]. The continuous improvement culture and process resulting from the *Kaizen* tool are typically very similar to those sought in environmental management systems, ISO 14001, and pollution prevention programs [1]. The advantage of Kaizen is that it involves workers from various areas who may have a role in improving a given process and they are encouraged to participate in reducing wasteful activities [21]. This is important because workers close to a specific process generally have suggestions and ideas that can be harnessed on ways to improve the process and reduce waste. In this way, *Kaizen* can be a powerful tool for discovering hidden wastes or waste generating activities and eliminating them, reducing environmental impact [11]. Like a *Kaizen* tool, the Total Productive Maintenance also brought benefits to companies in reducing wastes [22, 23]. Proper equipment and systems maintenance can reduce defects resulting from a process [24]. A reduction in defects can help eliminate waste of processes in three fundamental ways [22]:

- Fewer defects decrease the number of products that must be scrapped;
- Fewer defects also mean that the resulting raw materials, energy and waste associated with scrap are eliminated;
- Fewer defects decrease the amount of energy, raw material and waste that is used or generated to repair defective products that can be passed on.

From the waste identified above, the waste eliminated by the Total Productive Maintenance is the same waste as the sustainability strategies intend to eliminate. Also, the Just in Time (JIT) tool is indirectly linked to improving the environmental performance of organizations. The Just in Time system, aided by *Kanban*, enables organizations to eliminate overproduction, a substantial reduction in inventory levels, and reduce the floor space required for equal levels of production and, consequently, reduce transport [25].

3 Methodology

This case study is based on the action research methodology [26] applied in a Portuguese company that is dedicated to the manufacture of aluminum structures, where one of the main manufacturing processes is the welding process, supported by three welding robots, and where it was put the focus of this case study. The company in which the case study was applied initially had already applied a cell manufacturing concept although it contained some flaws in its processes. For the identification of these faults/wastes the route map shown in Fig. 1 was followed. With the wastes identification in the production and subsequent elimination of them, it is intended to quantify the consequent gains or losses in the processes and to assess whether it is possible through Lean, to increase the efficiency of a process and, at the same time, to reduce energy consumption, scrap and emissions to the environment, resulting in cost reductions for companies and contributing to environmental sustainability.

In the diagnostic phase of the process, the system was observed and quantified the cycle times, energy consumptions, percentage of defects that result in scrap and percentage of defects that result in rework before any implementation of Lean tools. During the diagnostic phase, it was possible to identify several wastes that besides raising the costs of the company, also increased their environmental impact.

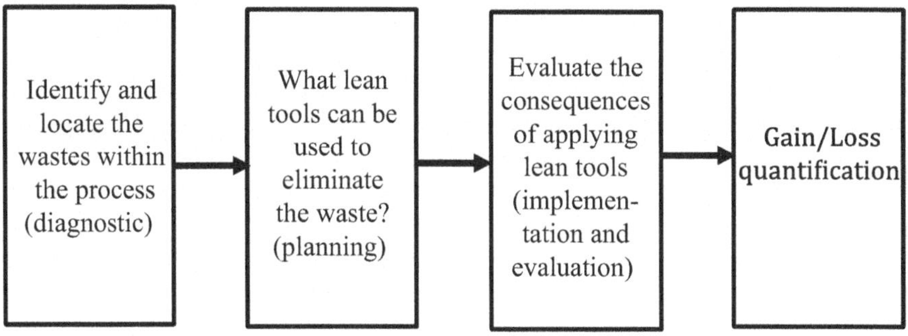

Fig. 1 Route map for wastes identification in manufacturing process

This analysis allowed to find wastes such as defects, high cycle times that led to waits and overproduction. Within the defects, there are defects that directly originate scrap and those in which there is the possibility of recovering the product through rework. The defects arose due to machine misalignment or due to the need for tuning some welding parameters due to changes in the ambient temperature or other factor influencing the welding process. Faced with these problems, the implementation of the Total Productive Maintenance and *Jidoka* tools was selected, which is able to detect the problem as soon as it arises, correcting it immediately. With the *Kaizen* technique, it was possible to reduce the cycle times thus reducing the waiting times because the welding process had been the process bottleneck.

4 Results

In this section, the lean and green gains obtained through the implementation of the lean tools addressed in the methodology are evaluated and quantified. In terms of defects, the scrap values before implementation were 600 kg/month. After the implementation of the Total Productive Maintenance and *Jidoka* tools, these values reduced considerably to 200 kg/month, a reduction of 66%. In terms of monetary value, this translates to approximately 1000 €/month (this value is calculated considering the value of the raw material to be subtracted by the value of the scrap material, in €/kg). On reworking, before implementation, it was 15% of total production and represented on average 1.5 h per day when only one worker is used on reworking. After the implementation of the Total Productive Maintenance and *Jidoka* tools, the percentage of rework fell to 4% and only needs an average of 0.5 h/day, thus reducing daily rework time by 1 h/day, 22 h/month. When we talk about welding, 22 monthly rework hours mean a significant reduction in environmental impact. The consumption of welding shield gas in hourly production is 0.75 m^3, so that in 22 monthly hours 16.5 m^3 is saved. Also, the emission of fumes resulting from the welding process causes environmental impact. The emission of these fumes into the air, considering a current intensity of 160 A and the shield gas 92% Ar^+ + 8% O_2, is 0.04 g/min

[27]. Thus, in 22 h, 52.8 g of fumes would have been emitted. Reducing rework time also reduces energy consumption. The average energy consumed monthly in the referred company is 1700 kWh/month working one single shift that corresponds to 10.6 kWh/hour. Thus, the 22 h correspond to an energy saving of 233.2 kWh/month. With the implementation of the *Kaizen* tool, a reduction of 25–40% in the cycle time of the different existing products was achieved, 1.5–3.2 min, and reduced the energy consumptions necessary to each product on average 0.42 kWh/part. This tool also eliminated waste from over processing. The collaboration of all the employees in the search for continuous improvement allowed the adjustment of suitable welding equipment to the process because of their interest in the process in which they participate and the knowledge of the problems on it, the employees have brought to the company the knowledge of a new welding technology that perfectly fits the needs of the process and the type of materials to be welded. After analysis, the expected cost reductions throughout the process would allow the return of the equipment exchange investment in the short to medium term beyond the many environmental benefits achieved. With this improvement, it was also possible to reduce the cleaning process by 70% and the over processing by 30% due to the ability to weld aluminium products with lower thickness, thus being able to resize the product according to the customer's needs and adding value to the product.

Table 1 provides a brief summary of the economic and environmental gains in the elimination of waste identified in the process using Lean tools. Thus, it is possible to conclude about the importance of Lean/Green relationship in reducing environmental impact and increasing economic performance of the organization.

5 Discussion

The application of Lean tools has proved to be an effective method in reducing waste in companies and in corporate sustainability. With the use of Lean techniques, the results of this research proved to be very positive for the company's environmental performance when these tools are used to eliminate waste identified in the manufacturing process. The application of the Lean tools allowed in total a reduction of 66% in scrap generation, a 38% reduction of the energy and welding shield gas consumption and corresponding welding fumes emission. An average reduction of 30% in the raw material used in each product and 70% of the chemicals used for cleaning was also achieved. These reductions are not only reflected in Green gains but also in economic gains that allowed the company to economize an average of 1786 €/month in scrap, labour, welding shield gas and energy costs. Also, product cycle times have been reduced by 25–40%. Other gains in reducing over-processing and non-value-added activities, although not reflected in a reduction in direct costs for the company, are reflected in a higher added value of the product for which the customer will pay. Although all the analysed tools have Lean and Green improvements, the combination of Total Productive Maintenance and *Jidoka* tools brought the most gains as they allowed the reduction of defects and all associated problems, such as the rework

Table 1 Lean and green gains quantification

Wastes (muda)	Lean tools	Lean gains	Green gains
Defects (scrap)	TPM and *Jidoka*	66% reduction in scrap = 400 kg monthly = 1000 €/month	400 kg of scrap reduction per month
Defects (rework)	TPM and *Jidoka*	Rework reduced in 22 h/month—Saving 198 €/month in labour considering 9 €/hour; Saving 16.5 m^3/month of welding shield gas = 143 €/month; Energy consumption reduction in 233.2 kWh/month = 69 €	Lower welding shield gas consumption; Reduction of 52.8 g/month in the welding fumes emission; Energy consumption reduction in 233.2 kWh
Waiting	*Kaizen*	Cycle time reduction of 1.5–3.2 min; Reduction of average energy consumption of 0.42 kWh per piece = 0.42 kWh × 1000 pieces/month = 420 kWh/month = 124 €/month; Waiting time reduction	Reduction of 24% in energy consumption, welding gas consumption and welding fumes emission
Over processing	*Kaizen*	Value added product;	Raw material reduced on average in each product =−3.5 kg
No value-added activities	VSM	Cleaning process reduced by 70%—7 min	Reducing the use of chemicals in 70%

cost and resources consumed. These results also are in line with others describe in the literature in other kinds of industries, such as the printing industry, where Lean tools were also applied, allowing a reduction of 32.9% in the defective products and a decrease in 8 min and 20 s in average the setup time, leading as well to a drastic reduction of the consumption of isopropyl alcohol in the fountain solution used in the printing machines [28], contributing by this way to a better/greener environment.

6 Conclusion

The results analysis of this paper demonstrates that the Lean tools implemented bring benefits to the environment by reducing waste and non-value-added activities contributing to the literature as a case study revealing that environmental concerns are not constraints to manufacturing operations and that integrating environment into Lean philosophy brings value to organizations. With these environmental gains, economic gains have also arisen as a result of Lean philosophy, so any expenses necessary for its implementation must be seen as an investment that in most cases has a return in the short or medium term. The demonstration of environmental gains coupled with economic gains in this case study can be a lever for companies to start using techniques that integrate the Lean and Green concepts and improve their environmental performance, thus contributing to sustainability through reductions in raw material consumption, energy consumptions, environment emissions, among others. As suggestions for future work, it is considered important to implement Lean tools in other sectors of activity so that their environmental impact can be measured. Within the same industry sector, it is considered important to implement other Lean tools in addition to those analysed in this study and it is suggested to study the implementation of the SMED tool in search of process improvements as it is a tool that has not demonstrated environmental gains in the literature mentioned in this paper.

References

1. Silva FJG, Gouveia RM (2019) Cleaner production—toward a better future. Springer Nature, UK. ISBN: 978-3-030-23164-4
2. Santos J, Gouveia RM, Silva FJG (2017) Designing a new sustainable approach to the change for lightweight materials in structural components used in truck industry. J Cleaner Prod 164:115–123
3. Rothenberg S, Pil F (2017) Lean, green, and the quest for superior environmental performance, September 2001
4. Rand G, Womack J, Jones DT (2006) Lean thinking-banish waste and create wealth in your corporation. J Oper Res Soc 48 (11):1148
5. Rosa C, Silva FJG, Ferreira LP (2017) Improving the quality and productivity of steel wire-rope assembly lines for the automotive industry. Procedia Manuf 11:1035–1042
6. Oliveira J, Sá, Fernandes A (2017) Continuous improvement through "Lean Tools": an application in a mechanical company. Procedia Manuf 13:1082–1089
7. Sousa E, Silva FJG, Ferreira LP, Pereira MT, Gouveia R, Silva RP (2018) Applying SMED methodology in cork stoppers production. Procedia Manuf 17:611–622
8. Ōno T (1988) Toyota production system: beyond large-scale production. Productivity Press
9. Jones DOB et al (2019) Existing environmental management approaches relevant to deep-sea mining. Mar Policy 103(February):172–181
10. Florida R (1996) Lean and green: the move to environmentally conscious manufacturing. Calif Manage Rev 39(1):80–105
11. Pampanelli AB, Found P, Bernardes AM (2014) A lean & green model for a production cell. J Clean Prod 85:19–30
12. Fercoq A, Lamouri S, Carbone V (2016) Lean/green integration focused on waste reduction techniques. J Clean Prod 137:567–578
13. Chiarini A (2014) Sustainable manufacturing-greening processes using specific lean production tools: an empirical observation from European motorcycle component manufacturers. J Clean Prod 85:226–233
14. Rosa C, Silva FJG, Ferreira LP, Campilho R (2017) SMED methodology: the reduction of setup times for steel wire-rope assembly lines in the automotive industry. Procedia Manuf 13:1034–1042

15. Leme RD, Nunes AO, Message Costa LB, Silva DAL (2018) Creating value with less impact: lean, green and eco-efficiency in a metalworking industry towards a cleaner production. J Clean Prod 196:517–534
16. Piercy N, Rich N (2015) The relationship between lean operations and sustainable operations. Int J Oper Prod Manag 35(2):282–315
17. Fahimnia B, Sarkis J, Eshragh A (2015) A tradeoff model for green supply chain planning: a leanness-versus-greenness analysis. Omega (United Kingdom) 54:173–190
18. Castro ACM, Carvalho JP, Ribeiro MCS, Meixedo JP, Silva FJG, Fiúza A, Dinis ML (2014) An integrated recycling approach for GFRP pultrusion wastes: recycling and reuse assessment into new composite materials using Fuzzy Boolean Nets. J Clean Prod 66:420–430
19. Silva FJG, Ferreira F, Ribeiro MCS, Castro ACM, Castro MRA, Dinis ML, Fiúza A (2014) Optimising the energy consumption on pultrusion process. Compos B 57:13–20
20. Kumar S, Dhingra AK, Singh B (2018) Process improvement through Lean-Kaizen using value stream map: a case study in India. Int J Adv Manuf Technol 96(5–8):2687–2698
21. O. US EPA. Lean thinking and methods—Kaizen. Available at: https://www.epa.gov/lean/lean-thinking-and-methods-kaizen [Acessed 21 May 2019]
22. O. US EPA, Lean thinking and methods—TPM. Available at: https://www.epa.gov/lean/lean-thinking-and-methods-tpm [Acessed 21 May 2019]
23. Guariente P, Antoniolli I, Ferreira LP, Pereira T, Silva FJG (2017) Implementing autonumous maintenance in an automotive components manufacturer. Procedia Manuf 13:1128–1134
24. Hammadi S, Herrou B (2018) Lean maintenance logistics management: the key to green and sustainable performance. In: 2018 4th IEEE International conference on logistics operations management, pp 1–6
25. Master KPP, Work T, Production I. Improving supply chain performance through lean and green. A study at Volvo Group India and Sweden
26. Susman GI, Evered RD, Susman G, Evered RD (2012) An assessment of the scientific merits of action research, vol 23, no 4, pp 582–603
27. Ladwig JON (2017) DANGER ZONE: control dust and fumes in metalworking and welding operations. Ind Saf Hyg News 51(12):24–25
28. Moreira A, Silva FJG, Correia AI, Pereira T, Ferreira LP, de Almeida F (2018) Cost reduction and quality improvements in the printing industry. Procedia Manuf 17:623–630

Towards Lean Ground Handling Processes at an Airport

Ana Paula Barroso, Virgínia Helena Machado and Helena Carvalho

Abstract Globalization of markets and the technological growth promote new business opportunities for companies, but also present major challenges in terms of competitiveness and sustainability. To meet these challenges, companies implement policies of continuous improvement and lean management reducing their waste. The case study presented belongs to a project regards to an airline company that lands at a main airport of a European city. It focuses on the local baggage flow on ground handling processes and aims to make lean its operations. Operational indicators and processes' waste identification were made on the gemba walk and a Waste Relation Matrix was developed allowing to quantify the relationship of the several types of waste identified on the baggage flow.

Keywords Lean management · Ground handling · Gemba walk · Waste relationship matrix

1 Introduction

Globalization of markets and the technological growth promote new business opportunities for companies, but they also present challenges. As company's competitiveness is a key factor to its success, they must implement innovation and continuous improvement policies to obtain flexibility, regardless of the economic, political and/or social environment in which they are inserted. Implementing Lean Management (LM) tools allow to improve the company's competitiveness and sustainability

A. P. Barroso (✉) · V. H. Machado · H. Carvalho
Department of Mechanical and Industrial Engineering, Faculdade de Ciências e Tecnologia, Universidade NOVA de Lisboa, Caparica, Portugal
e-mail: apb@fct.unl.pt

V. H. Machado
e-mail: vhm@fct.unl.pt

H. Carvalho
e-mail: hmlc@fct.unl.pt

M. Rossi et al. (eds.), *Proceedings of the 6th European Lean Educator Conference*, Lecture Notes in Networks and Systems 122,
https://doi.org/10.1007/978-3-030-41429-0_22

by reducing sources of waste and increasing value creation for customers [1]. Waste means something has no added value.

The current environment of liberalization and globalization of markets requires agile and efficient transportation systems. Fast connections between geographic points are indispensable, so air mode is one of the most competitive. Passenger air traffic in 2017 increased by 7.1% compared to 2016, which is higher than the average growth of the last 10 years [2]. Thus, in order to modernize and enhance the operations inherent to the flow of baggage and passenger, airports must be in line with airlines and handling companies. The baggage delivery time to the passenger is a relevant quality indicator of airports. Therefore, the ground handling process is critical for baggage handling, and the continuous improvement of the baggage unloading processes of the airplane and its delivery to the passenger may be relevant.

The European Parliament has defined ground handling indicators [3], such as the maximum time of delivery of the first and last baggage to the passenger. Thus, proper airport management and quality needs monitoring of service performance [4].

The present case study concerns a part of a vast project being developed at a European airport and aims to improve the performance of passenger services and meet European Union standards. It analyzes only the assistance provided by a ground handling company to a type of airplane (Narrow Body) in the Baggage Unloading (BU) and Baggage Delivery (BD) processes.

Some studies show that the application of LM strategies in ground handling processes improves the performance of baggage flow operations [5]. Thus, in case study the application of LM tools promotes the reduction of delays in the baggage flow time in BU and BD processes.

This paper has 6 sections. The current context of the ground handling sector and paper objectives were presented in Sect. 1. In Sect. 2 the Research Methodology is presented. Section 3 presents the Case Study. In Sect. 4 waste and its causes are analyzed. In Sect. 5 Waste Relationship Matrix is developed. Finally, Sect. 6 presents the conclusions and work to be carried out in the future.

2 Research Methodology

To develop the project a proposal of research methodology is presented in Fig. 1. This paper focuses on Waste Relationship Matrix development, included in phase 3.

Phase 1. The purpose of this project is to reduce baggage flow time at the airport. In this study only baggage flow time in the BU and BD processes and Narrow Body airplane are considered.

Phase 2. As is of baggage flow is analyzed in two steps: (1) Brainstorming among specialists to know how the baggage flow is managed and Ishikawa diagrams to identify the possible causes that generate delay in the baggage flow time; (2) In the gemba walk, possible causes of baggage flow time delay are identified and a sample with about 100 observations per process type is collected to quantify operational performance indicators.

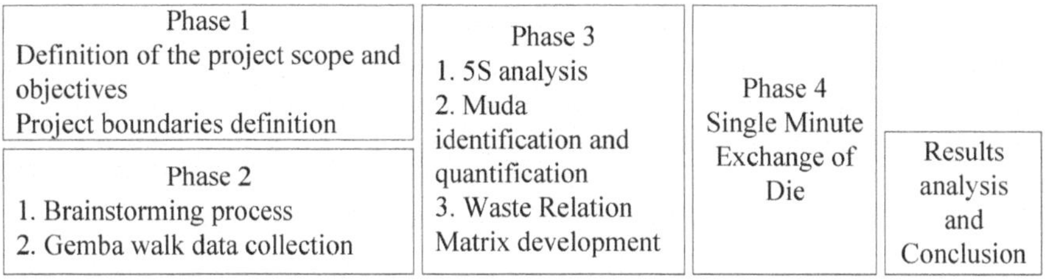

Fig. 1 Project research methodology proposal

Phase 3. An action plan is defined based on LM tools. To reduce waste, a 5S analysis is performed on the conveyor belt to unload the baggage from the airplane; Waste is characterized, and a Waste Relationship Matrix is developed to determine the relative importance of each type of waste.

Phase 4. The Single Minute Exchange of Die tool is then implemented in the system to reduce the time period between the arrival of the airplane and the delivery of the last baggage at the terminal.

3 Case Study

The case study concerns the ground handling of Narrow Body airplane of a major airline company that lands at the main airport of a European city. This type of airplane generates the largest number of flights from the airline company. The case study covers the handling processes performed by a service provider from the time the baggage is unloaded from the airplane until it is picked up by passengers in carousels at the terminal. Therefore, it focuses on local baggage flow.

Baggage flow encompasses three sequential processes (Fig. 2): BU from the airplane; Baggage transportation to the terminal; and BD at the terminal where the baggage is picked up by passengers. In this study BU and BD processes are analyzed.

In the BU process the baggage is unloaded from the airplane to the ground, either manually or on a conveyor belt. Then the baggage is moved to the terminal. The BD

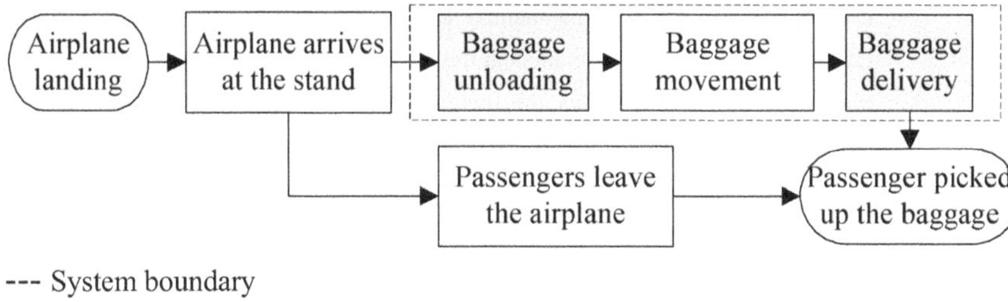

Fig. 2 Local baggage flow at the airport

process begins when the operator places the first baggage on the carousel and records the event on a Personal Data Assistant (PDA) fixed on the wall and ends when the operator places the last baggage on the carousel and records the event on the PDA. The BU and BD processes under study were analyzed in the gemba, and about 100 observations per process were carried out. Based on the collected data, performance indicators were quantified for the current operation of the service provider.

4 Baggage Flow. Waste and Causes

Waste identified in the BU and BD processes were classified according to Ohno classification [6]. An Ishikawa diagram per process was made applying a brainstorming to identify causes that lead to delays in baggage handling, which increase baggage flow time at the airport. Next, each cause was associated with a type of waste. Some causes and their associated type of waste are presented in Table 1.

- Motion
  In BU and BD processes many of the operator's movements cannot be eliminated for safety reasons and the equipment must be in specific places. However, activities contributing to unnecessary operator's movements have been identified. For example, in baggage carousel of BD process the proposed layout decreases, on average, 75% of the traveled distance (from 80 to 20 m) and 72% of the time spent (from 88 to 25 s).
  Figure 3 presents the Spaghetti diagram used to map the operator's movements in the as is layout (a) and the proposed layout (b).
- Process
  The lack of standardization procedures of the processes leads to waste. It was found that in 49% of the cases in which the processes were carried out there

Table 1 Type of waste and some causes identified in BU and BD processes

Waste	Cause
Motion	Unnecessary movements in BU process
	Inappropriate baggage terminal layout
	Lack of standardization of marshalling activity
Process	Lack of ground support equipment at the required place
	Placing the air power unit in a non-standard way
	Allocation of operators without a driving license
Defect	Placing the container on the stand with the wrong orientation
	Incomplete operator team at the arrival of the airplane to the stand
	Internal communication failures
Waiting	Pause due to over moving
	Pause due to the incomplete operator team at the stand after the airplane arrival

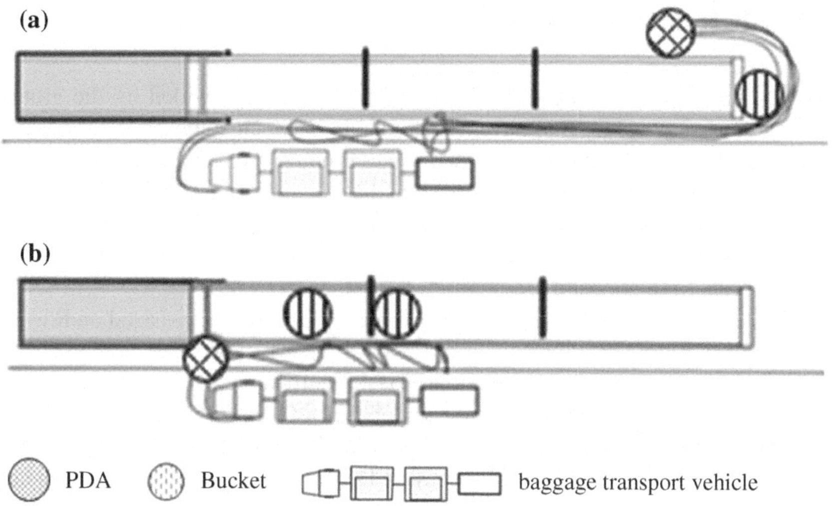

Fig. 3 Spaghetti diagram showing operators' movements in the baggage delivery process

were procedural errors. Also, on average, the performance of the activity with marshalling has a duration 39% lower than the one performed without marshalling.

- Defect
  A container with the wrong orientation on the stand must be set righted by the operator before the baggage be injected into the conveyor belt leading to rework. Also, the delay in recording on the PDA the time associated with the last baggage placed on the carousel constitutes a defect due to the lack of internal communication. It was found that 19% of the cases analyzed were performed with at least one defect.
- Waiting
  Figure 4 shown the unfulfilled rate for some activities of BU and BD processes related to activity time, which is internally defined by the airport.

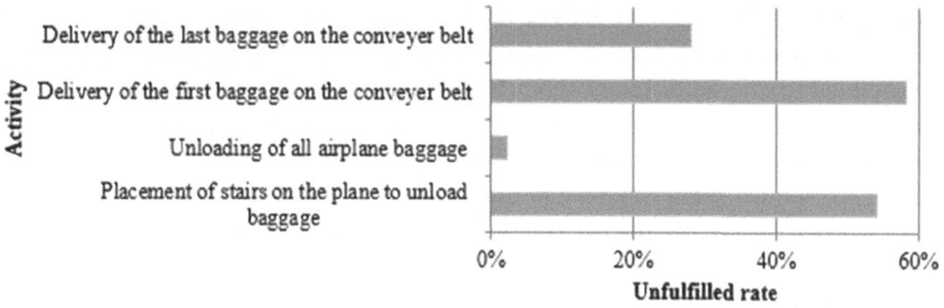

Fig. 4 Unfulfilled rate of some activities of BU and BD processes

5 Waste Relation Matrix

Based on the analysis of waste in BU and BD processes provided by the ground handling company, it seems to have a relationship between some types of waste in the baggage flow. The influence of each waste type on the others can be direct or indirect. The different types of relationships between them and its nature suggest that all these relationships have not the same contribution. To know which type of waste contributes more to the wasteful activities on the processes it is necessary to assign weights to relationships.

A Waste Relationship Matrix (WRM) was developed in 4 steps based on Rawabdeh model [7]. Direct relationships represented by pairs of types of waste (i, j), in which i $\neq$ j, are considered. The relationship (i, j) corresponds to the "effect caused by the type of waste i on type of waste j". Each type of waste was abbreviated using its initial (D: Defect; M: Motion; P: Process; W: Waiting). For instance, the pair (P,D) indicates the direct effect of Process on Defect, i.e. improper or non-standard processes lead to produce defects.

Step 1—The direct relationships must be identified. Some brainstorming sessions among specialists were conducted and the next ten direct relationships were identified: (M,W); (M,C); (P,M); (P,D); (P,W); (P,C); (D,M); (D,W); (D,C); (W,C).

Step 2—A measurement criterion based on a questionnaire was developed to quantify the strengths of waste relationships. It comprises 6 questions (Table 2) and each answer has a specific weight ranging from 0 to 4. The greater the weight, the stronger the strength of the relationship.

Step 3—Based on the gemba walk and several brainstorming sessions the questions were answered for each relationship previously identified. The weight was assigned to each answer of each relationship. The weighting of all answers of each relationship were added together, resulting in a score, which indicates the strength of each relationship (Table 3). Higher scores represent stronger relationships. The highest score corresponded to the pair of waste types (D,W), which leads to conclude that the Waiting is strongly influenced by Defects.

Step 4—The WRM is obtained from the assessment of the score of each relationship. Based on the range of score values (minimum = 1 and maximum = 20), the strength of each relationship was classified using symbols according to Table 4. To provide a more simplistic metric the score is converted into a percentage. A scale of 10 was chosen and divided into five identical subintervals allowing different weights to be defined for the six different symbols included in WRM (Table 4). The WRM is shown in Fig. 5 as shading. Each row indicates the effect of a certain type of waste on the other types. Each column indicates to what extent a certain type of waste will be affected by others. The diagonal of the matrix does not have values. Figure 5 also shows scores for each row and each column of the matrix, which are calculated based on the sum of the symbol weights of each row and each column, respectively.

The percentage of row i indicates the effect of the type of waste i on the other waste; the percentage of column j indicates how much the type of waste j is affected by the other waste. The waste type denoted by Waiting was found to be the most

Table 2 Questionnaire for evaluating the strengths of waste relationships

Question	Answer	Weight
Does i cause j?	**a.** Always	4
	b. Sometimes	2
	c. Rarely	0
What is the type of the relationship between i and j?	**a.** i increases j increases	2
	b. As i increases j reaches a constant level	1
	c. Random	0
The effect of j due to i:	**a.** Appears directly and clearly	4
	b. Needs time to appear	2
	c. Not often appears	0
Eliminating the effect of i on j is achieved by:	**a.** Engineering methods	2
	b. Standardization of processes	1
	c. Instructional solution	0
The effect of j due to i, mainly influences:	**a.** Service quality	1
	b. Productivity of resources	1
	c. Lead time of baggage	2
	d. Quality and productivity	2
	e. Lead time and operators' productivity	2
	f. Service quality and lead time of baggage	2
	g. Service quality, operators' productivity and lead time of baggage	4
In which degree does the effect of i on j increase baggage lead time?	**a.** High	4
	b. Medium	2
	c. Low	0

affected by the other waste (60%). The Process was found to be dominating the other waste with the highest percentage (48%). Process was the most serious waste type since it increases all the other types, W, M and D. Conversely, the other waste types do not influence Process, which can be seen by the percentage value (0%).

6 Conclusion

Recently, some airlines have begun to adopt LM principles for reducing waste and its sources. As the airline industry encompasses process, labor, and capital intensive, small waste reductions enable it to achieve better performance, improve customer service and engage employees to be more productive.

Table 3 Weight of the pairs of waste types identified

Question	1		2		3		4		5		6		Score
(i, j)	A	W	A	W	A	W	A	W	A	W	A	W	
(M,W)	a	4	a	2	a	4	a	4	e	2	b	2	18
(P,M)	a	4	a	2	b	4	b	1	e	2	b	2	15
(P,D)	b	2	a	2	b	2	c	0	f	2	g	4	12
(P,W)	a	4	a	2	a	4	b	1	g	4	b	2	17
(D,M)	b	2	a	2	b	2	b	1	c	2	e	2	11
(D,W)	a	4	a	2	a	4	b	1	g	4	g	4	19

A Answer; *W* Weight

Table 4 Range partition of strength of relationship

Range	Type of relationship	Symbol	Weight
–	Nonexistent	X	0
1–4	Unimportant	U	2
5–8	Ordinary closeness	O	4
9–12	Important	I	6
13–16	Especially important	E	8
17–20	Absolutely necessary	A	10

Fig. 5 Waste relationship matrix and respective values

(i,j)	M	P	D	W	Score	
M		X(0)	X(0)	A(10)	10	20%
P	E(8)		I(6)	A(10)	24	48%
D	I(6)	X(0)		A(10)	16	32%
W	X(0)	X(0)	X(0)		0	0%
Score	14	0	6	30	50	
	28%	0%	12%	60%		

The lack of resources and inadequate layouts at airports currently compromise the airport' performance, which can also be influenced by an inefficient management of baggage handling operations.

In the case study presented, the diagnosis of the current performance of the BU and BD handling processes that influence the flow time of the baggage allowed to identify improvement opportunities, which are the basis for implementing lean management tools. For this purpose, the processes were observed in the gemba walk and times from some activities were collected for quantify some relevant operational indicators. After identifying waste, a Waste Relationship Matrix was elaborated to quantify the direct relationship between the different types of waste and prioritize them with respect to their influence on the flow time of the baggage. It was found that of the four types of waste the Process waste is dominant.

In future work, in applying Phase 4 of the proposed research methodology to reduce baggage flow time, the matrix result will help in decision making when the Single-Minute Exchange of Die tool is implemented.

Acknowledgements The authors would like to thank the Unidade de Investigação e Desenvolvimento em Engenharia Mecânica e Industrial for its support (UNIDEMI—UID/EMS/0067/2019).

References

1. Asif M, Löwik SJA, Weusthof W, de Bruijn EJ (2010) Challenges in lean implementation in knowledge-intensive services. In: Proceedings of the 15th Cambridge international manufacturing symposium, pp 1–17. Cambridge

2. International Air Transport Association in https://www.iata.org/Pages/default.aspx
3. Burghouwt G, Poort J, Ritsema H (2014) Lessons learnt from the market for air freight ground handling at Amsterdam Airport Schiphol. J Air Transp Manag 41:56–63
4. Bezerra GCL, Gomes CF (2015) The effects of service quality dimensions and passenger characteristics on passenger's overall satisfaction with an airport. J Air Transp Manag 44–45:77–81
5. Nugroho IA, Riastuti UH, Iridiastadi H (2012) Performance improvement suggestions for ground handling using lean solutions approach. Procedia—Soc Behav Sci 65:462–467
6. Ohno T (1988) The Toyota production system: beyond large-scale production. Productivity Press, Portland
7. Rawabdeh H (2005) A model for the assessment of waste in job shop environments. Int J Oper Prod Manag 25(8):800–822

Reducing the Scrap Generation by Continuous Improvement: A Case Study in the Manufacture of Components for the Automotive Industry

J. Pereira, F. J. G. Silva, J. C. Sá and J. A. Bastos

Abstract The automotive industry is one of the most demanding sectors of the global market. The response capacity and flexibility of companies represent a key factor for their success. Applying Six Sigma, it was carried out an improvement project aiming at reducing the quantity of scrap on the most critical sector of a automotive components' manufacturer achieving a better comprehension of the flows, process characteristics and different variables associated to the scrap generation, identifying the equipment responsible for that scrap and its type. Brain- storming sessions were performed, as well as the application of 5 Why's and 5W2H techniques in order to fulfill the Ishikawa diagrams aiming at understanding possible root-causes for the scrap generation. A definition of the improvement actions has been developed. A reduction of 15% was achieve just in the machine identified as the main generator of scrap in these processes.

Keywords DMAIC · Continuous improvement · Six sigma · OEE · Automotive industry

---

J. Pereira · F. J. G. Silva · J. C. Sá (✉) · J. A. Bastos
ISEP—School of Engineering, Polytechnic of Porto, Rua Dr. António Bernardino de Almeida, 431, 4200-072 Porto, Portugal
e-mail: cvs@isep.ipp.pt

J. Pereira
e-mail: joaopereira.15@gmail.com

F. J. G. Silva
e-mail: fgs@isep.ipp.pt

J. A. Bastos
e-mail: jab@isep.ipp.pt

J. C. Sá
IPVC—School of Business Sciences, Polytechnic Institute of Viana Do Castelo, Av. Pinto da Mota, 4930-600 Valença, Portugal

M. Rossi et al. (eds.), *Proceedings of the 6th European Lean Educator Conference*, Lecture Notes in Networks and Systems 122,
https://doi.org/10.1007/978-3-030-41429-0_23

1 Introduction

In the automotive industry, and in particular the Portuguese automotive industry, that seeks to keep being competitive in the international panorama with increased processes' automatization and policies of low salaries, the fulfillment of the different customer delivery times and the quality requirements assume a key importance in order to keep the competitiveness of the national automotive industry [1–3]. The problem discussed in this case study had its origin on the evaluation of one of the main indicators of the factory: scrap generated. This indicator reflects the quality problems of some processes/equipment and can be evaluated under different perspectives. In this particular case, it was necessary to evaluate data from 2017 to understand which technologies or sub-modules were contributing the most to the scrap generation. After identifying the main source, it became necessary to understand the whole process to identify the possible causes of the scrap generation. DMAIC methodology has been used because it allows to fit the needs of reducing the scrap generated, as follows:

- Measure, identify and understand the causes, and the equipment associated to scrap generation on the most critical production module: comfort systems for car seats;
- Discuss the possible improvements and modifications that could be implemented on the most critical equipment;
- Follow the implementation, verify and control the results of the improvement actions.

The global goal of this work is to reduce the scrap generation in a sustainable way, acting mainly in two different equipment: LBT and Tecnogial machines, which are responsible for the metallic wire bending and assembly of comfort systems for car seats.

This work is divided into five main sections: the first is the introduction that presents the main motivations behind this work, the second is the Literature Review which de- scribes some developments and applications carried out in other works, the third designates the methodology used in this work, the forth describes the DMAIC application and the fifth discusses the results and summarize the main achievements.

2 Literature Review

Keeping productivity at truly high levels will have to be one of the main objectives of the automotive components industry. Many efforts have been made in order to improve the productivity in the manufacturing of components for the automotive industry, which also is strongly related to the reduction of scrap generation. Rosa et al. [4] has studied how to reduce setup time on production lines of Bowden cables used in motor vehicles. The change of some internal tasks to external ones, as well as the implementation of visual aids and the increase of the workers training, allowed a setup time reduction of 58.3%. A parallel study based on the same type of product [5] also allowed excellent results using some Lean tools such as Value Stream Mapping (VSM), achieving productivity gains of 41%. Moreira et al. [6] also improved the operation costs in a printing company. Changing some products used as consumables in the equipment and taking care more accurately of the equipment maintenance, the number of nonconformities was drastically reduced (−32.9%). Antoniolli et al. [7] studied production lines devoted to air-conditioning tubes for motor vehicles. Furthermore, in this case the indicator used was OEE, which allowed to detect that the failures were essentially in the productivity factor. Thus, Lean and Kaizen tools were applied, such as Standard Work and Production Line Balancing. By improving some of the factors, the OEE was increased from 70 to 86%. After that work, the capacity of the line was matched with the demand. The generation of scrap is a common problem in many industries and it is very common to hold Brainstorming sessions, through which an Ishikawa's diagram is generated, which intends to highlight the problems root causes [8, 9]. However, the reduction in scrap generation can also be performed directly, through actions performed on the hardware responsible for the positioning, tooling or work on the parts. Costa et al. [2] designed a new equipment able to produce complex sets of components used on the transmission system of automotive windshield wipers, avoiding the initial problems of scrap generation and ensuring an increased productivity. Pinto et al. [10] investigated the moulds geometry in order to avoid the scrap production of small components injected in Bowden cables used in motor vehicles. Castro et al. [11] developed a new tool concept for the crimping process of electrical terminals, reducing by this way the scrap generation. Costa et al. [12] developed a new clamping system regarding the machining process of parts with complex shape, avoiding the generation of scrap or the need for rework. Barbosa et al. [13] and Costa et al. [14] studied quality problems at different stages of tire production for the automotive industry. Barbosa et al. [13] studied the bead APEX production process. The application of the DMAIC methodology and the statistical control of the process were crucial for a 41% increase in the quality rate due to a rigorous study and refinement of the process parameters. On the other hand, Costa et al. [14] developed a study to improve the extrusion process of sub-products used in tire production. DMAIC methodology was also used. A decrease of 0.89% in the scrap generated was achieved, which represents around € 165,000 per year.

3 Methodology

Being the excessive quantity of scrap a complex problem, with known and unknown variables responsible for its existence, it is necessary to understand all these variables and how they relate with each other. This led to a need for a deeper analysis phase that would allow defining the adequate improvement actions to eliminate these problems. In this case study, as it refers to a Six Sigma approach, it was used a traditional DMAIC methodology, which is more adequate for more complex problems with more associated variables. However, it is possible to refine it in more specific steps, summarized as follows: (a) Observation and identification of the major source for the scrap production; (b) Observation of the processes associated with the respective source; (c) Identification of the possible variables associated with the scrap production along the corresponding processes; (d) Identification and analysis of possible causes for the scrap production; (e) Possible discussion improvements and modifications to equipment and procedures; (f) Validation of the effectiveness and results after actions implementation.

4 Results: DMAIC Cycle Application

Define step: In this initial stage, a complete definition of the project was made trying to identify the problem, and to establish the goals to be achieved, and corresponding benefits. The different team members, their roles and responsibilities along the project were also defined. A process mapping was carried out in order to understand, the different operations and activities under analysis and the respective stakeholders.

The product discussed in this work is a part of the automobile seat. It basically consists in a wire structure responsible for the comfort offered in the seats for their users. The manufacturing process of comfort system products varies according to the product. Some projects can only be produced in certain machines, having its own features. Thus, the manufacturing process follows the path 1, 2 or 3 (Fig. 1), according to the type of machines that can produce them. The production sequence also is Fig. 1.

Measure step: with the analysis of the scrap data of the previous year, it was possible to identify and refine which equipment, shifts, type of defects and specific operators that contribute more to the scrap value. It was necessary to undergo a specific data collection plan in order to reach the root causes of the scrap production (Table 1).

This data collection was created according to the following segmentation factors: (a) Equipment; (b) Shift; (c) Adjustment operator; (d) Type of defect; and (e) Raw material traction effort. Through this new data collection plan it was possible to identify the LBT, Tecnogial 1 and Tecnogial 8 as the equipment with a bigger ratio between the amount of scrap produced and the quantity produced (Fig. 2). The LBT stands out with a much higher percentage than the other equipment. This percentage

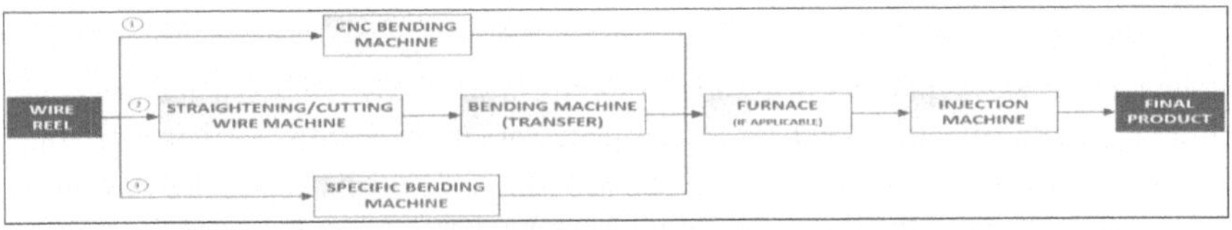

Fig. 1 Different variations of the manufacturing process of comfort system products

Table 1 Parameters of data collection plan implemented

Data collection plan parameter	Description
What	Weight of scrap (kg)
How	Quantity of scrap placed in the scrap container
Time	From January 2018 until the final of project
Frequency	Every day

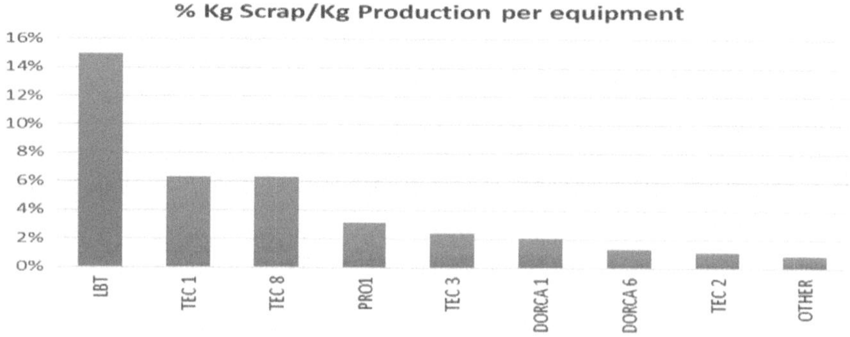

Fig. 2 Distribution of the ratio of scrap produced/quantity produced by equipment

allows visualizing a more real and accurate scenario in terms of scrap production, as it extrapolates the difference in quantity produced between the different equipment.

The evolution of the data related to the percentage of the ratio between the amount of scrap produced and the total quantity produced is visible in the control chart present in Fig. 3. There is a high amount of values above the upper control limit, that is automatically calculated according to the mean (3σ from the mean). This control chart shows that the process is not under control over the time. According to the scrap source, it was made a box plot graph in order to understand the distribution and dispersion of data for each source (Fig. 4), considering the main sources of problems previously identified (AF—Adjustments; AV—Breakdowns; BNC—Internal

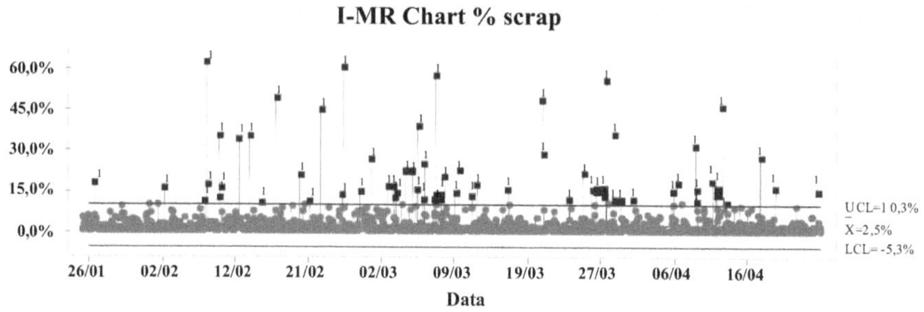

Fig. 3 Control chart with evolution of ratio of scrap produced/quantity produced

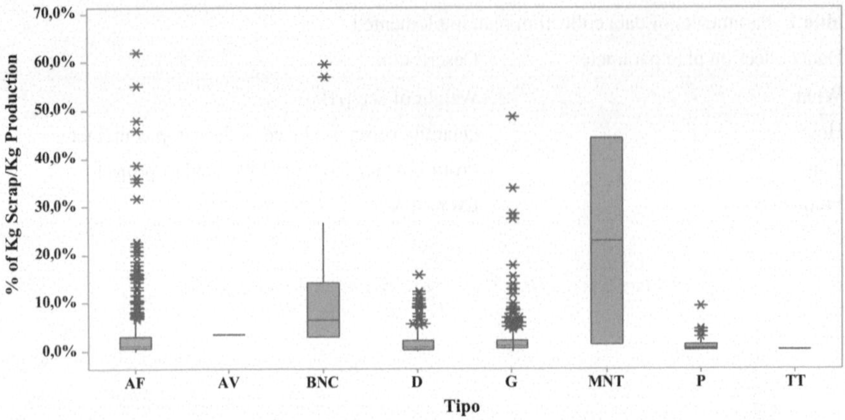

Fig. 4 Boxplot with distribution of ratio of scrap produced/quantity produced by scrap

complaints; D—Bends; G—Non-conform hooks; MNT—Maintenance; P—Production; TT—Heat treatment). In terms of breakdowns, BNCs, maintenance interventions, and heat treatment, the values could be disregarded as they represent one-off situations.

Analyse step: The initial data collection process allowed to identify the most critical equipment in terms of quantity of scrap produced and in terms of the ratio scrap produced/total quantity produced. The most critical equipment was identified as the LBT and the Tecnogial 1 and Tecnogial 8. Also, the most frequent source of scrap identified happens during the adjustment process. Brainstorming sessions were carried out sup- ported by the six sigma team members and adjustment operators specialized in the respective equipment. These brainstorm sessions had the aim to reach the root causes through Ishikawa technique. For the root causes identified, it was further defined im- provement actions plans regarding each equipment.

LBT Equipment

The LBT equipment was the second major responsible for the scrap production in the period between January and April of 2018, with 748 kg of scrap. This value corresponds to 15% of the total quantity produced of this equipment and for the same period, being the machine with the highest ratio quantity of scrap/quantity of production. The major scrap cause is the operators' adjustments and setup operations (42.7%). Brain—storming sessions allowed to draw a Ishikawa's diagram pointing out the main causes (Fig. 5).

Tecnogial 8 (TEC 8) Equipment

For the same period, the Tecnogial 8 is the second biggest contributor in terms of scrap ratio calculated with 6% (alongside with Tecnogial 1), but it is the major contributor in terms of kg of scrap produced, with 778 kg. The major cause identified in this

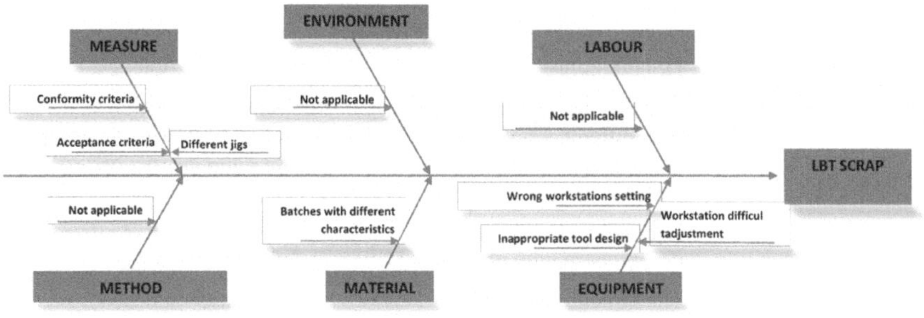

Fig. 5 Ishikawa diagram related to the scrap produced on the LBT equipment

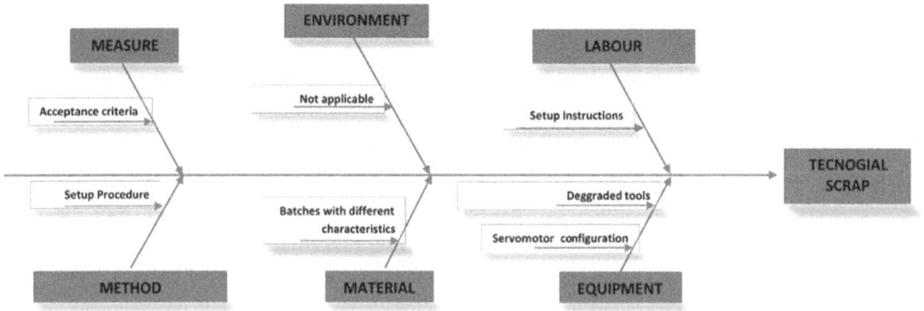

Fig. 6 Ishikawa's diagram related to the scrap produced during the adjustment process on the Tecnogial 8 equipment

equipment was the adjustment process/setup performed by the adjustment operator (Fig. 6).

Improve step: With the support of the data collection plan that was implemented, the experience of workers as operators, adjustment operators, the maintenance team, the processes team and the production director, and based on the developed Ishikawa's diagram, it was possible to define improvement actions with the aim to reach the main goal of this project: reduce the quantity of scrap on the comfort systems module. Based on the previous Ishikawa's diagrams, which is frequently used on six sigma projects, it was possible to identify the elements that contribute to the variation in the process and define a set of improvement actions. In the Table 2 are summarized, for the critical equipment, the main potential causes to produce scrap and the respective improvement actions.

Control step: LBT equipment was one of the most critical asset regarding the scrap production, presenting the highest ratio percentage of scrap/quantity of production and the second most representative in terms of weight of scrap generated. Also, in terms of improvement actions finished, the LBT was the equipment that could receive the most complete and deep intervention, replacing the most important tools. Figure 7 depicts a control chart regarding the ratio quantity of scrap produced/quantity of production. After completing this main intervention, the values

Table 2 Potential causes for scrap production and respective improvement actions defined

Equipment	6Ms	Potential cause	Improvement action
LBT	Method	Complex adjustment process and with differences depending on the adjustment operator who performs it	Replacing tools to allow a faster setup and creation of detailed and precise setup instruction
		Equipment parameters not defined or incorrectly defined	Accurate definition of all equipment parameters
	Machine	The difficulty with the adjustment process ("zero" of the equipment)	Second phase—it depends on the improved tools implementation
		Degradation tools: presser, sticks, workstation, pins and wire guide	Replace tables, pressers, pins, bend tools, and wire guides
		Deficient workstations fixation	Repair old workstation welding's
		Damaged controller	Replace parameters and equipment movements controller
TEC 8	Measure	Lack of visualization of scrap on this equipment by type of defect	Divide scrap containers according to the type of defect
	Method	Adjustment process requires the change of all workstations	Budget approval to "lock the center of the equipment" and to change the headers
	Machine	Degraded tools of the transfer mechanism: cars; tweezers; springs; chains; rolls; bushings	Evaluate tools replacement

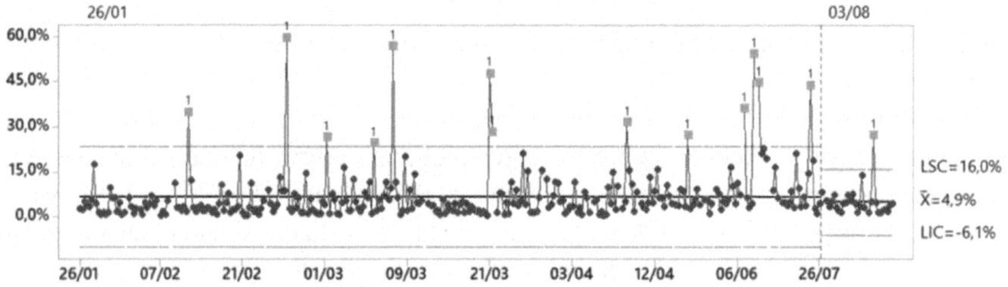

Fig. 7 Control chart of quantity of scrap per quantity produced (LBT)

show a lower dispersion between them, comparing to the period before the intervention. The control limits are tighter for the period after the intervention, presenting an average value of 4.9%, with the superior value moving from 23.8 to 16.0% and the inferior from −8.9 to −6.1%.

5 Discussion and Conclusions

Since the start of the project, at the beginning of 2018, the scrap cost value per month has been gradually reducing as it is visible in the graph of Fig. 8. In terms of equipment, for the LBT equipment, where it was implemented the major improvement actions, it is already visible a positive effect. This equipment was identified as the top producer of scrap for the first four months of the year. In the weeks before the intervention, it was produced between 40 and 60 kg of scrap per week, representing, on average, about 15% of the production. After two weeks of the intervention in this equipment, it was noticed that the defined improvements allowed to reduce the quantity of scrap produced for values between 25 and 35 kg, on the first four weeks of activity (Fig. 9). However, it is visible that, relatively to the quantity produced, the amount of scrap represents a lower portion, confirming that the implemented modifications led to the expectable results.

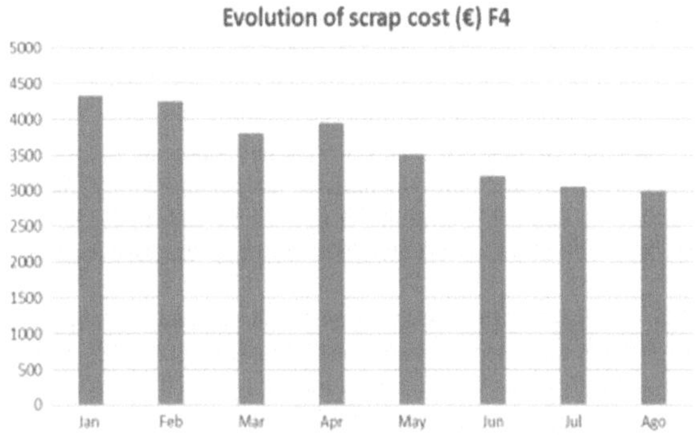

Fig. 8 Evolution of scrap cost (€) during the beginning of the project

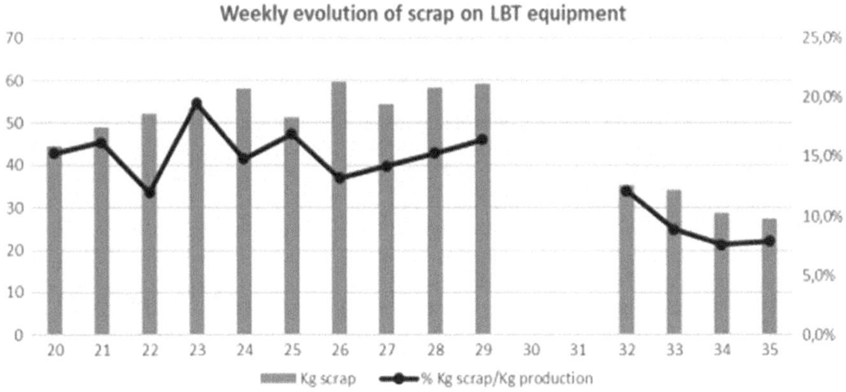

Fig. 9 Weekly evolution of scrap on LBT equipment

Based on the goals initially defined, it can be concluded that using the DMAIC tool it is possible to improve processes and increase its quality. If the Measure step is important to quantify and identify the problems, the Analyse step is crucial and depends on the experience and insight of the teams involved. Here, Ishikawa's diagram and Brainstorming sessions are crucial to accurately point out the root causes. When correctly defined the root-causes, the ideas on how to improve the processes are vital, and depend again on the team expertise. After that, control will help the team to keep under control the improvement actions implemented. Now, the company management is much more aware about the difficulties related to the equipment adjustment and corresponding tools. The final results show the successful application of the DMAIC methodology, allowing to reach the desired results.

References

1. Costa RJS, Silva FJG, Campilho RDSG (2017) A novel concept of agile assembly machine for sets applied in the automotive industry. Int J Adv Manuf Technol 91:4043–4054
2. Costa MJR, Gouveia RM, Silva FJG, Campilho RDSG (2018) How to solve quality problems by advanced fully-automated manufacturing systems. Int J Adv Manuf Technol 94:3041–3063
3. Santos J, Gouveia RM, Silva FJG (2017) Designing a new sustainable approach to the change for lightweight materials in structural components used in truck industry. J Clean Prod 164:115–123
4. Rosa C, Silva FJG, Ferreira LP, Campilho R (2017) SMED methodology: the reduction of setup times for steel wire-ropes assembly-lines in the automotive industry. Procedia Manuf 13:1034–1042
5. Rosa C, Silva FJG, Ferreira LP (2017) Improving the quality and productivity of steel wire-ropes assembly-lines for the automotive industry. Procedia Manuf 11:1035–1042
6. Moreira A, Silva FJG, Correia AI, Pereira T, Ferreira LP (2018) Cost reduction and quality improvement in the printing industry. Procedia Manuf 17:623–630
7. Antoniolli I, Guariente P, Pereira MT, Ferreira LP, Silva FJG (2017) Standardization and optimization of an automotive components production line. Procedia Manuf 13:1120–1127
8. Neves P, Silva FJG, Ferreira LP, Pereira T, Gouveia A, Pimentel C (2018) Implementing lean tools in the manufacturing process of trimmings products. Procedia Manuf 17:696–704
9. Silva FJG, Campilho RDSG, Ferreira LP, Pereira MT (2018) Establishing guidelines to improve the high-pressure die casting process of complex aesthetics parts. In: M. Peruzzini et al (eds) Proceedings of the transdisciplinary engineering methods for social innovation of industry 4.0. IOS Press. https://doi.org/10.3233/978-1-61499-898-3-887
10. Pinto H, Silva FJG (2017) Optimisation of die casting process in Zamak alloys. Procedia Manuf 11:517–525
11. Castro TAM, Silva FJG, Campilho RDSG (2017) Optimising a specific tool for electrical terminals crimping process. Procedia Manuf 11:1438–1447
12. Costa C, Silva FJG, Gouveia RM, Martinho RP (2018) Development of hydraulic clamping tools for the machining of complex shape mechanical components. Procedia Manuf 17:563–570
13. Barbosa B, Pereira MT, Silva FJG, Campilho RDSG (2017) Solving quality problems in tyre production preparation process: a practical approach. Procedia Manuf 11:1239–1246
14. Costa T, Silva FJG, Ferreira LP (2017) Improve the extrusion process in tire production using six sigma methodology. Procedia Manuf 11:1104–1111

Material Flow Cost Accounting as a Way to Apply Lean Manufacturing

Helena Cecílio, Paulo Peças, Inês Ribeiro and Diogo Jorge

Abstract The industrial sector is under an increasing pressure to achieve quality products with the lowest possible cost and environmental impacts, leading to the necessity of developing methods to support management decision. The Material Flow Cost Accounting (MFCA) and the Lean Management tools usually provide that support, separately. The MFCA assess the economic and environmental performance of the production system, evaluating the physical flows in monetary units, presenting the results in terms of costs of the product itself and the waste. The Lean philosophy analyses, mainly, the physical flows and physical waste. Lean methods/tools allow the diagnosis of different systems and uses problem solving strategies towards continuous improving. This paper present an approach combining both methods/tools, MFCA and Lean, supported by its application to a real case study. This approach allows a clear and quick identification of critical cost and energy related problems, allowing an early problem-solving phase.

Keywords Lean tools · MFCA (Material flow cost accounting) · Cost · KPIs

H. Cecílio · D. Jorge
Erising, Efficiency Rising, 1600-491 Lisboa, Portugal
e-mail: helenacecilio@erising.pt

D. Jorge
e-mail: diogojorge@erising.pt

P. Peças (✉) · I. Ribeiro
IDMEC, Instituto Superior Técnico, Universidade de Lisboa, 1049-001 Lisboa, Portugal
e-mail: ppecas@tecnico.ulisboa.pt

I. Ribeiro
e-mail: ines.ribeiro@tecnico.ulisboa.pt

M. Rossi et al. (eds.), *Proceedings of the 6th European Lean Educator Conference*, Lecture Notes in Networks and Systems 122,
https://doi.org/10.1007/978-3-030-41429-0_24

1 Introduction

Nowadays, a modern, competitive and environmental concerned society is pressuring companies to achieve higher productivities with the lowest possible environmental impact [1]. Thus, few alternative methods have been emerging, to support management decisions in terms of economic performances and, simultaneously, considering the environmental impact and production volumes [1, 2].

The Material Flow Cost Accounting (MFCA) is considered as one of the main tools for Environmental Management Accounting [1]. MFCA, ISO standard 14051 [3], is a management tool which fosters the transparency of energy and material flows and consumptions. This method has been developed to support industrial companies to increase the efficiency of the material and energy, and to support management decisions by presenting the effective value of the company's waste.

Lean Management is recognized as a solution for waste elimination. Its main goal is the identification and elimination of several types of waste allowing companies to achieve an efficient customer demand [4]. Lean is supported by a comprehensive set of performance indicators regarding flow and effectiveness allowing the diagnosis of the production system, moreover it is also supported by a several tools/methods for problem-solving and continuous improvement.

Despite the potential of these two approaches, they are often used separately, nevertheless its complementarity is evident. In this paper an approach is proposed combining these two methods. Firstly, the identification of differences and complementarities is discussed, then the proposed combination methodology is presented. Finally, the methodology is applied to a case study and the results analysed.

2 Material Flow Cost Accounting and Lean Complementarities

The MFCA is a method to diagnose production systems based on the quantification of the material flows splitting the material used to manufacture the products from the material losses (waste) [5]. It allows the identification of inefficiencies through the production system and presents the results regarding the product and waste cost flows separately. Once, the cost of waste is visible, it can drive managers to re-plan their strategy. As soon as their strategy is implemented, the resources reduction can be achieved and consequently a reduction of the overall production cost as well as the environmental impact [5]. Thus, MFCA aims to support companies to enhance its environmental and economic performance through the reduction of resources usage [6].

The MFCA method divides the entire production system into Quantity Centres (QC) [7, 8]. The QCs are parts or sub-divisions of the manufacturing system where the inputs and outputs must be quantified in physical and further in monetary units. Usually, these areas corresponds to places where materials are transformed, or stocked [3]. The QC is the starting point for data collection in physical units in terms of resources measurements.

The base concept of the MFCA, is the conservation law of material and energy. Considering this principle, and in order to guarantee that all flows are accounted, a mass balance should be performed to the production system per QC individually. MFCA considers the production of goods as a system of material's flow, based on the mass balance. It distinguish the movements of materials in [7]: (i) Desired material flow—Movement of material that intend to become part of the final product; (ii) Undesired material flow—movement of unintended materials output.

The MFCA maps and quantifies the places (QCs) and the amount of resources consumed, as well as the material (and energy) losses. MFCA divides the several types of cost into: (i) Material cost, (ii) Energy cost; (iii) System cost; (iv) Waste management Cost. The system cost includes the cost of 'all expenses incurred in the course of in-house handling of the material flows, except the material, energy and waste management cost' [3]. Thereafter, a calculation model should be developed to compile all the information required resulting in an MFCA flow map. So, the flow map should be presented and analysed by the company's managers aiming to be used for seeking improvement opportunities [3].

Concluding, MFCA shows the performance of each QC, through the mapping of information related to each QC. This information allows the stakeholders to identify critical processes only based on the monetary aspect but it does not allow the analysis of the criticality level of the QC neither its correspondent QC-Section. This is due to the lack of indicators (within MFCA indicators) able to identify single QC's and QC—Section's contributions for the Total Cost, or any parameter above the expected or the desired value. From the MFCA output the stakeholders are able to analyse different cases from that information, however the MFCA do not present directly those indicators [6, 8, 9]. The literature on MFCA largely neglects strategies for taking advantage of its detailed and monetary based diagnosis, as procedures of critical aspects' identification and strategies to develop solutions [9].

Lean has a different approach for diagnosis and critical aspects identification. While MFCA is essentially a diagnostic tool and is concerned to make "visible" the monetary value of the production waste, Lean has a more incisive diagnostic, supporting the identification of critical processes/tasks, as well as the root causes including problem-solving tools/approaches [4]. The diagnostic tools of Lean approach is divided in Value Stream Mapping (VSM) and Gemba Walk (and the use of Key Performance Indicators) which analyse the production system in a macro perspective without a direct relation to the monetary value of the processes and/or the production cost [4, 10]. Nevertheless, these analysis aim to identify the critical processes/tasks to launch continuous improvement projects (based on continuous improvement principles—Plan-Do-Check-Act). In fact, the main goal of Lean diagnosis tools is to identify non-value added activities and analyse them.—Plan phase—aiming to eliminate waste regardless of its nature (8 MUDA [4]). The action-plan application accomplishes the effective waste elimination—Do phase—where the problem-solving tools as Kaizen events, A3 Problem Solving and 8D method are used to define cooperatively between the company collaborators. The intrinsic characteristic of these tools leads to the need of data collection tasks in physical units to analyse the results (sometimes with consecutive Gemba Walks). These results are then shown regarding non-added value time (inefficiencies). During these problem-solving methods, supporting tools for root-cause identification are used. Namely, 5Whys, 5W+1H, Is/Is not, among others [10]. Then counter-measures are generated aiming to eliminate the root-causes, usually using good-practices of Lean tools like 5Ss, SMED, Kanban, Mizusumashi, among others [10]. The continuous improve- ment process continues by assessing the impact of the counter-measures—Check phase—and after a few iterations a final procedure/solution is implemented (visual management is often used as a Lean tool) and standardized—Act phase. The process continues for the next problem to solve.

So, MFCA and Lean tools use different approaches to identify wastes and achieve better performance of production system, but they have the same aim and starting point: both analyse the production flow in physical units and present the actual production performance status. However, Lean is mainly concerned about reducing MUDA, and MFCA is concerned about the waste economic impact and its reduction based on its cost. (Fig. 1). On the one hand, MFCA's goal is to reveal the improvement opportunity by showing the waste cost but it is not primarily designed for problem-solving, nor to present specific solutions. On the other hand, Lean management aims to reduce all types of waste and uses the problem-solving methods and Lean tools to identify the root causes and to provide solutions. Nevertheless, it is not designed to present the results in monetary units—such as MFCA.

Consequently, there is a potential to combine these two approaches (Fig. 1), taking advantage of their complementary aspects. On the one hand, MFCA aims to present to the managers the real monetary value of the waste and the QC within the manufacturing system that has the highest contribution to the waste value. On the other hand, Lean Management tools goals concern to physical flow analysis and problem-solving solutions. From this, the research question of the present paper arises: How to combine MFCA and Lean management tools for a continuous improvement system? To

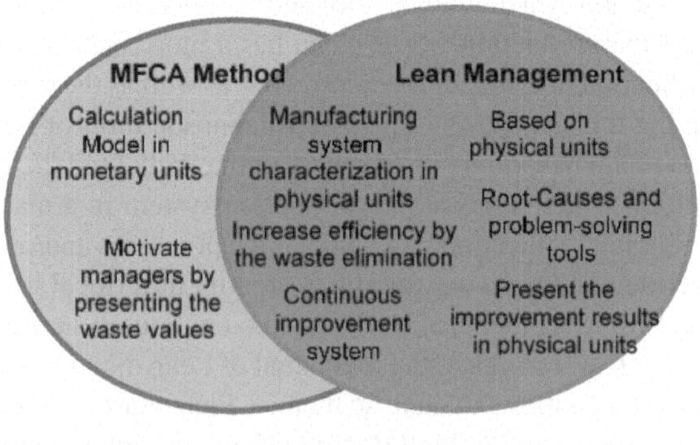

Fig. 1 Complementary aspects and combination opportunity

answer this question, an integration methodology is proposed and further validated through its application to a case study.

3 MFCA-Lean Proposed Methodology

The proposed methodology (Fig. 2a) combines MFCA structured phases with an adaptable application logic of Lean tools, i.e., the tools should be selected according to the production issues. It also incorporates a very important rationale for the effective success of its implementation: the continuous improvement foundations. The PDCA cycle [10] is imbibed in the MFCA-Lean methodology although is not explicitly mentioned in the methodology sequential phases.

The first step is the objectives definition that should be aligned with the company's strategical plan. These "macro objectives" should be a translation of the strategic objectives in operational performance figures. Furthermore, the scope definition will delimit the production system or part of it where the methodology will be applied. The second step is the selection of the MFCA and Lean-based operational performance indicators considering the objectives and scope. The KPIs derived from MFCA are "mandatory", whereas the ones from Lean are more dependent on the objectives and scope. Moreover, the Target Values (TV) for each KPI should be defined, according to the company strategy. The third step is the application of the MFCA method, namely related to QC definition and data gathering related to QC and to Lean-based KPIs. The

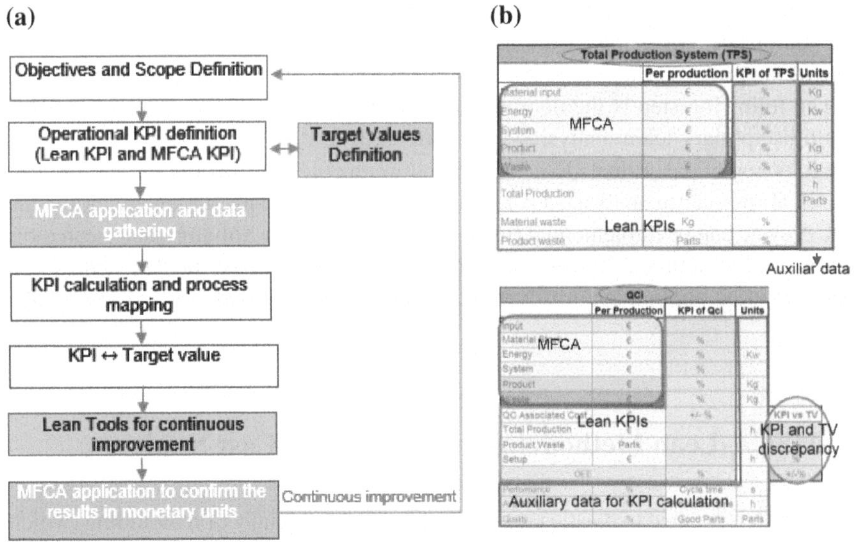

Fig. 2 **a** Overview of MFCA and Lean combination approach, **b** A proposal of QC and total production system dashboards output data, the comparison with target values and the performance indicators

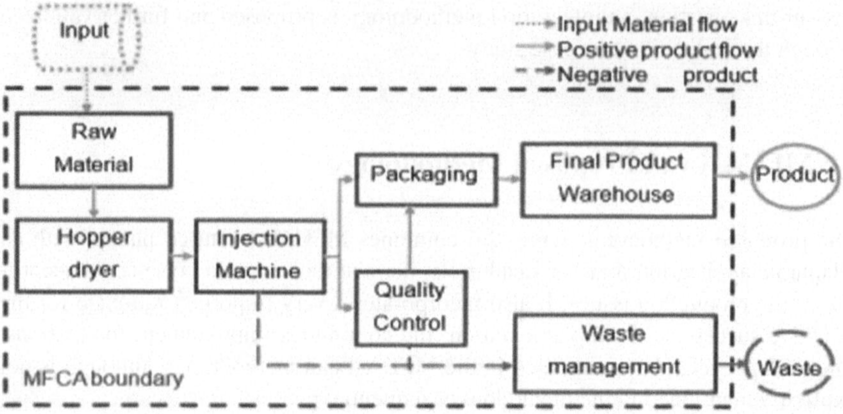

Fig. 3 Material flow model

next two steps are the KPI calculation (vis a vis TV) and process mapping. Following the visual management benefits, two types of dashboards are proposed (Fig. 3b) one for each QC and the other showing the total performance of the system. Both dashboards suggested two main areas, one dedicated to the MFCA indicators, and the other to the Lean indicators. They also have a column that connects each KPI with the Target Value. The QC dashboard has more detailed information related to specific operation or tasks, if existent. This dashboard also shows the contribution of each QC to the over-all production cost. The dashboard related to the total performance has the final MFCA typical indicators of performance as well as the total cost involved. The contrast be- tween the KPIs observed, and the Target Values indicates the current state of the process where improvement opportunities might be visible.

The following step is the determination of the critical QC and the critical KPIs (multidisciplinary team analysing the dashboard, in PDCA quality circles logics). The selection of the critical QCs and KPIs allow an efficient and effective subsequent phases of root-cause analysis, a solutions development and its implementation—the PDCA cycle can be implemented (using A3 or 8D Problem-Solving), and Lean tools can be used to improve performance and reduce waste. Finally, the data acquisition task should be applied again to verify the improvement in MFCA and Lean KPIs results.

4 MFCA-Lean Methodology Application—Case Study

The MFCA is applied to plastic part production system by injection moulding (which follows a make-to-stock manufacturing strategy), to appraise its current economic performance. Due to the lack of space in this paper, only relevant information will be given for the sake of results understanding. The company defined as its main goal, the increase of gain margin and the reduction of material, energy and human resources. Then, the scope was defined as the entire process from the material supply until the product delivered. The appropriate KPIs to analyse the current state of the production system considering the company's objectives were to reduce total production system (TPS) cost, the percentage of material waste (TV: <3%), the setup-time (TV <90 min) and increase OEE (TV: >65%). To perform MFCA analysis, the steps presented in Sect. 3 were followed. Firstly, the data collection period was established as one production bunch, i.e. the time required to produce the total order (36,000 good parts). Then, the production system was sub-divided in QC, and the material flow was analysed (Fig. 3). The KPIs were properly defined, as well as each TV, and data was acquired accordingly in each QC (Figs. 4 and 5).

The proposed dashboards map presents not only the costs per QC (input; output, product and waste) and per QC-section as the original MFCA, but also, the actual

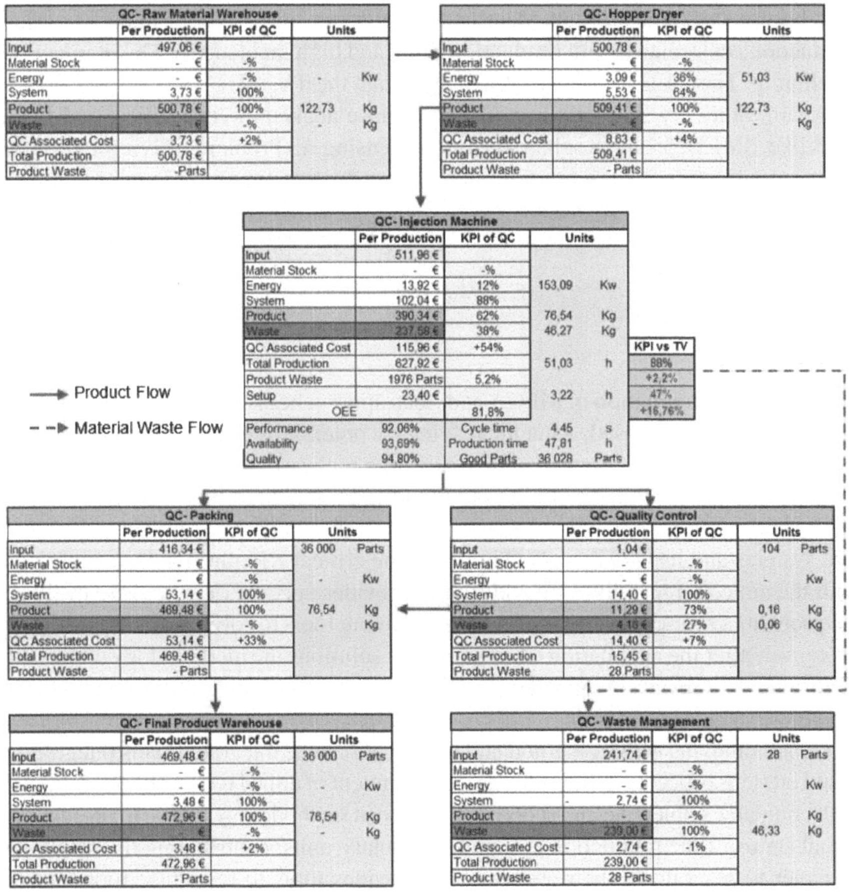

Fig. 4 Methodology dashboard for the production system per QC

Fig. 5 Methodology general dashboard of total production system (TPS)

Total Production System (TPS)				
	Per production	KPI of TPS	Units	
Material input	515,37 €	72,4%	122,73	Kg
Energy	17,02 €	2,4%	204,12	Kw
System	179,58 €	25,2%		
Product	472,96 €	66%	76,54	Kg
Waste	239,00 €	34%	46,19	Kg
Total Production	711,96 €		51,03 38 004	h Parts
Material waste	46,19 Kg	38%		
Product waste	2004 Parts	5%		

process performance through the KPIs presentation. Moreover, this MFCA-Lean dash-board also presents the discrepancy between the KPI and the Target Value. Allowing the user to evaluate if the pre-defined plan was being fulfilled as planned. These properties make the MFCA-Lean dashboard more extensive than the original MFCA flow map, and more detailed and objective than Lean. In particular, and very briefly, the critical points identified were: setup time, level of material waste, production cost (related with production time) and high incidence of QC of injection machining. The OEE showed a value higher than the TV.

Again, in a very synthetic way, the root-cause analysis were applied, and Lean tools like, SMED, 5S and Kanban were applied, using A3 Problem-Solving approach, being possible to reduce setup time in 51%, production time in 3% and an overall production cost reduction of 3%.

5 Conclusions

Fostering the combination of MFCA and Lean approaches is the aim of this paper, so a methodology is proposed. After identifying the potential benefits of this integration, the methodology was presented and applied to a real case.

From the application results it was possible to conclude that the MFCA-Lean methodology allows, not only, the understanding of the costs incurred in its production systems and its flow, but also highlights the critical KPI through its comparison with the aimed target values. In addition, it provides specific Lean tools to evaluate the problem's root-cause and uses problem-solving tools to solve the existent issues. Moreover, after the application of the proposed solutions the methodology allows the confirmation of results in monetary units due to the performance of the improvement activities.

Concluding, the MFCA method and Lean tools can be integrated. This integration based on steps procedure allows the accomplishment of aimed results directly aligned with company's objective and scope. MFCA-Lean methodology is able to present the actual status of the production system in monetary units, contributing to encourage manager to re-evaluate their strategy and provides tools to recognise root-causes, to support and improve employees' activities guiding their work efficiently. This methodology should be implemented as a continuous improvement cycle so the production process moves closer to the ideal optimized process.

As future developments, the authors intend to create a guide to implement the propose approach with practical recommendations, including also guidelines to model the energy flows independently of the material flows. The allocation of the energy under the material flow usually neglects several types of energy waste as, vibrations or heat transfer.

Acknowledgements This work was supported by FCT, through IDMEC, under LAETA, project UID/EMS/50022/2019. Authors gratefully acknowledge the funding of Project POCI—01-0247-FEDER-017637, cofinanced by Programa Operacional Competitividade e Internacionalização and Programa Operacional Regional de Lisboa, through Fundo Europeu de Desenvolvimento Regional (FEDER) and by National Funds through FCT—Fundação para a Ciência e Tecnologia. Authors gratefully acknowledge the funding of Project POCI-01-0247-FEDER-024541, cofinanced by Programa Operacional Competitividade e Internacionalização and Programa Operacional Regional de Lisboa, through Fundo Europeu de Desenvolvimento Regional (FEDER) and by National Funds through FCT—Fundação para a Ciência e Tecnologia.

References

1. Kokubu K, Tachikawa H (2013) Material flow cost accounting: significance and practical approach. In: Handbook of sustainable engineering
2. Sygulla R, Bierer A, Gotze U (2011) Material flow cost accounting—proposals for improving the evaluation of monetary effects of resource saving process designs. In: Proceeding of the 44th CIRP conference on manufacturing system
3. DIN-EN-ISO (2011) Environmental management—material flow cost accounting—general framework (ISO 14051:2011)
4. Spear SJ (2004) Learning to lead at Toyota. Harv Bus Rev
5. Schmidt A, Götze U, Sygulla R (2015) Extending the scope of material flow cost accounting—methodical refinements and use case. J Clean Prod
6. Christ KL, Burritt RL (2016) ISO 14051: a new era for MFCA implementation and research. Rev Contab
7. Sygulla R, Gotze U, Bierer A (2013) Material flow cost accounting: a tool for designing economically and ecologically sustainable production processes. In: Pecas P et al (ed) Technology and manufacturing process selection, 1st edn
8. Guenther E, Jasch C, Schmidt M, Wagner B, Ilg P (2015) Material flow cost accounting—looking back and ahead. J Clean Prod
9. Cecflio H (2017) Material flow cost accounting application and its integration with lean tools Universidade de Lisboa Instituto, Superior Técnico
10. Basu R (2009) Implementing six sigma and lean: a practical guide to tools and techniques

On the Way of a Factory 4.0: The Lean Role in a Real Company Project

Federica Costa and Alberto Portioli-Staudacher

Abstract In the last three decades, Lean Production has been largely adopted, while, more recently Industry 4.0's technologies have been raised new management paradigms. Literature has been investigating the role played by Lean Production and Industry 4.0 and the interrelations between the two on the operations performances of companies. In this conference paper, we want to show a part of the results of an improvement projects, conducted in an Italian manufacturing company, in which Company's first goal was to digitalize the workflow in order to enhance traceability. The goal of the project has been achieved, however the Lean Production played a preliminary important role to reduce the inefficiencies. This would be an example of an implemented improvement project in a company that looks at digitalization and it represented an interesting learning point for the company.

Keywords Lean production · Digitalization · Improvement project · Manufacturing

1 Introduction

The wide adoption of Lean Production (LP) practices and principles has consistently occurred throughout different industries and contexts during the last three decades [1, 2]. Such intensive adoption is due to the expected benefits that LP implementation can entail, such as cost reduction, quality and productivity enhancement, delivery and customer satisfaction improvement, etc. [3]. In this sense, a diversity of organizations has been investing a lot of effort to adapt and implement LP in their processes and systems [4].

F. Costa (✉) · A. Portioli-Staudacher
Department of Management, Economics and Industrial Engineering, Politecnico di Milano, Via Lambruschini 4/B, 20156 Milan, Italy
e-mail: federica.costa@polimi.it

A. Portioli-Staudacher
e-mail: alberto.portioli@polimi.it

M. Rossi et al. (eds.), *Proceedings of the 6th European Lean Educator Conference*, Lecture Notes in Networks and Systems 122,
https://doi.org/10.1007/978-3-030-41429-0_25

With the advent of Industry 4.0, new management paradigms have been raised through novel technologies adoption [5]. As Industry 4.0 is characterized by modern information and communication technologies (ICT), products, machines and processes can become interconnected, allowing the establishment of the 'smart factory' concept [6]. In this sense, many authors spotlighted the potential benefits of adopting technologies such as 3D printers, Internet of Things (IoT) and augmented reality models (i.e. [7]), generating great expectations and enthusiasm about the theme.

However, literature evidence regarding Industry 4.0's integration with other management approaches, such as LP, is still scarce. Some previous studies (e.g. [8]) attempted to examine how some LP practices could benefit from the incorporation of a certain set of technologies. Additionally, other researchers (e.g. [9, 10]) have suggested a positive relationship between LP and Industry 4.0, but literature falls short of empirical validation of such synergy.

In this sense, the paper presents a real case of a manufacturing company that started a digitalization project with the Lean Excellence Centre of Politecnico di Milano and shows how LP practices supported and played a role.

The rest of this article is structured as follows. Section 2 presents in the Background the recent literature on the studied topics. Section 3 describes the company and the goal of the project, whose implementation is detailed in Sect. 4, before Sect. 5 that concludes the article indicating the main take aways.

2 Background

Industry 4.0 represents the integration of automation technologies, such as the Cyber Physical Systems (CPS), Internet of Things (IoT) and cloud computing, in the manufacturing industry. In this ICT driven industrial context, prominent technological frameworks for manufacturing processes at both intra- and inter-organizational levels have been proposed, entailing an array of solutions to the growing needs of informatization in manufacturing industries [11]. Hence, there has been a growing demand for research regarding Industry 4.0 in order to provide insights into the issues, challenges, and solutions related to the design, implementation and management of Industry 4.0 [12].

However, for many manufacturers the current ICT infrastructures may not be entirely ready to support the transformation into Industry 4.0, which aims at integrating operations horizontally and vertically, as well as end-to-end [13]. Further, Industry 4.0 adoption may impact other key aspects of an organizational structure, such as human resources development [14] and customer relationship management.

LP aims at streamlining the flow of value while continually seeking to reduce the resources required to produce a given set of products [1]. It was conceived as an evolutional detachment from the precepts of traditional mass-production manufacturing [15]. Although the adoption of LP is not a new concept, few organizations fully understood the philosophy underlying its practices and principles. Based on a people-centric system where people are directly involved in the process of continuous improvements, LP practices are deployed to allow employees solving problems at their own workplace [16].

There is a general agreement that LP is positively associated with operational performance, in both developed [17] and developing economies' context [18]. The relationship between Industry 4.0 and LP has been increasingly highlighted in operations management research [19].

Lean Management challenged successfully mass production practices [3]. However, nowadays, Lean Management and Industry 4.0 must be linked to compete in the market through Cyber Physical Production Systems adoption, providing IT integration between production and planning, between customers and suppliers [20].

As a matter of fact, Industry 4.0 does not make Lean obsolete, instead the two are mutual dependent and they have their particular domain of application and combination in product variability and production volume [10]. Nowadays, the combination of Lean Management and I4.0 (i.e. *Lean 4.0*) seems essential to boost productivity [21].

Another shared point is that Industry 4.0 and Lean Management are completely compatible [9, 22].

The current literature seems to believe that Lean management could be the most suitable underlying strategy for a digital transformation. So, Industry 4.0 has been proposed as "the new boost" for the hunt of wastes, because it supports coordination and alignment of systems and organizations. Hence, Lean and Industry 4.0 are not alternatives but rather they complement each other.

For example, I4.0 needs stable and standard, simple and clear processes in order to exploit digitalization benefits, but at the same time, the proper exploitation of Industry 4.0 technologies on stable processes straight improvement and innovations in the systems. Industry 4.0 technologies are considered as enablers useful to support improvements in operations, much more in a well-organized Lean environment: this condition is necessary to assure and sustain the success and sustainability of disruptive technologies [23].

In this sense, Lean is a prerequisite before automating a process to avoid costs and wastes, achieving a better quality and avoid digitization of inefficiencies. According to Netland et al. [24], "Lean will not fade with Industry 4.0": digital revolution incorporate Lean and, at the same time, Lean principles become more important. As a matter of fact, Industry 4.0 may permit to generate the true Lean Enterprise [1] allowing a much richer understanding of the customer demand and, at the same time, the immediate sharing of the demand data throughout complex supply chains and networks.

However, according to Maguire [25], relationship between Lean Management and I4.0 did not lead necessarily to success. Piszczalski [26] defined them as "two opposing camps", referring to Lean Management as "almost anti information systems in its stance". Lean is associated to an idea of simplicity, while IT solutions provide opportunities to introduce complexity and automation [27]. Moreover, sometimes automation generates lower flexibility: some processes, once automated, become much more difficult to change and hence improve. According to this issue, a perceived risk is the interruption of learning by doing, thereby obstructing operators from effectively understanding how the factory works.

The acknowledgement of the relevant integration of new technologies into LP has been evidenced in early 1990s and denoted as Lean Automation (LA). Lean Automation was defined by Dulchlnos and Massaro [28] as "technique which applies the right amount of automation to a given task. It stresses robust, reliable components and minimizes overly complicated solutions". Hence, its principal goal is the improvement of productivity, focusing on lean production to reduce waste, make the system less complex and increase efficiency, while relying on automation to cope with increased quality demand and complex product design. Lean system allows companies to deal with complex production processes and robotic automation integrates robots and conveyance system to improve flexibility in material handling.

Despite its potentialities, the academic world did not give much attention to Lean Automation for over two decades. However, thanks to the new opportunities offered by Industry 4.0 technologies it has regained its importance.

More recently, much attention has been given to LA with the advent of Industry 4.0 [29].

In essence, while there are authors advocating that Industry 4.0 can conflict with the ground principles of simplicity, continuous and small improvements from LP, others might claim that Lean and Industry 4.0 may be positively related.

3 Company's Introduction and Project's Goal

Fifty years of experience and know-how has allowed Gefran to become a leader in the automation components and industrial process control fields. The production portfolio of Gefran is characterized by a wide variety of products and here the focus in on one product family that is produced in Gerenzano. The production process is, thus, starting from the motherboards, work in progress components and, then, final goods.

Each production order begins with the processing of the motherboard, called SMD, through two automatized and parallel machines, which deal with the positioning of small components and their welding through an oven. After the SMD order is cooled, it is processed through a visual check machine in order to detect all possible defects. The aim is to deliver final products to customers with the highest quality level, since the beginning of its production process. Once all the motherboards are inspected, they are ready for the manual positioning of components, which delimits the starting of the PTH area, and their consequent proceeding into the wave soldering machine. Boards are completed with some final procedures that can change in accordance with their different characteristics. In particular, the most common phases are the boards review, checks through programmed workstations and the manual or automatized coating. As soon as all the related work in progress materials are processed, the assembly of the final products begins with a consequent execution of pre-tests, insulation, burn-in—resistance test for inverters—or loading test for converters and final checks, which

Fig. 1 Product evolution from SMD, PTH to final product

are performed on all the final goods. Finally, the last production phase consists in the packaging of the client' order in the production department or in the warehouse, ready to be delivered.

Since the beginning, it has constantly invested in technological innovation businesses, which have offered the opportunity of increasing the performances of the current machines in terms of accuracy and efficiency thanks, also, to research towards equipment more in the vanguard. The research towards a higher productivity level has empowered the company to make the customer-based approach as its core and successful business strategy. A relevant aspect to highlight is that innovations in the technology field are predominantly incremental; in particular, they regard upgrades that can extend the life expectancy of machines. Nevertheless, as a technology become obsolete, incompatibilities appear evident between the existing machines and their related costs.

Since the beginning of a six-months project, Lean Excellence Centre of Politecnico di Milano together with Gefran focused their efforts in defining and analysing any relevant tools aimed at improving the productivity and enhancing the traceability of the main product families within the production department. Gefran's first objective was to improve the traceability of the products in order to control and monitor real-time the workflow on the production lines (Fig. 1).

The main benefit of the former is the possibility of shortening the lead time, which is a critical variable for a process that covers a wide diversified production mix. Instead, the latter allows to reach a higher level of control, which is essential in order to exploit all the resources and provide a flexible and efficient service level to customers. Another relevant benefit in the development of a system based on a higher traceability, is the possibility to extend the control not only on the Italian Motion and Control department, but also to its worldwide sales subsidiaries and, in particular, to the Indian one.

4 Project's Implementations

Gefran manufacturing department, which is specialized in the Motion and Control market, is divided into two main areas: boards and final products assemblies. In particular, the former is subdivided between SMD and PTH, or work in progress, production. Before developing the root cause analysis, the team has considered necessary to make a detailed study in order to identify the most critical production phase in terms of impact on the overall lead time, and so to better address the stakeholders' target in the short term. For each of the above-mentioned manufacturing areas, the value-added time (VAT) and non-value-added time (NVAT) are computed. The NVAT considered are the queueing time—QT, the transportation time—TT and the waiting time—WT. The analysis revealed that NVATs of the PTH area is the one with the highest incidence compares with the other ones. So, it makes that production sector as the most critical one and focus of our project. Once having identified the PTH area as the most critical one, the Ishikawa diagram appears as a useful tool in order to highlight the most meaningful root causes that have as final effect a high PTH NVATs.

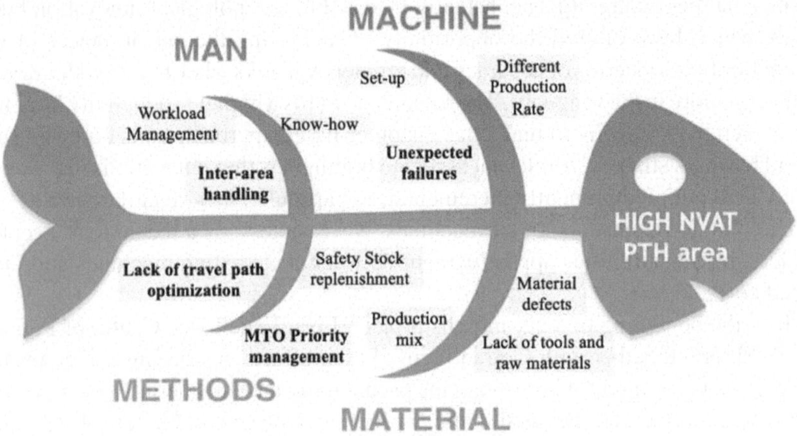

The team decided to focus its efforts in developing a deeper analysis only on the most significant causes, which are identified based on surveys done in the PTH area, in order to select the most critical one. More precisely, the four main areas object of studying are: rack handling, failure time, priority management and lack of travel path optimization and in this paper, we present our proposals implementation for priority management and travel path improvement.

The first proposal consists in a review of the policy planning methodology, which is also linked to the stock management of final products. A Make to Order policy is the current methodology applied in the PTH area for the management of the orders. In particular, it consists in the prioritization of boards managed through a make to order policy over the ones regulated by a make to stock plan. In accordance with a context of high customization level in combination with low production lead time, the assembly to order—ATO—stock management policy of final products represents the most suitable policy in accordance with the company. So, starting from the analysis and the classification of the stock management of final goods based on data gathered from the previous and current year, the purpose of the developed countermeasure is to optimise the policy to be used for different final products, based on the frequency of requested product. In particular, the first solution, called "Reduction of noise", was developed in order to optimise the current stock management of the final products, aiming at reducing the impact of make to order goods on the PTH production area and, so, improving the current management of the production complexity.

The second solution implemented consists in a workload controlled release of the actual work orders in the PTH area in order to obtain a production flow as linear as possible. Indeed, the actual company strategy is to release board orders by considering an infinite capacity level. The main benefits would be the following ones:

Optimization of work order path and processing time—the freezing of the consequent production day. It implies a higher visibility on the processing time and a better control of the flows in the PTH area.
Workload balance—planning orders in order to release the right workload to each workstation and employee. It allows to saturate downstream phases, avoiding, at the same time, the potential creation of bottlenecks or line stoppages.
Compliance with the theoretical due date and reduction of the non-value-added time—a better planning management allows to optimize the production paths and reduce the NVATs.

Lastly, a solution called "Visual shop floor monitoring system", was proposed with the purpose of developing a virtual interface, aiming at centralising the production monitoring of the boards area, leading to a consequent improvement in the awareness of the current flows to allow a real-time quick decision-making process based on real time information.

5 Conclusions and Take Aways

Considering the achieved results through the implementation of the developed solutions, it is possible to confirm that the target was successfully achieved by developing cutting-edge proposals towards the enhancement of the company's productivity and traceability. In particular, the team was able to reduce the production lead time by 40%, which leaded to an overall non-value-added time reduction of about 22% compared to the As-Is scenario. The estimated benefit of the implementation of the real-time visual platform is about 20% in terms of an optimisation of VAT and an effective bottlenecks individualisation.

The learning point of the project that came out brainstorming with Gefran is that the implementation of the real-time monitoring platform to enhance the traceability of product's workflow has to be done consequently to the streamlining of the product's flow. Firstly, the reduction of NVAT activities has been achieved in order to set and prepare the ground for the visual real time platform. Gefran's first objective has been achieved—visual monitoring real-time platform—however the company touched with hands that, before digitalizing, Lean's approach of reducing NVAT and waste has to be undertaken and applied.

References

1. Womack J, Jones D (1997) Lean thinking—banish waste and create wealth in your corporation. J Oper Res Soc 48(11):1148
2. Soliman M, Saurin TA (2017) Lean production in complex socio-technical systems: a systematic literature review. J Manuf Syst 45:135–148
3. Portioli-Staudacher A, Tantardini M (2012) Lean implementation in non-repetitive companies: a survey and analysis. Int J Serv Oper Manag 11(4):385–406
4. Holweg M (2007) The genealogy of lean production. J Oper Manag 25(2):420–437
5. Lasi H, Fettke P, Kemper H, Feld T, Hoffmann M (2014) Industry 4.0. Bus Inf Syst Eng 6(4):239–242
6. Kagermann H, Helbig J, Hellinger A, Wahlster W (2013) Recommendations for implementing the strategic initiative Industrie 4.0: securing the future of german manufacturing industry. Final report of the Industrie 4.0 Working Group. Forschungsunion, Berlin
7. Lu Y, Xu X (2018) Resource virtualization: a core technology for developing cyber-physical production systems. J Manuf Syst 47:128–140
8. Kolberg D, Knobloch J, Zühlke D (2017) Towards a lean automation interface for workstations. Int J Prod Res 55(10):2845–2856
9. Sanders A, Elangeswaran C, Wulfsberg J (2016) Industry 4.0 implies lean manufacturing: research activities in Industry 4.0 function as enablers for lean manufacturing. J Ind Eng Manag 9(3):811–833
10. Sanders A, Subramanian K, Redlich T, Wulfsberg J (2017) Industry 4.0 and lean management—synergy or contradiction? In: IFIP international conference on advances in production management systems. Springer, Cham, pp 341–349
11. Mourtzis D, Vlachou E (2018) A cloud-based cyber-physical system for adaptive shop-floor scheduling and condition-based maintenance. J Manuf Syst 47:179–198
12. Xu L, Xu E, Li L (2018) Industry 4.0: state of the art and future trends. Int J Prod Res 56(8):2941–2962
13. Liao Y, Deschamps F, Loures E, Ramos L (2017) Past, present and future of Industry 4.0—a systematic literature review and research agenda proposal. Int J Prod Res 55(12):3609–3629
14. Dworschak B, Zaiser H (2014) Competences for cyber-physical systems in manufacturing—first findings and scenarios. Procedia CIRP 25:3–8
15. Marodin G, Frank A, Tortorella G, Saurin T (2016) Contextual factors and lean production implementation in the Brazilian automotive supply chain. Supply Chain Manag Int J 21(4):417–432
16. Liker J (2004) The Toyota way: 14 management principles from the world's greatest manufacturer. McGraw Hill Professional, New York
17. Shah R, Ward P (2003) Lean manufacturing: context, practice bundles, and performance. J Oper Manag 21(2):129–149
18. Jasti N, Kodali R (2016) An empirical study for implementation of lean principles in Indian manufacturing industry. Benchmarking Int J 23(1):183–207
19. Mourtzis D, Fotia S, Vlachou E (2017) Lean rules extraction methodology for lean PSS design via key performance indicators monitoring. J Manuf Syst 42:223–233
20. Mrugalska B, Wyrwicka MK (2017) Towards lean production in Industry 4.0. Procedia Eng 182:466–473. https://doi.org/10.1016/j.proeng.2017.03.135
21. Rüttimann BG, Stöckli MT (2016) Lean and Industry 4.0—twins, partners, or contenders? A due clarification regarding the supposed clash of two production systems. J Serv Sci Manag 9(6):485–500
22. Wagner T, Herrmann C, Thiede S (2017) Industry 4.0 impacts on lean production systems. Procedia CIRP 63:125–131
23. Secchi R, Camuffo A (2016) Rolling out lean production systems: a knowledge-based perspective. Int J Oper Prod Manag 36(1):61–85
24. Netland T (2015) Industry 4.0: where does it leave lean? Available at www.leanmj.com. Apr 2015, pp 22–23

25. Maguire K (2017) Lean and IT—working together? An exploratory study of the potential conflicts between lean thinking and the use of information technology in organisations today, understanding the lean enterprise. Springer, Berlin, pp 31–60
26. Piszczalski M (2000) Lean versus information systems. Automot Manuf Prod 112(8):26–28
27. Jones D (2012) How can IT support a lean transformation. Presented at European lean IT summit, 22–23 Nov 2012. Available at https://www.slideshare.net/LeanUK/lean-and-it
28. Dulchlnos J, Massaro P (2005) The time is right for labs to embrace the principles of industrial automation. Drug World Discovery 2006(Winter-Issue):25–28
29. Rossini M, Costa F, Tortorella GL, Staudacher-Portioli A (2019) The interrelation between Industry 4.0 and lean production: an empirical study on European manufacturers. Int J Adv Manuf Technol 102:3963–3976. https://doi.org/10.1007/s00170-019-03441-7

Lean Performance Evaluation: Models and Application

Margarida Barros, Helena Cecílio, Diogo Jorge and Paulo Peças

Abstract Lean Manufacturing is based on continuous improvement, aiming at optimizing the production system, eliminating waste and minimizing resources use. In this study, an in-depth analysis regarding the Lean performance evaluation methods present in the literature, as well as already implemented in the industry, is performed, focusing on the gaps of current models and needs for future ones. This paper proposes a method for Lean performance evaluation, following the robust theoretical approaches of existent assessment methods and keeping simplicity of application. The proposed method is divided into two distinct models: The Lean Assessment model and the Lean Maturity model. The Fuzzy logic is used in the models to attenuate the ambiguity and vagueness of the participants' responses. The models' validation was accomplished by comparing the method application in an industrial company with *in loco* Lean assessment.

Keywords Lean manufacturing · Lean performance evaluation · Lean assessment · Lean maturity · Fuzzy logic

1 Introduction

For companies to keep up with the changing market demands, it is necessary to focus on the rapid and efficient production of high quality and low-cost products. Therefore, organizations want to identify ways of reducing costs and time, while

M. Barros · P. Peças (✉)
IDMEC, Instituto Superior Técnico, Universidade de Lisboa, Lisbon, Portugal
e-mail: ppecas@tecnico.ulisboa.pt

M. Barros
e-mail: margarida.barros@tecnico.ulisboa.pt

H. Cecílio · D. Jorge
Erising, Efficiency Rising, Lisbon, Portugal
e-mail: helenacecilio@erising.pt

D. Jorge
e-mail: diogojorge@erising.pt

M. Rossi et al. (eds.), *Proceedings of the 6th European Lean Educator Conference*, Lecture Notes in Networks and Systems 122,
https://doi.org/10.1007/978-3-030-41429-0_26

increasing quality. A good solution for such an objective lies in Lean Manufacturing. This philosophy consists on the continuous reduction of waste (non-added value), producing according to market demand, while optimizing quality [1].

In this context, there is a need to know how to evaluate a company's Lean performance, allowing for identification of Lean training and implementation requirements, as well as for unbalanced knowledge among persons in the company. There were developed Lean evaluation methods by several authors, focusing on measuring the performance and progress of Lean organizations, helping to choose the most appropriate improvement actions and highlighting the areas with potential application [2, 3].

As presented in Sect. 2 of this paper, there are different kinds of Lean evaluation methods. Some of those methods are very simple to implement by companies, usually through a list of questions. Nevertheless, they do not cover all the Lean thinking dimensions, or the output is very limited, just showing direct results from the questions. There are also complex methods with dozens of questions, which analysis requires external consultant elements to do the assessment and the evaluation of results. Based on the limitations of the existent methods, this paper proposes a new Lean evaluation method aiming to be self-applied by the companies, giving comprehensive and meaningful results. The output of the method is visual based, allowing an intuitive identification of the areas which need further improvement. The method was applied in a mould making company, where a two months *inloco* full Lean assessment was also performed. This allowed the assessment of the robustness of the proposed method: the knowledge and implementation gaps regarding Lean Manufacturing identified by the method, corresponded to the ones identified in the *inloco* assessment.

2 Lean Performance Evaluation Methods

Prior to read the following section, the reader should be aware of Lean Manufacturing principles and methods, as well as the origins of Lean Manufacturing: The Toyota Production System. The authors of this paper opted to emphasize the results of the literature review about Lean evaluation methods, recommending for the ones not familiarized with Lean Manufacturing to read [1, 4, 5].

A literature survey was performed by searching Google Scholar, ResearchGate and ScienceDirect databases, using the keywords: Lean Manufacturing, Lean Production, Lean Assessment Review, Lean Maturity Model, Lean Production Audit, and Lean Evaluation.

A common feature identified among the several Lean evaluation models found [2, 3, 6–24] is the use of two definitions: Lean Maturity and Lean Assessment. According to [6–9], Lean Maturity models aim to find the flaws between the company's current practices and the optimal level, with the establishment of improvement strategies and the assessment of competence, capacity and innovation level of organizations through specific criteria while comparing the current state with other levels or organizations. In addition, those authors claim Lean Maturity models aim to indicate logical paths for the improvement of the organization and to evaluate the impact of Lean principles and tools implementation. The Lean Assessment models, as claimed by [2, 3, 10], are more direct about the practices and the tools applied by the company, and are usually qualitative. In addition, those authors refer the Lean Assessment models are a necessary tool throughout the continuous improvement process, while they help to identify the practices and areas to be developed.

From the analysis of the two central concepts found, it is clear that there is some difficulty in the correspondence of definitions, as some of the characteristics found for both terms represent the same concept. Although, both definitions allow the Lean evaluation and are essential to define the best practices and strategies to be followed.

Despite the unclear differentiation between these two terms, the analyzed publications converge in several characteristics that should be present in Lean evaluation models. An evaluation model must be measurable and aligned with the strategic objectives. It should allow a performance evaluation and the understanding of the current state, assisting in the selection of improvement opportunities, while seeking the balance between the detail of evaluation and its simplicity. Some of the publications refer the self-assessed evaluation process by company elements [8, 12], others proclaim the need for external evaluation by partners or specialists in the area [10, 13].

Regarding the model's structure, there are several alternatives. The use of questions is the most common structure in the analyzed models [14, 16]. Others structures are based in statements or indicators [17–19]. Some of the methods assess directly the company about Lean methods and tools, while others ask questions without mentioning the Lean-related themes [20, 21]. Concerning the number of analysis fields, there are models using uniform number of fields per category, others have a non-uniform field number [15, 18]. Some of the models propose a long list of items to analyze [17] with 162 fields to fill, others present a very short list [19] with only with 8 fields. Of course, some of them are in between, having a balanced number of topics [15, 22] ranging between 20 and 50 fields to fill.

Among the existing models some use qualitative approach other use quantitative approach. There are also many models which use both types of analysis, especially in the form of fuzzy logic [3, 10, 12, 17]. The Likert Scale is the most commonly used evaluation scale [18].

Some of the models aim to cover all the Lean related areas [12, 20], other proposed a specific focus evaluation [16]. Finally, the results are presented as a single value, as a list of best-performing categories or as a visual scheme. The different models' features analyzed are listed in Table 1.

As a conclusion of survey of Lean evaluation models published in literature, there is no typical structure for a Lean evaluation method, neither a clear definition of the difference between Lean Maturity and Lean Assessment models. This aspect is also referred by other authors [11, 14]. The complexity of some models is high, hence the authors propose the existence of an external assistant for its implementation [14]. The use of explicit Lean-related terms differs among models, which may have a negative influence on the results, in case the company elements are not aware of those terms

Table 1 Differentiating features of lean performance evaluation models. Based on [2, 3, 6–24]

Evaluation process	Self-assessment		External assessment
Participants	Single		Multiple
Structure	Questions	Statements	Indicators
Approach	Direct		Indriect
Number of fields	Uniform		Non-uniform
Extension	Extended	Balanced	Short
Analysis	Qualitative	Quantitative	Qualitative and quantitative
Evaluation scale	Likert scale		Other scale
Focus	Integral		Specific
Result	Single value	List	Visual scheme

[20]. Some models do not cover all Lean areas, therefore they do not accurately represent the overall level of Lean performance, as mentioned by Zanjirchi et al. [15]. Other critical aspects were found through this analysis and were pointed by some of the models' authors: there are quite extensive models, although they are complete [11]; there are unbalanced models in terms of fields per category [12, 24]; some models are not cleared about the actions to be taken after the analysis [11].

Hence, there is a need to develop a Lean self-evaluation model that allows a clear, simple, direct and comprehensive analysis of the results, covering all the Lean essential areas.

3 The Proposed Lean Evaluation Method

The proposed evaluation aims to assess the company's regarding Lean Manufacturing knowledge and implementation, and to be useful by selecting the action and training action required towards performance improvement. It takes into account that there are companies with Lean knowledge, although they have difficulties implementing the best practices; companies which implement good practices, although they have a low formal knowledge concerning the Lean principles and tools; and companies whose lack of knowledge in Lean hinders the implementation of continuous improvement actions. For the proposed method's design several requirements were taken into account, having in mind the published work and gaps identified in Sect. 2. In summary, the method must (i) be simple enough to allow self-assessment by the company, therefore the output should easy to interpret; (ii) include all the hierarchy levels in the assessment; (iii) have a quantitative evaluation (not exclusively); (iv) use terms which are easily understood by the participants so the filling is consistent and the final result realistic; (v) enable a comprehensive assessment, encompassing key fundamentals and Lean methods, but have no more than 50 questions; (vi) identify Lean areas with potential opportunities for improvement.

In order to have no influence from the ambiguity on the participants' responses and considerations, together with a numerical result, the simultaneity of the quantitative and qualitative analysis, a fuzzy logic data treatment is used—this data treatment in not visible to the user. Regarding the evaluation scale, a seven-level Likert Scale with equidistant intervals was selected for the method. One of the challenges regarding the method's design was to cover both dimensions of evaluation: the level of knowledge and the level of implementation. To deal with these two dimensions, the method is composed by two distinct parts, with different objectives: The Lean Assessment model and the Lean Maturity model. The two models can be used separately, since they have different objectives. However, when used in the same company, it makes the analysis more complete.

The Lean Assessment model aims to measure each employee's perception of Lean methods implementation in the company, as well as the personal knowledge of these methods and Lean fundamentals. This model does not aim to assess the implementation level, but only the individual perception about the level of implementation. In addition, it intends to assess the individual perception about Lean principles and tools. This model's output allows the comparison of perceptions among different elements of the company. The Lean implementation's level is not the main subject, but the focus on the individuals' willing to acquire more knowledge about Lean and/or to ask for a higher implementation level, despite the "real" one. The Lean fundamentals, vocabulary, methods, and problem-solving actions are the ones assessed in this model (Table 2). Each individual must classify each item according to a seven-level Likert Scale (Table 2). The inputs (answers) are computed using the fuzzy logic approach, in order to attenuate the ambiguity and imprecision which is characteristic from numerical values or exact linguistic expressions [12]. In this paper there is not enough space to explain the fuzzy logic parameters and equations, however they are explained in [25]. Finally, a visual representation is outputted where the results of different elements from different areas are plotted. In Fig. 1 the output from the model's application in a mold making company is presented. There are various conclusions after this model's application. However, the most obvious one is the distinguished perception of Lean implementation among the different elements. Matching this analysis *vis-à-vis*, the assumed knowledge about each topic (not presented), allows the identification of which actions should be taken regarding improvement and training practices.

The Lean Maturity model seeks to assess the importance associated to the practice of each Lean's fundament or method, along with the level of implementation. This model's questions approach the type of procedures and practices used in the company, with no direct question using Lean terms. Naturally, all the questions are linked with a Lean fundament or method, aiming to allow a comprehensive assessment. This model complements the Lean Assessment one, as it was designed to: (i) have an accessible vocabulary which facilitates its filling; (ii) evaluate all the organization's essential areas as well as the Lean related practices; (iii) permit to identify differences between distinct hierarchy levels; (iv) obtain a result that clarifies the areas in which to act.

Table 2 Lean concepts in the Lean Assessment model (a) and Knowledge and Implementation weights scale for Lean Assessment model (b)

(a)

Fundamentals	Just In Time (JIT)
	Jidoka
	Heijunka
	Standardized Work
	Kaizen
Vocabulary	Muda, Muri e Mura
	Work In Process (WIP)
	Lead Time
	Takt Time
	Overall Equipment Effectiveness (OEE)
Methods	5S
	Single Minute Exchange of Die (SMED)
	Total Productive Maintenance (TPM)
	Poka-Yoke
	Kanban
Problem-solving	PDCA (Plan, Do, Check, Act)
	Go to Gemba
	Kaizen Events
	Value Stream Mapping (VSM)

(b)

Knowledge weight		Implementation weight	
Very low	1	Very low	1
Low	2	Low	2
Relatively low	3	Relatively low	3
Medium	4	Medium	4
Relatively high	5	Relatively high	5
High	6	High	6
Very high	7	Very high	7

The Lean Maturity model is divided into importance weight (scale similar to Lean Assessment model) and implementation weight (seven-level scale based on frequency—never to always). The importance weight allows employees to express their point of view regarding the importance of each Lean foundation and method. The implementation weight expresses the company's performance in the described good practices. Thus, it is possible to understand the most relevant fundamentals and tools according to the company elements, as well as which wastes type and principles

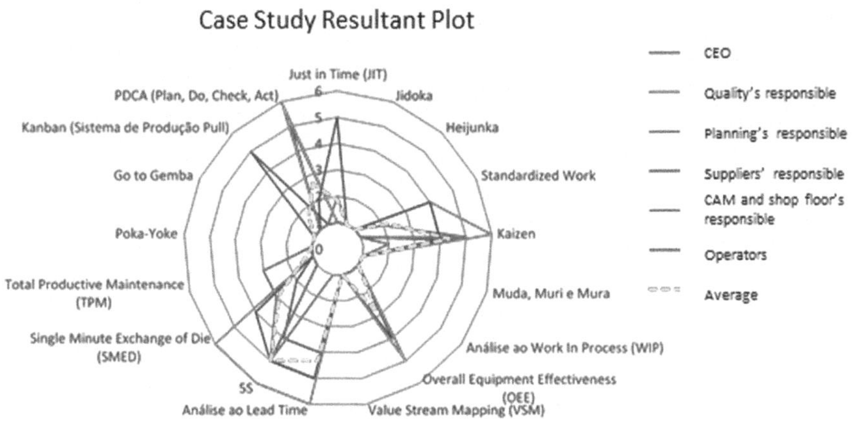

Fig. 1 Final output of Lean Assessment model for the case study company regarding the implementation level

are covered by these fundamentals and tools implementation. Fuzzy logic was also used to compute this model's inputs.

The model's questions were based on Lean principles, following a coherent logic and assigning a fair weight to each of the fundamentals. The TPS House [6, 8] was used as one of the foundations to assure comprehensiveness, since it is an organized way of presenting the fundamentals. In addition, the types of waste (*Muda*, *Mura* and *Muri*) were used for the questions' categorization, to ensure that all of them were addressed in the same way. The questions' organization were based on the traditional 5Ps model for Lean Manufacturing. The full questionnaire is presented in Table 3, where is possible to check the balance of all covered areas. Finally, the results are showed using the TPS House aiming to allow an easy and intuitive interpretation of the results. In Fig. 2 the application of the Maturity model to the case study company is presented. The result is showed using colors, complementing the quantitative value obtained from fuzzy logic, allowing the observation of how far is the performance to change for the next level (not represented in this paper). In this case, the CEO filled out the complete questionnaire and the heads of each department fill out only the questions related to their area. The obtained results matched the *in-loco* analysis where it was found that most of the Lean areas were considered with a high importance (good), although there were flaws in the implementation of people's empowerment strategies. Besides, collaborators focus on internal planning and technical/quality aspects, but they do not give enough attention to customer specific needs and to the right strategies for keeping the design and production activities stable during the weeks.

Table 3 Lean Maturity model questions and categories

5P	Question	3M	TPS House
Product	Do you evaluate the quality of the final product?	Defects/rework	Customer
	Do you use mechanisms which avoid the error?	Defects/rework	Jidoka
	Do you use quick response strategies to nonconformities?	Talent	Jidoka
	Do you use standardization strategies for the exchange and availability of information?	MURA	Standardization
	Do you ensure employees' alignment with the strategic objectives of the company?	–	Stability
	Are there mechanisms to disseminate the company's mission?	–	Stability
	Do you use systems which prevent error propagation (defects)?	Defects/rework	Jidoka
Plant	Are the safety aspects reviewed?	–	Customer
	Do you apply day-to-day visual management?	MURI	Involvement
	Do you promote the use of standardized procedures for the management of operational spaces?	Motion	Standardization
	Do you have internal transportation routes established?	Transportation	Standardization
	Do you check the equipment's operational performance?	Waiting	Jidoka
	Are the workstation's organization and tidiness audited?	Motion	Stability
	Do you evaluate the impact of component and product transportation?	Transportation	JIT
Processes	Are production costs evaluated?	MUDA	Customer
	Do you measure setup times?	Waiting	JIT
	Do you use formal problem -solving approaches?	MUDA	Jidoka
	Do you control process variability?	MURA	Jidoka
	Do you develop internal projects for continuous improvement?	MUDA	Involvement
	Do you apply Standardized Work formally?	MURA	Standardization
	Do you use production levelling methods (*Heijunka*, "one-piece-flow")?	Overproduction	Standardization
Planning	Do you evaluate operations' Lead Time?	Waiting	Customer
	Do you confirm Pull production logics' application for all products?	Overproduction	JIT

(continued)

Table 3 (continued)

5P	Question	3M	TPS House
	Do you analyse the product flow?	Inventory	JIT
	Do you confirm the Takt time accomplishment?	Waiting	JIT
	Do you analyse the WIP (Work in Progress)?	Inventory	JIT
	Do you check the existence of unnecessary operations/processes from the customer's perspective?	Over processing	Customer
	Do you check if there is equipment overload, and there is another without load?	MURI	Stability
People	Are employees' levels of motivation measured?	Talent	Customer
	Do you develop strategies for employees' involvement and decision autonomy?	Talent	Involvement
	Do you support teamwork in projects?	Talent	Involvement
	Do you promote employee's flexibility in terms of skills?	MURI	Involvement
	Do you use strategies which promote "doing well at first"?	Defects/rework	Stability
	Do you promote cross-audits to identify redundancies?	Over processing	Involvement
	Do you verify the levels of exhaustion and pressure on employees?	MURI	Stability

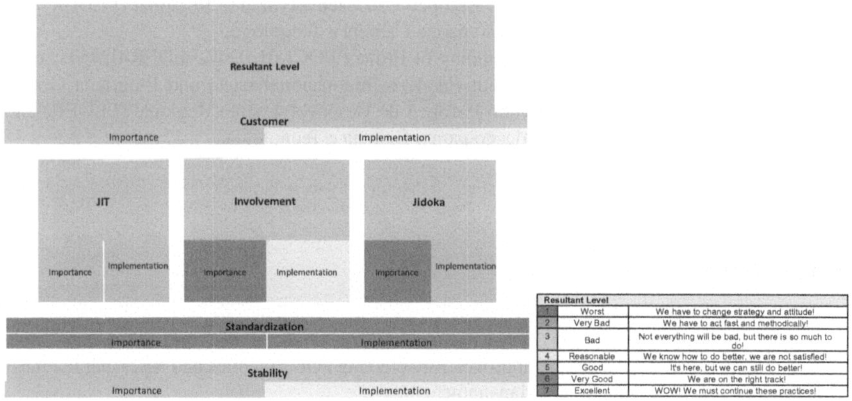

Fig. 2 Final output of Lean Maturity model for the case study company

4 Conclusions

This work aims to propose a new Lean evaluation method, based on the identified gaps on the existing Lean performance evaluation methods published. Comparing with the reported models, the one proposed guarantees a consistent and measurable evaluation, a balance between the model extension and the evaluation detail, and an easy filling. The several aspects identified during the literature review were covered by the application of the two different models. They address all the essential Lean fundamentals and tools allowing the company a complete analysis. Regarding the models' results, both helped choosing the most appropriate improvement actions to be applied in the company, clearly exposing the most needed areas. The developed Lean Assessment model's goal is to measure each employee's perception of personal knowledge and the company's implementation of Lean fundamentals and tools. Compared to the existing ones, this model focus on people and compares perceptions and knowledge among the various participants. It confirms to be a model of rapid application and results. The developed Lean Maturity model's purpose is to assess the importance assigned to each good practice, together with the frequency of implementation in the organization.

The method was applied in a company, as the *in-loco* assessment confirm the method results.

Acknowledgements This work was supported by FCT, through IDMEC, under LAETA, project UID/EMS/50022/2019.

Authors gratefully acknowledge the funding of Project POCI-01-0247-FEDER-017637, cofinanced by Programa Operacional Competitividade e Internacionalização and Programa Operacional Regional de Lisboa, through Fundo Europeu de Desenvolvimento Regional (FEDER) and by National Funds through FCT—Fundação para a Ciência e Tecnologia.

Authors gratefully acknowledge the funding of Project POCI-01-0247-FEDER-024541, cofinanced by Programa Operacional Competitividade e Internacionalização and Programa Operacional Regional de Lisboa, through Fundo Europeu de Desenvolvimento Regional (FEDER) and by National Funds through FCT—Fundação para a Ciência e Tecnologia.

References

1. Womack JP, Jones DT, Roos D (1992) The machine that changed the world. Bus Horiz
2. Al Amin M (2013) A systematic approach for selecting lean strategies and assessing leanness in manufacturing organizations. Management
3. Susilawati A, Tan J, Bell D, Sarwar M (2015) Fuzzy logic based method to measure degree of lean activity in manufacturing industry. J Manuf Syst
4. Womack JP, Jones DT (1997) Lean thinking—Banish waste and create wealth in your corporation. J Oper Res Soc
5. Yazdani B (2003) Toyota production system: an integrated approach to Just-In-Time. Comput Integr Manuf Syst
6. Nesensohn C, Bryde D, Ochieng E, Fearon D, Hackett V (2014) Assessing lean construction maturity. In: 22nd annual conference of the international group for lean construction: understanding and improving project based production, IGLC 2014

7. Becker J, Niehaves B, Pöppelbuß J, Simons A (2010) Maturity models in IS research. In 18th European conference on information systems, ECIS 2010
8. Urban W (2015) The lean management maturity self-assessment tool based on organizational culture diagnosis. Procedia Soc Behav Sci
9. Cronemyr P, Danielsson M (2013) Process management 1-2-3—a maturity model and diagnostics tool. Total Qual Manag Bus Excell
10. Vinodh S, Balaji SR (2011) Fuzzy logic based leanness assessment and its decision support system. Int J Prod Res
11. Srinivasaraghavan J, Allada V (2006) Application of Mahalanobis distance as a lean assessment metric. Int J Adv Manuf Technol
12. Vinodh S, Vimal KEK (2012) Thirty criteria based leanness assessment using Fuzzy logic approach. Int J Adv Manuf Technol
13. Hosseini Nasab H, Aliheidari Bioki T, Khademi Zare H (2012) Finding a probabilistic approach to analyze lean manufacturing. J Clean Prod
14. Da Wan H, Frank Chen F (2008) A leanness measure of manufacturing systems for quantifying impacts of lean initiatives. Int J Prod Res
15. Zanjirchi SM, Tooranlo HS, Nejad LZ (2010) Measuring organizational leanness using Fuzzy approach. In: Proceedings of the 2010 international conference on industrial engineering and operations management
16. Gonçalves MT, Salonitis K (2017) Lean assessment tool for workstation design of assembly lines. Procedia CIRP
17. Bayou ME, de Korvin A (2008) Measuring the leanness of manufacturing systems-a case study of Ford Motor Company and General Motors. J Eng Technol Manag JET-M
18. Sânchez AM, Pérez MP (2001) Lean indicators and manufacturing strategies. Int J Oper Prod Manag
19. Behrouzi F, Wong KY (2011) Lean performance evaluation of manufacturing systems: a dynamic and innovative approach. Procedia Comput Sci
20. Doolen TL, Hacker ME (2005) A review of lean assessment in organizations: an exploratory study of lean practices by electronics manufacturers. J Manuf Syst
21. Vinodh S, Chintha SK (2011) Leanness assessment using multi-grade fuzzy approach. Int J Prod Res
22. Chan FTS, Qi HJ (2003) An innovative performance measurement method for supply chain management. Supply Chain Manag
23. Pakdil F, Leonard KM (2014) Criteria for a lean organisation: development of a lean assessment tool. Int J Prod Res
24. Soriano-Meier H, Forrester PL (2002) A model for evaluating the degree of leanness of manufacturing firms. Integr Manuf Syst
25. Barros M (2019) Lean performance evaluation: models and application. Instituto Superior Técnico, Universidade de Lisboa, Portugal

JIT Implementation in Manufacturing: The Case of Giacomini SPA

Bassel Kassem, Federica Costa and Alberto Portioli-Staudacher

Abstract *Purpose*: The paper aims at exploring a JIT implementation framework integrating the hard and the soft pillars of Lean Production in a manufacturing company. *Design/methodology/approach*: A JIT cell creation project is analyzed in Giacomini using a Learn, Use, Teach and Inspect, LUTI JIT framework. *Findings*: Any continuous improvement project should embrace its actors in a lean culture for it to succeed, that is grasping the "who" and the "why" aspect. *Practical implications*: The main contribution of the article is showing empirically, the importance of the "who" and the "why" aspect in successfully transforming a batch-and-queue production system into a Just-In-Time system. *Limitations*: The framework is applied in one company and one sector which may hinder generalizability issues.

Keywords Lean production · Lean culture · Lean education · JIT · LUTI

1 Introduction

In 1950, Toyota Chief Engineer Taiichi Ohno was entrusted to improve the Toyota production system to reach the same productivity level of American competitors, such as Ford. Ohno overcame the expectations through the creation of Toyota Production System (TPS) otherwise known as "Lean Production" (LP) [8]. LP has two main pillars: "Respect for people" and "Continuous improvement". The former pillar was less emphasized by managers until the publication of "The Toyota way" by Toyota corporation in 2004 [7] that showcased that the first pillar intrinsically enables the second one [3]. "Respect for people" advocates people's engagement and active

B. Kassem (✉) · F. Costa · A. Portioli-Staudacher
Dipartimento di Ingegneria Gestionale, Politecnico di Milano, Via lambruschini 4, 20156 Milano, Italy
e-mail: bassel.kassem@polimi.it

F. Costa
e-mail: federica.costa@polimi.it

A. Portioli-Staudacher
e-mail: alberto.portioli@polimi.it

M. Rossi et al. (eds.), *Proceedings of the 6th European Lean Educator Conference*, Lecture Notes in Networks and Systems 122,
https://doi.org/10.1007/978-3-030-41429-0_27

participation in value creation through a culture of learning and embracing change. The role of management in instilling trust between the different actors of the organization is crucial for continuously improving company's operations, hence the second LP pillar [7].

Just in Time JIT is one of LP hard practices. It is focused on having the right product at the right time in the right quantity to the right customer [8, 9] and requires people's involvement in its implementation. The literature review presents frameworks that deal with Lean transformation, when it comes to JIT implementation frameworks, those are not JIT specific but rather of generalized nature that authors tend to apply it to JIT. The main contribution of the article is the creation of an adhoc framework for transforming a batch-and-queue production system into a Just-In-Time system, by coupling LP's two pillars in a Lean project at Giacomini SPA. The "LUTI JIT framework" is a JIT-specific framework, where LUTI stands for Learn, Use, Teach and Inspect, an acronym which explains that the model comes from the experience of a Lean expert and then it is turned into a standardized, replicable model.

This paper is structured as follows: after the introduction, Sect. 2 discusses the literature regarding the JIT implementation frameworks followed by Sect. 3 that introduces the project at Giacomini and its results using the framework suggested. The article concludes with important considerations and implications.

2 Literature Review

This section highlights the most relevant articles in the literature discussing JIT implementation frameworks found in Scopus that are available to download and their access is not restricted by their authors.

Harvey [6] proposed eight implementation phases for JIT: (1) establish a focused factory; (2) streamline the order flow; (3) adopt good manufacturing techniques; (4) improve material availability; (5) plant-adopted standards, including information, design, process, and equipment standards; (6) concern with the quality improvement; (7) attention focused on improving skills and awareness, and upgrading the responsibilities of the employees, and (8) moving closer to customers. However, it is done with no explanation of these steps in an exhaustive way but rather describes them in generic terms.

Gélinas [4] believes that project management is a suitable tool to properly implement JIT: planning, controlling and monitoring can be also applied for JIT as shown in Fig. 1.

The framework constitutes of development and realization. Development starts by defining the problem, followed by evaluating the situation and possible solution to the problem. Then feasibility is performed, an evaluation of physical, managerial and financial constraints followed by the implementation audit. It enables the verification of the company's position according to the drivers required for a successful JIT implementation. The result of this step is the description of the strengths and weaknesses of the company according to a list of managerial and operational key

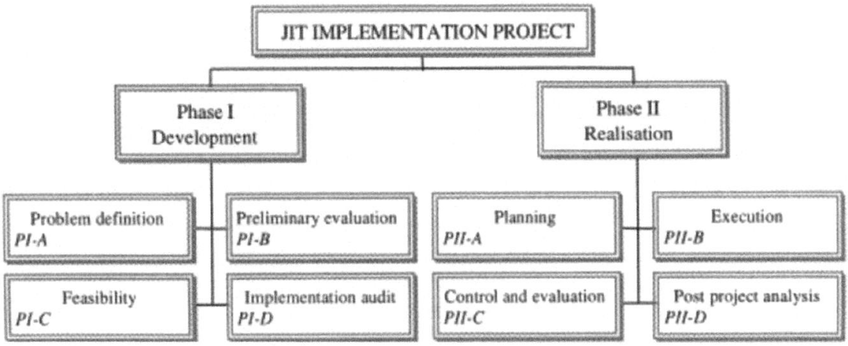

Fig. 1 Project management framework for JIT implementation

factors. The realization phase, on the other hand, deals mainly with the planning and the execution. The planning part defines the activities to be implemented, their sequence, and the related resources needed—both financial and human. Then, after the execution of the planned activities, a control and evaluation of the ongoing activities is done. Finally, a post project analysis is performed. This project management is advantageous; it forces a deep, structured analysis about the prerequisites of each phase and defines the operations sequence in a systematic way, without losing energy and time in unnecessary activities.

Van Aken et al. [11] developed a model for designing, managing and improving Kaizen event programs. This differs from the former, in the sense that it is a cycle rather than a process and can be used for any kind of LP activity according to the steps to follow:

- Plan: identification of the stakeholders involved, the selection of the Kaizen event to be improved and the definition of the project charter.
- Implement: event preparation and execution in addition to follow-up after event and full-scale deployment.
- Sustain: reviewing the results and share them, through a standardization of the best practices.
- Support: education and motivation of employees as they are critical activities which should be done alongside the whole project.

Wong et al. [13] instead, creates a very comprehensive model which, starting from the 5 principles of Lean mentioned in [12], defines three action areas (Think Lean, Act Lean and Human) which leads to the implementation of Lean into the whole company. The most relevant paper is that of [2]. The paper discusses four main variables: the "what" (e.g. what are the techniques to be used), the "how" (e.g. how to implement them), the "who" (e.g. who is going to perform a specific task) and the "why" (e.g. the reason beyond a specific activity). These four elements are discarded by the surveyed papers described previously, but they shall be relied upon for the project conducted in Giacomini SPA.

Fig. 2 LUTI JIT framework

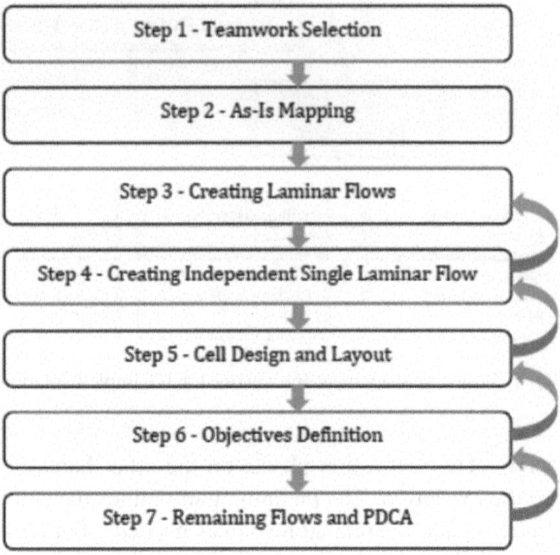

The LUTI framework, which is JIT specific, follows seven main steps (see Fig. 2) preceded by a preliminary phase, indicated as Step 0. This phase is the preparation of the activities, which includes: (i) the selection of the project, based on the Gantt diagram listing all the improvement projects forecasted for the year, (ii) the analysis of the product codes involved and their volumes and (iii) the preparation of the module for the time registering.

Step 1 is the Teamwork selection. According to the typology of project and the related skills level required, a team is created. Indeed, the lean specialist should not work alone. He will require the support of the maintenance technician, the programmer, the line responsible, the department responsible and the operators.

Step 2 is the as-is mapping and a detailed definition of the process map, registering all the transformation phases of the products. In this step, it is fundamental working directly on the shop floor, in Gemba.

The data to record are the production volumes, the characteristics of each working stations (C/T, C/O, machine availability, OEE, process phase description), human resources per machine and the product travel distance.

Step 3 consists of creating laminar flows for each product family, laminar in the sense that the flow is smooth just like particles. This is a theoretical step: what will be the ideal flow of the products with totally dedicated and independent resources?

Step 4: the laminar flows created in the Step 3 should be grouped together in a single laminar flow, which include all the families' products. Then, the model requires to account the Takt Time and to compare it with the C/T of the new line.

Step 5 is the cell design and layout. It is composed by six sections: (i) collection of the scaled layout of the department, (ii) realization of the shapes for machines and equipment, (iii) creation of a possible layout per person considering physical (e.g. presence of pillars), ergonomic (e.g. considering the golden zone rule in the working stations) and technical (e.g. feasibility of the electric current provisioning), (iv) comparison of the layouts and selection of the best alternative, (v) share of the final layout with the whole teamwork and (vi) testing the layout in a simulated environment.

The Step 6 where a list of objectives should be prepared to assess the main differences: lead time, WIPs, travel distance, number of resources, and economic evaluation.

Step 7 is related to the remaining flows—the ones have been removed from the JIT cell due to low production volumes—and how to treat them. An ad hoc solution may be necessary. After this activity, the project can be presented to the top management. Once the approval is received, the team should prepare a detailed PDCA for the cell realization and kick off the project. The process is shown above.

Most of the frameworks seen, focus only on the "what" (e.g. the techniques of TPS).

(1) The majority of the analyzed frameworks do not describe the sequences and the phases to transform a company in a Lean organization [1, 2].
(2) Most of the models do not indicate the "who" of a specific step, the specific internal stakeholders to which assign a task [1, 2].

Anand and Kodali [1] tackled the "how" as well. They deeply explain what to do and how to do it, describing the specific steps to follow. Chay et al. [2] build on their work, adding another dimension of analysis, which is missing in [1]: the "why" factor. They review 30 frameworks in the literature and believe that blindly following certain steps does not mean transforming a company into a Lean enterprise. Otherwise, every company could do it perfectly. Since each company has specific external and internal contexts, the main point is to understand the reasons of a certain technique and model, before implementing it.

The last consideration taken from [2] is related to the bottom-up versus top-down approach. All frameworks starting from a management input increase the risk of fail due to the poor workers' participation.

3 Methodology

The importance of the "who" and "why" part in any JIT implementation process is evident in a project conducted at Giacomini. This section introduces the company and the project conducted on its site applying LUTI JIT framework comparing it with the works of [2].

Giacomini S.p.A. is a world leader in the production of components and systems that use water to produce comfort, to manage water usage "in situ" and to extinguish fire. The Group has a well-established international presence, with new subsidiaries and sales offices opening around the world. Brazil, Thailand, Canada, India and Turkey are the most recent examples. Giving a few numbers, about 200 million in revenue, more than 80% of it coming from foreign markets, with 3 Italian production

Fig. 3 Giacomini house of lean

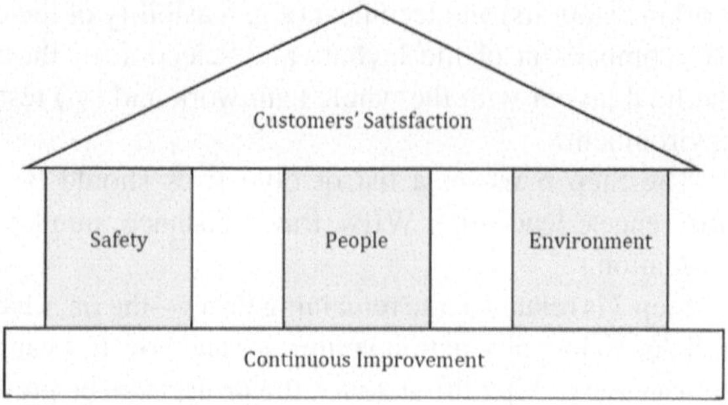

facilities, 18 international subsidiaries, 900 employees and 70 tons of brass processed each day. The company implemented Lean as both a need and a vision [5]. On one side, they pursue efficiency to get costs reduction. On the other, they believed that Lean could be more than a simple set of rules: a source of competitive advantage and a source of innovation that led Giacomini towards a next step in its evolutionary process. The company created for itself a "house" (see Fig. 3) embodying everything that they stand for: Safety, People and Environment, with "Continuous improvement" as its essence and customer satisfaction as its objective.

One of the crucial steps adopted by the company in the Lean implementation is the communication between top management and workers. Communicating means creating a better sense of belonging. Another tool used to improve the relationship among management and workers is the "Plant of ideas" or as it is called in Italian "Fabbrica delle idee", a document which can be filled by the operators with suggestions, ideas and comments about their job. The proposal of interesting ideas will be rewarded with a certain amount of points. The points collected can be spent in recreational activities.

Students of Polimi participated in the project conducted by Giacomini with the support of Lean Excellence Center of Polimi by creating JIT cells using "LUTI JIT framework", one of the pillars of World Class Manufacturing WCM [10].

The first aspect emerged by the LUTI JIT framework implementation is the presence of a strong iterative component. However, the implementation on the ground reveals the need to go back to the previous steps, according to the discovery of a specific constraint. In particular, the final steps require to consider physical, economical and managerial limitations, that could require moving back to the previous phases.

A main issue is highlighted: the LUTI JIT framework theoretically does not explain the "why". Comparing the LUTI JIT model with the considerations done by [1] and [2], the framework can be analyzed according to four main variables: the "what" (e.g. what are the techniques to be used), the "how" (e.g. how to implement them), the "who" (e.g. who is going to perform a specific task) and the "why" (e.g. what the reason beyond a specific activity). The LUTI model deeply defines the techniques, their implementation and describes the tasks' responsibilities throughout its steps as well as in the PDCA procedure used in the Step 7. However, it shows a lack of "why". Of course, this missing element is not present in the top management. They are perfectly aware of the reasons, advantages and limitations of a JIT production system. The workers may see the framework as a simple change in the operations and in their sequence. Their involvement is mainly in the layout phase, not in the conceptual one. They may understand the idea of the cell manufacturing but will miss the connection with the comprehensive picture: Why a JIT system is needed, why Lean improves the AS-IS situation and, most importantly, why working in a JIT line improves their working conditions. The main reason is the instillation of the Lean culture. If Lean was a simple set of rules to be blindly followed, all the companies could reach the objectives of costs reduction, higher quality and shortest lead times. Instead, there is more. The main difficulty is inside the company culture. Obtaining the proper mind-set, both in the management and in the workers, is harder than instructing an operator about a specific task. The Lean methodology requires a complete mentality change.

In Giacomini, this cultural evolution started few years ago, and, as every long-term process, it started producing some results, despite the presence of some constraints. Another consideration to have in mind in comparison with the work of [2] is the "who" aspect. This driver of analysis examines who is going to perform a specific task. Two main opposite approaches can be found among the possible ways to assign tasks to the workers. The first is the "top-down" approach: the assignment of tasks to each person is driven by the top management. Thus, the operative workers simply receive the instructions about the new operating standards, without being active actors in the transformation process. They receive orders from the managers, without having a clear idea about the context where they are working. On the other side, the "bottom-up" approach relies on the operative workers from the initial phases of a transformation process. Sometimes, they could be the initiator of the project itself. The managers' role is to coordinate the project with a high degree of delegation to the workers.

The LUTI JIT framework shows a good trade-off between a top-down and a bottom up approach. The model initially receives a strong input from the management. Indeed, the first four phases do not involve so much the operators while Step 5 is almost centered on the workers' participation.

Figure 4 shows the qualitative involvement of the operators against the managerial support (accounted dividing the number of hours dedicated by the people involved in each step of the framework). The communication among management and operators follows a similar path, increasing in the fifth step. Some possible suggestions about how to improve the operative workers' involvement are: increasing the communication between the two categories since the initial phases (e.g. weekly meeting could be useful for this scope); increasing the communication among the whole company's workforce (e.g. company journals address this point) and constantly providing training to the operative workers about the Lean techniques and principles, with a specific focus to the models and tools used in Giacomini. The proportion among the two typologies of workers is 60 h spent by a manager against 300 h spent by the operative workers, meaning less than one third of the time spent.

Since the nature of the LUTI is to be improved and adjusted every time it is applied, the previous analysis about the "who" driver leads to the following elaboration.

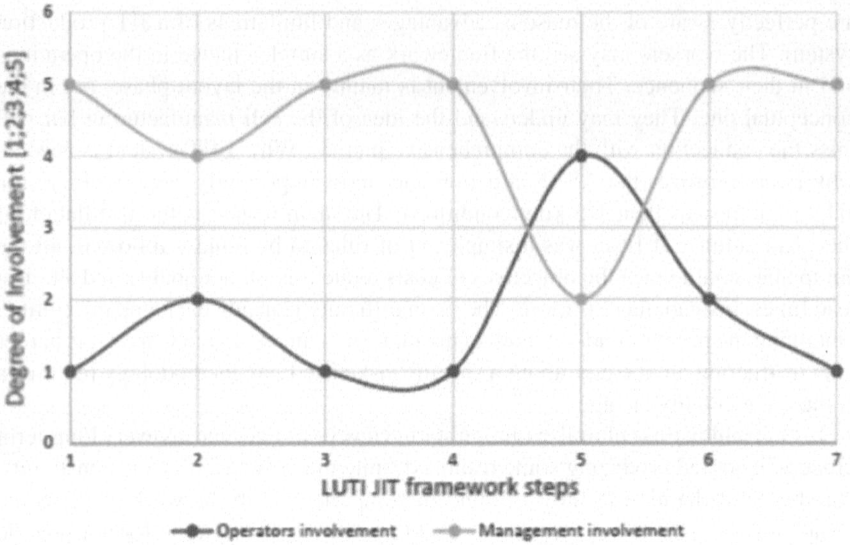

Fig. 4 Distribution of workload for the project

The Giacomini application suggests dividing the "who" driver into two parties: the manager, who is leading the process, and the operative workers, who are effectively implementing the transformation tasks. In particular, in the PDCA documents the distinction between these two typologies of workers should be evident. The main benefit of this improvement is the time saving achieved by the manager, who can dedicate time to other projects as well as the improvement of workers' performances and motivation.

Finally, among the variables analyzed by the paper of [2], the time dimension was not considered. The reason is related to the characteristics of each project: the complexity of the process directly affects the timeline. However, some considerations about the Giacomini case can be done. The financial savings brought by the JIT cells months has surpassed the expectations. Apart from the financial side, the operational advantages of the lines included: reduction of lead time, reduction of WIP along the line, reduction of the human resources and improving of material handling. In only 2 months, the Lean Manager was able to achieve incredible results. The main reason of this success is having a well-working model to reach the objectives is a facilitator element.

4 Conclusion

The presence of a fixed schema to follow, with a very high level of specificity and details, enables the Lean Manager to delegate to the Lean Assistants the less strategic activities, without anyway losing the control on the project development. The LUTI JIT framework proves how the usage of models is fundamental in the realization of a Lean transformation with the necessary involvement of both white and blue collars in all phases of the project. The Giacomini Group took the right way, which is the creation of ad-hoc transformation structure, perfectly suitable for the company organization.

This article is a testimony of the importance of including both the hard and the soft part of Lean whenever a project of continuous improvement is to be held in a manufacturing context. On the one hand it is essential to have a lean culture instilled in the company and on the other hand people are the core of LP. Their contribution would bring success when involved actively in the processes and the contribution of the workers have been increasing ever since in Giacomini leading to greater results.

The main limitation of the research is that it is based on one case study, but the team of LEC is continuously performing improvement projects in various Italian companies and there is the possibility of generalizing the findings with other projects that are ongoing as well.

Acknowledgements We would like to thank Giacomini SPA represented by Mr. Emanuele Peraldi along with their management and their operators for their support of this project. The authors would like also to thank the Lean Excellence Center team and the students of Politecnico.

References

1. Anand G, Kodali R (2010) Analysis of lean manufacturing frameworks. J Adv Manuf Syst 9(1):1–30
2. Chay T, Xy Y, Tiwari A, Chay F (2015) Towards lean transformation: the analysis of lean implementation frameworks. J Manuf Technol Manage 26(7):1031–1052
3. Emiliani ML, Stec DJ (2005) Leaders lost in transformation. Leadersh Organ Dev J 26(5):370–387
4. Gélinas R (1999) The Just-In-Time implementation project. Int J Project Manage 17(3):171–179
5. Giacomini SPA, https://it.giacomini.com/ (visited on 10/06/2019)
6. Harvey RE (1986) How to deliver Just-in-Time. Manuf Manage 229(18):36–40
7. Liker JK (2004) The Toyota way: 14 management principles from the world's greatest manufacturer. McGraw-Hill
8. Shah R, Ward PT (2003) Lean manufacturing: context, practice bundles, and performance. J Manage Oper, 129–149
9. Shah R, Ward PT (2007) Defining and developing measures of lean production. J Oper Manage, 785–805
10. Sukarma L, Azmi H, Abdullah NL (2014) The impact of world class manufacturing practices on company performance: a critical review. Appl Mech Mater 564:727–732
11. Van Aken EM, Farris JA, Glover WJ, Letens G (2010) A framework for designing, managing, and improving Kaizen event programs. Int J Prod Perform Manage 59(7):641–667
12. Womack J, Jones D (1996) Lean thinking. Banish waste and create wealth in your corporation. Free Press, Simon & Schuster, Inc., New York, NY
13. Wong YC, Wong KY, Ali A (2010) A study on lean manufacturing implementation in the Malaysian electrical and electronics industry. Eur J Sci Res (2009) 38(4):521–535

Lean and Sustainable Continuous Improvement: Assessment of People Potential Contribution

Matteo Rossini, Alberto Portioli-Staudacher, Fabiana Dafne Cifone, Federica Costa, Fabio Esposito and Bassel Kassem

Abstract Many industries all over the world have been increasingly adopting Lean Management practices, from manufacturing to service sectors. Notwithstanding the demonstrated astonishing benefits in operational performance of Lean, many firms strive to successfully sustain Lean and its gains in the long-term. Many authors in the literature relate these failures to an exasperate effort to correctly implement the so called "practices" of Lean. This excessive focus on practical aspects led companies to completely neglect the human side of developing people in order to create a base for sustainable continuous improvement system that Lean implementation entails. For that reasons, increasing attention has been placed by academic researchers towards the importance of the human side in successfully sustaining Lean implementation and Continuous Improvement Programs. Despite this increased attention, only few studies have tried to analyze this aspect from a broader and comprehensive perspective. Indeed, before the contribution of this dissertation, no studies in the literature have been able to identify a comprehensive model for assessing the development of people's potential to contribute at the sustainable continuous improvement. This paper represents a preliminary research attempting to fill this gap, proposing a new assessment model for monitoring people development.

Keywords Lean · Sustainable continuous improvement · Assessment model

1 Introduction

Even if Lean is recognized as one of the major managerial approach for leading companies to success [1], some author argue that wrong application of the tools and

M. Rossini (✉) · A. Portioli-Staudacher · F. D. Cifone · F. Costa · B. Kassem
Department of Management, Economics and Industrial Engineering, Politecnico di Milano, Via Raffaele Lambruschini, 20156 Milan, Italy
e-mail: matteo.rossini@polimi.it

F. Esposito
Ghelfi Ondulati Spa, via dei Lavoratori n. 10, 23010 Buglio in Monte, Sondrio, Italy

M. Rossi et al. (eds.), *Proceedings of the 6th European Lean Educator Conference*, Lecture Notes in Networks and Systems 122,
https://doi.org/10.1007/978-3-030-41429-0_28

techniques inherent to lean philosophy may also entail wasting resources up to 30% more than before [2].

These outcomes partially lie in the nature of lean implementation, which is complex, time-consuming, and needs a substantial amount of human resources and effort [3, 4]. In the literature, failures in implementing lean are commonly related to the complexity of implementation, lack of considering other operations management practices, not coherent planning and strategy, and difficulty in adapting the concepts to the specific context [5]. Companies usually do not really understand lean concepts, thus making critical mistakes like focusing on reducing some forms of wastes and disregarding others [6]. Tortorella and Fogliatto [7] reports that sometimes a superficial approach leads lean programs to fail, while [8] shows the existence of inconsistencies between what companies really feel important about lean and how these factors are executed. However, the real problem is that companies fail in sustaining the change lean brings in the organization, believing that the first results achieved will last forever, and ever better ones will come without improvements or with unsustainable ones. Thus, two major problems affect lean implementation: (i) not improving what already done or failing in continuous improvement, which is one of the key principles of lean; (ii) not creating a robust and large base of people capable and devoted to continuous improvement process.

Then, this preliminary research focuses on the latter point, aiming at creating an assessment model for evaluating the real strength of a Lean organization that is composed by the employee participating in the continuous improvement process. In order to do that, a case study has been carry out and preliminary results are presented in this paper.

2 Background

Despite the clear evidence highlighting the benefits related to Lean Manufacturing, the literature suggests that many companies fail in the implementation of such a managerial system, thus failing in gaining sustainable competitive advantages against competitors. Indeed, the adoption of Lean tools and techniques does not automatically bring to a successful implementation of Sustainable Continuous Improvement (SCI) system that is sustainable over the long-term. Many companies all around the world struggle with the implementation of Lean Management practices, meaning that simple implementation of the different tools does not ensure sustainable increased performance. The evidence suggests that two out of three organizational change processes fail [9] and in many cases companies that fail in implementing Lean return back producing according to their traditional means. The isolated use of tools such as 5S, SMED, JIT and other techniques, could bring strong improvement of performances that cannot be lead to SCI system if the company is not able to change the organizational culture [10].

The performance advantages that lean can enable are strictly related to a strong commitment to continuous improvement enabled by people development [11]. Continuous improvement consists of highly frequent minor changes carry out by many that, added up, may entail a revolution, and result in a positive impact on performance [12]. Some authors consider it a culture concerned with quality as an integral part of the processes within the organization [13]. Similarly, Rossini et al. [14] state that continuous improvement is a pervasive culture that focuses on eliminating waste in the processes within all organization's levels, requiring everyone to look for problem root causes and sources of variation, and try to get rid of them.

The benefits of implementing continuous improvement programs are many and involve the company from a broad perspective [15]. As mentioned above the organizational learning is the basis for continuous improvement, but the relationship is still valid if observed at the opposite. CI programs encourage organization to undertake a process of learning through which continuously revising assumptions and values, and triggering new problem-solving schemas [16]. Employees come to be more positive and satisfied, showing a greater sense of responsibility and desire to learn and develop new skills, as well as a deeper understanding of the impacts of their actions on the process as result of CI [17]. A strong decrease of turnover and absenteeism can be observed, in addition to a higher productivity and better quality [18].

However this process of change is not simple, and nowadays it has been recognized that measuring level of leanness only looking at practices is not enough. Companies must be able to measure their strength in doing continuous improvement.

Though Continuous improvement is extremely important and has a huge impact on companies from a very wide perspective, its implementation results difficult and particularly hard to sustain in the long term [19]. What is really missed is a sustainable implementation of lean and continuous improvement. Sustainability is a quite ambiguous term, which is defined in several different ways in the literature. As an example, Flumerfelt et al. [20] argue that sustainability is nothing other than an employee based process improvement. Other authors affirm that sustainability is the company's ability of developing new knowledge to cope with ever coming problems and inefficiencies. The company must be capable of involving employees to successfully sustain a change, and pushing them to think and learn so to increase knowledge, thus reaching what is usually referred to as learning organization [21]. From a completely different perspective.

Therefore, there is not a clear pattern in the literature on how to measure leanness level and SCI in the company and this paper proposes a new point of view, which is focused on SCI and, more in detail, on the people and their potential impact in the creation of a SCI system.

3 Methodology

Case studies present a fundamental feature in respect to the actual research to perform. They can rely on a variety of sources of evidence that usually are not available for other explanatory methods [22]. In fact, case studies can leverage information coming from documents, artefacts, interviews of people part of the study and direct observation of the studied events [22]. This characteristic is fundamental for studying a current phenomenon like Industry 4.0 that is still characterized by a lot of uncertainty. Lastly, differently from what happens with experiments, in which the investigator has full or partial control over variables, case studies do not require control of behavioural events [22]. Then, the case study is the method that best fits the previously mentioned criteria and, hence the final aim of the research.

4 Case Study

4.1 The Company

Ghelfi Ondulati S.p.A. (Ghelfi) is an Italian SME (small-medium enterprise) operating in the corrugated cardboard industry since 1952. Ghelfi mainly operates in Italy in fruit & vegetables and food industries, but it has also an important customer base in Europe (in particular in France) and some clients in northern Africa. The company represents the stereotype story of today's Italian companies: a successful SME, created by scratch of an entrepreneur after the Second World War and now led by the third generation of the family. In this context, the governance (composed by the family and few other employees) developed a great vision for the company: developing people capabilities, at all levels, in order to create the base of management for the future and so to prepare the exit of the family from the operative management of the company.

This vision led Ghelfi bet on Lean journey for the growing of its employee.

4.2 The Case

A Lean program has been set and many people have been involved. The company did not focus on the operational performance results (let's say, not only), the company wanted to measure and evaluate how people were becoming stronger in company working life, how they were proactive and successful in improving Ghelfi processes. Within this point of view, traditional measurement, based on operational performances, appeared not sufficient anymore.

The necessity of a new measurement system model that includes the growth of employees showed-up. This research is the preliminary representation of a new measurement system model developed by the Lean Excellence Centre of Politecnico di Milano that assesses employee's capability in pursuing continuous improvement.

4.3 *The New Measurement System Model*

The growth of the people has been measured along two dimensions: a vertical dimension, named "people strength", which considers the capability of people in problem solving and in leading, and a horizontal dimensions, named "people view", which considers the extension of an employee experience in different areas of the company.

4.3.1 People "Strength" Dimension

This dimension evaluates employee growth in three main topics: Lean techniques (LT), A3-PDCA (A3), Leadership capability (LE).

LT's sphere assesses employee's capability in the knowledge and the correct use of traditional Lean techniques (i.e. 5S, SMED, Pareto analysis …).

A3's sphere assesses employee's capability of problem setting and of pursuing correctly the Deming cycle, i.e. passing through a robust analysis instead of jumping to solution.

LE's sphere assesses employee's capability of leadership: the emphasis of this dimension is the growth of a person that changes from being an agent of change to be a facilitator and a coacher for new agents of change, from being a person that is led to be a person that is a leader of the continuous improvement process. Figure 1 depicts an assessment for one experienced employee in Ghelfi.

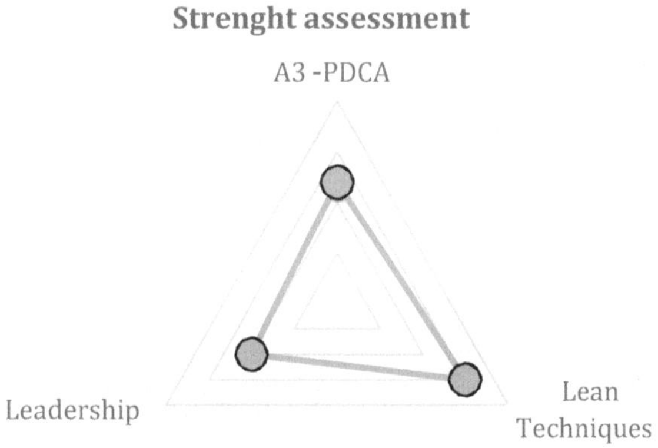

Fig. 1 Single person assessment model for the "strength" dimension

Table 1 People view assessment for 10 employees of Ghelfi

People	Company functions						
	Sales-indust.	Sales-fruit	Logistics	Purchasing	Production	Tech. design	R&D
AB				O	X	O	
CD					X	O	O
EF							X
GH				X	O	O	
IL				O	O	X	
MN	O	X	O				
OP	O	O	X				
QR	O	O	O	O	X	O	O
ST	X	O	O				
UV	O	X	O				

X: in charge of that function; O: in contact through improvement projects

4.3.2 People "View" Dimension

This dimension evaluates employee's growth in overall knowledge of the company. With the overall knowledge we mean the consciousness of people of what happens (procedures, people, processes) in company's functions different from the one where they are used to work. The importance of this dimensions comes-up from the experience that the biggest opportunity of innovation and of improvement lay in the interrelation between different departments. So, the wider the employee's knowledge of the company businesses, the bigger the capability of the employee of proposing and pushing continuous improvement. Table 1 represents the assessment of a Ghlefi's employee view.

5 Conclusions

Although more than two decades have passed since the publication of Womack's book, both academic and practitioners' worlds are still in strong consensus with these argumentations. In their literature review [23] state that the continuous improvement process changes the role of employees and the separation between the so called white-collar and blue-collar is no longer possible. Employees are expected to conduct the continuous improvement themselves, embracing and internalizing the company's problem-solving procedures and should be motivated to approach the daily improvement as a natural behavior, driven by a wish of personal development and the sense of achievement. Hence, employees' engagement and long-term development is a fundamental characteristic for the sustainability of Lean Production Systems. According to the centric view of people development, it is clear the necessity to have a measurement system which support this view. This paper attempts to fill this lack in the literature and propose a preliminary research conducted in a real case. An evaluation framework based on two dimensions has been proposed and the aim of this research is to continue to investigate in order to create a complete and more robust framework that practitioners could use for assessing their SCI systems.

References

1. Costa F, Lispi L, Staudacher AP et al (2018) How to foster sustainable continuous improvement: a cause-effect relations map of Lean soft practices. Oper Res Perspect. https://doi.org/10.1016/j.orp.2018.100091
2. Pavnaskar SJ, Gershenson JK, Jambekar AB (2003) Classification scheme for lean manufacturing tools. Int J Prod Res 41:3075–3090. https://doi.org/10.1080/0020754021000049817
3. Tortorella GL, Marodin GA, Miorando R, Seidel A (2015) The impact of contextual variables on learning organization in firms that are implementing lean: a study in Southern Brazil. Int J Adv Manuf Technol. https://doi.org/10.1007/s00170-015-6791-1
4. Costa F, Denis Granja A, Fregola A et al (2019) Understanding relative importance of barriers to improving the customer-supplier relationship within construction supply chains using DEMATEL technique. J Manag Eng. https://doi.org/10.1061/(asce)me.19435479.0000680
5. Mackelprang A, Nair A (2010) Relationship between just-in-time manufacturing practices and performance: a meta-analytic investigation. J Oper Manag 28:283–302. https://doi.org/10.1016/j.jom.2009.10.002
6. Jasti NVK, Kodali R (2014) Lean production: literature review and trends. Int J Prod Res 53:1–19. https://doi.org/10.1080/00207543.2014.937508
7. Tortorella G, Fogliatto F (2013) Assessment of organizational maturity for lean change. Sist Gestão. https://doi.org/10.7177/sg.2013.v8.n4.a10
8. Kundu K, Rossini M, Portioli-Staudacher A (2019) A study of a kanban based assembly line feeding system through integration of simulation and particle swarm optimization. Int J Ind Eng Comput. https://doi.org/10.5267/j.ijiec.2018.12.001
9. Schipper T (2018) Innovative lean development
10. Bortolotti T, Boscari S, Danese P (2015) Successful lean implementation: organizational culture and soft lean practices. Int J Prod Econ. https://doi.org/10.1016/j.ijpe.2014.10.013
11. Tortorella GL, de Castro Fettermann D, Frank A, Marodin G (2018) Lean manufacturing implementation: leadership styles and contextual variables. Int J Oper Prod Manag. https://doi.org/10.1108/IJOPM-08-2016-0453
12. Thürer M (2017) Hoshin Kanri for the lean enterprise: developing competitive capabilities and managing profit. Qual Manag J. https://doi.org/10.1080/10686967.2013.11918356
13. Wilkinson A, Redman T, Snape E et al (2015) TQM, organisational change and human resource management. In: Managing with total quality management
14. Rossini M, Costa F, Tortorella GL, Portioli-Staudacher A (2019) The interrelation between Industry 4.0 and lean production: an empirical study on European manufacturers. Int J Adv Manuf Technol. https://doi.org/10.1007/s00170-019-03441-7
15. Thürer M, Tomašević I, Stevenson M et al (2018) On the meaning and use of excellence in the operations literature: a systematic review. Total Qual Manag Bus Excell. https://doi.org/10.1080/14783363.2018.1434770
16. Netland TH, Schloetzer JD, Ferdows K (2015) Implementing corporate lean programs: the effect of management control practices. J Oper Manag. https://doi.org/10.1016/j.jom.2015.03.005

17. Netland TH, Ferdows K (2014) What to expect from a corporate lean program. MIT Sloan Manag Rev
18. Wiengarten F, Gimenez C, Fynes B, Ferdows K (2015) Exploring the importance of cultural collectivism on the efficacy of lean practices. Int J Oper Prod Manag. https://doi.org/10.1108/ijopm-09-2012-0357
19. Netland TH, Ferdows K (2016) The S-curve effect of lean implementation. Prod Oper Manag. https://doi.org/10.1111/poms.12539
20. Flumerfelt S, Siriban-Manalang AB, Kahlen FJ (2012) Are agile and lean manufacturing systems employing sustainability, complexity and organizational learning? Learn Organ
21. Lucey JJ (2009) The concept of a lean sustainability zone. Manag Serv
22. Yin RK (2009) Case study research: design and methods. Sage Publications Inc, Thousand Oaks, CA
23. Dombrowski U, Mielke T (2014) Lean leadership – 15 rules for a sustainable lean implementation. Procedia CIRP 17:565–570. https://doi.org/10.1016/j.procir.2014.01.146

Evaluating Manufacturers' Smart Readiness and Maturity Exploiting Lean Product Development Approach

Claudio Sassanelli and Sergio Terzi

Abstract Manufacturers need to address Industry 4.0 to enhance their competitiveness on the market supported by the gradual adoption of digital technologies. However, devoting investments to these technologies should first be coupled by the need to gauge both their own digital status quo and the mechanisms grounding their support to perform product development process. In this context, the importance of lean approach, leading under a continuous improvement and value focused perspective how the process of product development is organized and managed throughout the functions involved, is paramount and strategic. Methods and maturity models have been introduced in the extant literature to evaluate the readiness and maturity of either their digital level or their design and engineering process. A coalescence of these tools still needs to be employed to evaluate a company under the twofold level of lean and digital: to do this, a case study has been implemented.

Keywords Industry 4.0 · Readiness model · Maturity model · Lean Thinking · Case study

1 Introduction

The fourth industrial revolution is driving companies towards a significant enhancement of their systems to remain or become more competitive. Indeed, Industry 4.0 is a new production system applying new technologies to manufacturing: it represents a breakthrough affecting all the activities bonded to manufacturing, from planning to processes, involving also products and work organisation [1]. These technologies, grouped in 11 different types (among which IoT platforms, 3D printing, smart sensors, augmented reality), can lead to a heterogeneous set of practical applications for

C. Sassanelli (✉) · S. Terzi
Department of Management, Economics and Industrial Engineering, Politecnico di Milano, Piazza Leonardo da Vinci 32, 20133 Milan, Italy
e-mail: claudio.sassanelli@polimi.it

S. Terzi
e-mail: sergio.terzi@polimi.it

M. Rossi et al. (eds.), *Proceedings of the 6th European Lean Educator Conference*, Lecture Notes in Networks and Systems 122,
https://doi.org/10.1007/978-3-030-41429-0_29

manufacturers [2]. However, companies need to know their status quo before knowing how to invest in this kind of technologies: first of all they need to understand which is their actual digital level. Looking at the extant literature, DREAMY (Digital REadiness Assessment MaturitY model) [3], can support manufacturing companies to assess their readiness level for initiating their digital evolution process and also to create a roadmap for investments in digitalization and switching to smart manufacturing. Furthermore, manufacturing companies have to realize and gauge how digital technologies can enhance their product development process alongside the whole company. This can be done selecting the most suitable best practices to apply in the development process through CLIMB (Chaos-Low-Intermediate-Mature-Best practice) model [4], measuring maturity in product development activities. With the aim of obtaining a more rapid and effective process, lean techniques can be used in product development: MyWaste and MyTime [5, 6] are methods to detect wastes and to guide designers to address continuous improvement. A systematic combination of these tools needs to be implemented to fully measure and evaluate a company in a complete way: to do this, a case study has been conducted. The aim of this paper is to analyse the AS-IS situation of the order development process of Company A, particularly focusing on the digitization level of the functions involved. The main result of this study is represented by a coalescence of the models and methods extant in literature and presented in Sect. 2. Section 3 describes the methodology adopted, reporting the case study approach criteria and an overview of the assessed company. Section 4 explains the results of the study. Finally, Sect. 5 concludes the paper, discussing results, further researches and limitations.

2 Research Context

So far different methods and maturity models have been proposed in literature to support practitioners to appraise the readiness and maturity of either their smart manufacturing level or of their design and engineering process and analyse the existing wastes along the development process. Before presenting the models and methods adopted in the case study, it is necessary to clarify the research context they refer. DREAMY and CLIMB are assessment models, thus based on a comparison with consolidated best practices codified according an advancement/maturity degree. While, MyWaste and MyTime are analysis methods gathering data regarding the AS-IS company situation according to a referring format. In order to detect the touching points among them, some heterogeneous concepts need to be defined:

1. 'smart manufacturing readiness' is the capability of a manufacturing company to deploy smart manufacturing concepts;
2. 'smart manufacturing maturity' is the capability of a manufacturing company to employ smart manufacturing concepts;
3. Lean Thinking is based on the idea to give the customer what he wants, when he wants it and at the right amount he desires. It means companies must create

Table 1 DREAMY maturity model [3]: aim, structure and output

Digital REadiness Assessment MaturitY model (DREAMY)	
Aim	1. To assess manufacturing companies' readiness level for starting the digital transitioning process 2. To identify a manufacturing company's strengths, weaknesses, opportunities, creating a roadmap for investments in digitalization and transitioning to smart manufacturing
Structure	4 dimensions mutually involving 5 areas, composing together the digital backbone of the company and affecting the value generation along the key processes
Output	A 5-levels maturity scale and a radar chart composed at its angles by the four dimensions considered

value, intended as everything the customer is willing to pay for, and banish waste, which literature classifies in 8 macro-classes (Overproduction, Inventory, Waiting, Motion, Transportation, Defects, Processing, Unused employees creativity) [7–9]. The entire process is based on and guided by the continuous improvement concept [10].

In the following, the single approaches are presented.

2.1 DREAMY

DREAMY is an assessment model based on the idea that, to be able to invest in digital technologies, companies should know their status quo. The model structures the manufacturing company's digital readiness and maturity along four dimensions: Process, Monitoring and Control, Technology, and Organization [3]. These dimensions can mutually involve the digital backbone of the company, affecting the value generation along the key processes five areas (A1. Design & Engineering; A2. Production Management; A3. Quality Management; A4. Maintenance Management; A5. Logistics Management) (Table 1).

2.2 CLIMB

CLIMB is a maturity assessment model based on a selection of the prevalent product development best practices from literature. It is structured into a semi-structured questionnaire, based on the proposed product development best practice framework and each of the questions investigates one of the best practices. The framework

Table 2 CLIMB maturity model [4]: aim, structure and output

Chaos-Low-Intermediate-Mature-Best practice (CLIMB)	
Aim	1. To create awareness and understanding in academia and industry on the extant best practices in product development 2. To retrieve an AS-IS picture of the practices usage in the industrial context, providing practitioners with the possibility of self-assessing their processes and benchmarking with both literature and other industrial cases
Structure	Based on the prevalent product development best practices from literature structured into a questionnaire (categorizing 107 product development best practices, across 8 areas grouped into 4 categories)
Output	An evaluation scale made of 5 maturity levels and a radar chart

categorizes 107 product development best practices across eight areas: 1 activities and flow, 2 decision making, 3 training, 4 roles and collaboration, 5 knowledge management (KM) process, 6 KM techniques, 7 methods, 8 computerization and software. These areas are grouped into four categories. Process concerns how the flow of the development process is managed, the activities performed, the decision-making methods, the orientation to the client and the value, the methods used. People refers to how the development process is structured, which are the actors involved, the respective roles and competences. KM deals with methods and tools used in the company to support knowledge management, accumulation, maintenance and reuse of the same. Tools refers to support tools for the product development process and to improve the data integration level along it.

Results are deployed on a five maturity levels evaluation scale and a radar chart (Table 2).

2.3 *MyWaste*

MyWaste is a methodology that companies can adopt to enhance their New Product Development (NPD) processes in a continuous way. It consents to analyse and rank a given library of product development wastes. Five steps compose the method: companies are led through NPD process waste and criticalities analysis and removal. The method is applied to improve an existing process under lean perspective, reflecting lean principle of pursue perfection through progressive improvement actions (Table 3).

Table 3 MyWaste method [5, 6]: aim, structure and output

MyWaste method	
Aim	To improve companies' NPD processes in a continuous fashion, evaluating and ranking a given library of product development wastes
Structure	Questionnaire. Composed of 5 steps (grouped into 3 macro-activities: MyWaste Analysis, Map-it Process and Change Implementation) which can be recursively applied in order to continuously improve NPD process
Output	NPD process waste and criticalities analysis and removal: better performances of the whole development process

Table 4 MyTime method [5, 6]: aim, structure and output

MyTime method	
Aim	To measure how time along the NPD process is spent
Structure	Questionnaire to evaluate time of the actors involved: main activities (design and test), knowledge recovery (through traditional sources or informatics ones), specification and documentation development, data input in informative tools, coordination with other colleagues or partners and other activities
Output	Detection and ranking of the main problems faced by actors working along NPD process, unveiling how time allocation is perceived by each of them

2.4 MyTime

MyTime is aimed at measuring through a questionnaire how time along the NPD process is spent. It assesses how time of the players involved in this process is split among main activities (design and test), knowledge recovery (through traditional sources or informatics ones), specification and documentation development, data input in informative tools, coordination with other colleagues or partners and other activities. In the meanwhile, it also detects and ranks the main problems coped by users along this process, showing how time allocation is perceived by each of them (Table 4).

3 Research Methodology and the Case Study

The study conducted followed the embedded single-case study design approach and encompassed semi-structured interviews to gather data [11]. The choice of the unit of analysis fell on two main embedded macro-areas of Company A: production and ICT for the digital readiness and maturity of the company, technical department and R&D for the readiness evaluation of the design and engineering process. In particular, Company A is an Italian engineering to order company: it is the international reference point in manufacturing of innovative technologically advanced machinery and

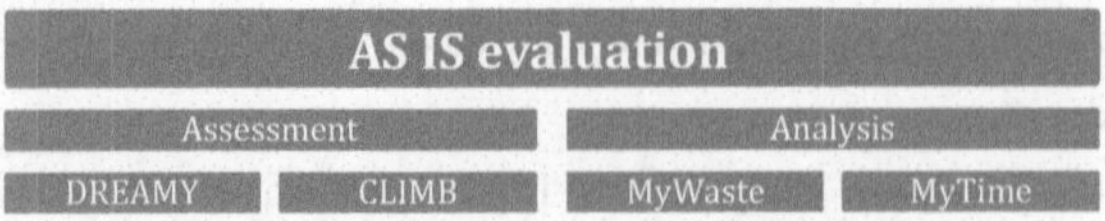

Fig. 1 AS-IS evaluation research approach

in engineering and service solutions of customized machines. It operates in various industrial sectors such as Converting (e.g. primary and converter first-class machinery in the paper, film and aluminium foil industries), Automotive (e.g. semi-automatic and fully automatic machinery for the thermoforming of automotive interiors soft trims) and other Industries (e.g. customized vertical injection moulding machineries for the over-moulding of components with complicated shapes and dimensions and the possibility of manual or automatic loading). The more demanding customers and highest standards required has contributed to the increasing complexity of projects. To remain competitive and operate in an effective and efficient way, offering to their customers the most valuable solutions and simultaneously optimizing development steps and workflow management, Company A wants to analyse its order development processes and to identify critical factors with a particular attention to the digitalization level of the areas involved. Finally, the company would like to find the correct solutions to intervene on them under a continuous improvement fashion. Ten interviews and one workshops were conducted in the company, with a total duration of about 45 h. Interviews were related to DREAMY with actors from production (and lean production), operations and ICT. CLIMB was applied with technical director, system engineering and R&D, service/after sales, SW BU and mechanical BU. Finally, MyWaste and MyTime were conducted in a workshop with all of them. Therefore, the research approach adopted (Fig. 1), composed by the integration of DREAMY and CLIMB models with MyWaste and MyTime methods, presented in Sect. 2, was used to conduct the case study, providing the following results.

4 Results

The first main result of this research is represented by the successful application in Company A of the approach reported in Fig. 2. The joint use of the models and methods selected resulted suitable to obtain the data needed to gauge the maturity level of manufacturing industries to integrate digital technologies along their value-adding processes, providing a special focus on NPD.

First, DREAMY and CLIMB resulted to be structured on the same four pillars: process, monitoring and control (supported by KM practices), technology (comprising tools, methods and computerization level) and organization (dealing with roles, collaboration and training of people). Thanks to this compatibility, results coming

Fig. 2 AS-IS assessment of Company A: DREAMY and CLIMB results

from the two models were crossed effectively. A complete evaluation of each of the four dimensions and the related main issues is provided.

Dealing with process, a process culture does not exist, strengthened on a strong division among functions and a lack of continuous improvement strategy. A macro-perspective in the management of processes is quite diffused but not completely shared. It is not clear where the process starts and finishes, as well the value for the market and how it should be considered internally. Formal methods able to manage the process and facilitate value-added activities along it are not adopted. Regarding people, the order development process is not well structured and key roles for coordination and management (e.g. project managers, industrialization, continuous improvement) are lacking. Roles allocation along the process is not clear (who is in charge of controlling and their responsibilities are not clear). Training is not performed (also for technical departments) and the majority of the existing knowledge is grounded on the personal experience of employees. Concerning tools, basic tools of virtual prototyping are used and there is also a basic support of PDM (only for 3D modelling) but they are used in a basic way. Some attempts have been done to digitalize but result obsolete and low integrated to each other, limiting the interoperability and leading to several low value-added activities (e.g. transcoding) almost ever not supported by suitable methods and tools. In terms of monitoring and control & KM, there is a widespread issue with knowledge formalization, strongly de-structured. Apart from the basic IT tools, there are not methods supporting knowledge protocolling and sharing. Indeed, knowledge sharing results inefficient and based on the physical interactions among employees. There is a lack of dedicated resources to knowledge management and continuous improvement of the organization and KPIs to monitor critical processes are not used.

Moreover, MyWaste and MyTime methods contributed to further strengthen these results. The main wastes along the NPD process (grouped in the 8 macro-categories according to the lean theory) were detected. The two most relevant wastes are related to knowledge management and inappropriate processes (that are often not clear, not defined or not well managed). These kind of issues are mainly due to mistakes or lacks related to organization activities, e.g. lack of roles allocation, of knowing who does what or who codifies the voice of the customer. It has to be said that, independently from the type of waste, there is a relevant difference among functions of their perception: this is maybe due to a shortage of a common perspective of the process involving the different actors. As a result, it emerges that knowledge represents the major waste (waits to process information and not efficient information and knowledge management), followed by misalignment with the market (over-engineering), waste of resources (inappropriate process and activities not needed) and finally time wasted (in reworks and revisions). The wastes detected were also linked, per each of the 8 macro-categories, to specific causes and effects. In general, the majority of them were triggered by organizational factors (34%), followed by process (25%), knowledge (23%), resources and tool (15%) and other (3%) causes. It was also found that wastes mostly lead to an increase in generation of delays (24%), of development costs (15%) and of product costs (13%), of development time (12%), reduction of productivity (17%) and other effects (19%). Moreover, MyTime method revealed that the main issues in the NPD process are linked to the excessive time spent in codification, management, sharing and retrieval of knowledge (56%), not neglecting the time dedicated to coordination and organization (21%). Indeed, the majority of daily time (70%) resulted to be quite squeezable through the implementation of a major level of automation and digitization and a more efficient operability.

5 Conclusions and Further Researches

The aim of this paper has been to practically study the maturity level of manufacturing companies to coalesce digital technologies along their value-adding processes, through the lenses of NPD process. An explanatory embedded single-case design case study has been presented in Company A, an engineering to order Italian company. Looking at the extant literature, DREAMY and CLIMB assessment models and MyWaste and MyTime analysis methods have been selected and combined: they analyze different aspects in the company but seemed complementary for the final objective of this research. The main output of the study has been the positive coalescence of these approaches in a more comprehensive and integrated one, providing a broad assessment of the adding-value processes and raising the main problems and wastes occurring along them: concepts as readiness and maturity, dealing with digitization and design practices, but also as value, waste and continuous improvement, related to the Lean theory, have been used. Both company path towards industrial digitization and the order development process effectiveness dramatically depend on the wastes detected. In the meanwhile, wastes, if adequately managed, can be the key for the company to head, according to a continuous improvement approach, towards a more proficient product centric knowledge management. Here the capacity of the joint method adopted and presented is unveiled in its role of delineating a plan of possible interventions through which the existing situation could be improved: first, the organizational structure should be re-modelled (based on people's training and empowerment) to support the changes to be enacted. Then, the order micro-processes' streamlining should be the key to solve macro-problems, opening room to finally evaluate the adoption of informative systems still unattended in the organization. However, this research has some limitations: the method obtained with this research brings with itself the limitations of the single approaches that need to be considered and coped during the analysis, requiring to systematically combine different aspects to get a more effective and holistic approach.

References

1. De Santis S, Monetti M (2017) Industry 4.0 for the future of manufacturing in the EU. Roma
2. PwC and GMIS (2016) Industry 4.0: building the digital industrial enterprise
3. De Carolis A, Macchi M, Negri E, Terzi S (2017) A maturity model for assessing the digital readiness of manufacturing companies. In: Advances in production management systems. The path to intelligent, collaborative and sustainable manufacturing, APMS 2017. IFIP advances in information and communication technology, vol 513, pp 13–20
4. Rossi M, Terzi S (2017) CLIMB: maturity assessment model for design and engineering processes. Int J Prod Lifecycle Manag 10(1):20
5. Rossi M, Taisch M, Terzi S (2012) Lean product development: a five-steps methodology for continuous improvement. In: 2012 18th international ICE conference on engineering, technology and innovation (ICE)
6. Rossi M, Kerga ET, Taisch M, Terzi S (2011) Proposal of a method to systematically identify wastes in new product development process. In: ICE 2011
7. Liker JK, Morgan JM (2006) The Toyota way in services: the case of lean product development. Acad Manag Perspect 20(2):5–20
8. Ohno T (1988) Toyota production system: beyond large-scale production, vol 15, no 2
9. Womack JP, Jones DT (1996) Lean thinking, vol 55, no 1
10. Womack JP, Jones DT, Roos D (1990) The machine that changed the world: the story of lean production
11. Yin RK (2009) Case study research design and methods, vol 5, 4th edn

Employees' Participation and Involvement in Lean Management: The Experience of a Training Program of Assembly Lines Workers

Luigi Campagna, Luciano Pero and Margherita Pero

Abstract This paper tackles the problem of how to train employees to foster involvement and participation to support the implementation of Toyota Production System (TPS). To this aim, it presents the case of a training program aimed at involving workers in a TPS project, increasing their knowledge of the lean management model, and starting a discussion with them, to fine-tune the new organization, especially on the themes of versatility, knowledge exchange, cooperation and teamwork. The training program involved 450 workers of a manufacturing company, it lasted three days per participant, where activities in the classroom and in the shop floor were combined. Different didactic methods were used: theoretical lessons, group discussion and teamwork, didactic games and practical experience in the production line. Results show that the approach was successful and allowed the company to fully reap the benefits of TPS.

Keywords Lean management · Training · Employee involvement · Participation

1 Introduction and Background

Lean management, also known as Lean Thinking or Toyota Production System (TPS), is a production strategy that was developed in the 1950s by Taichi Ohno. The aim of lean management is reaching operational excellence by continuously improve

L. Campagna · L. Pero
Meta Governance & Innovation Studio,
Via Mancini 3, 20129 Milan, Italy
e-mail: luigi.campagna@meta-innovation.com

L. Pero
e-mail: luciano.pero@meta-innovation.com

M. Pero (✉)
Department of Management, Economics and Industrial Engineering, Politecnico di Milano, Via Lambruschini 4B, 20156 Milan, Italy
e-mail: margherita.pero@polimi.it

M. Rossi et al. (eds.), *Proceedings of the 6th European Lean Educator Conference*, Lecture Notes in Networks and Systems 122,
https://doi.org/10.1007/978-3-030-41429-0_30

production process and procedures, leveraging on standardization and a culture oriented to workers involvement and participation [1]. The main focus of most of the lean management projects is waste (in Japanese "muda") elimination. To this aim, a plethora of tools can be used effectively, ranging from e.g. Just-In-Time (JIT), to value stream mapping (VSM) and 5S [2].

However, TPS is not simply a set of tools and techniques that can be used to improve operative performance, but it is a "philosophy", that encompasses some principles on which to build the organizational model. Liker [3] presents the 14 principles of "Lean thinking", which summarize the main guidelines for a successful implementation of TPS in a company. Employees' empowerment and training are at the core of the TPS philosophy. In fact, among the others, principle six stresses the importance of employee empowerment to support growth and improvement, while principle nine underlines the importance of training and education of employees to build a learning organization. Finally, principle thirteen suggests the importance of building consensus of the solution found to problems. Therefore, to fully reap the benefits of TPS implementation, employees' participation and involvement are fundamental.

However, despite the "theory" supports a bottom-up and participated approach to TPS introduction, Campagna et al. [4] shows that, especially in Small and Medium Enterprises (SMEs), there are still many cases of unsuccessful application of lean management. These are due to the use of a "top down" approach, aimed at simply imposing new organizational models and processes without properly involving and training employees. Thus, the need for proposing successful stories of employees' involvement and training formats centered on employees and based on employees participation.

This paper aims to fill this gap, by providing answer to the following research question: *How can workers and employees be trained and involved when a company moves from a traditional production strategy to TPS?*

To provide answer to the research question, this paper discusses a case study of a company that underwent a change in the organization of the production process aimed at implementing TPS. In a first phase (phase 1), the company used a top-down approach, imposing the "new way of working" to the workers on the assembly lines. The benefits were limited. Higher benefits in terms of productivity were gained when the company changed its approach and moved to a phase 2 where employees were trained and involved. This paper, therefore, building on [4], presents the training approach and tools used to increase workers understanding of TPS and their willingness to support the "new way of working".

The case study presented is based on the personal experience of the authors. In fact, the authors were involved by the company in the design and the delivery of a training program to the 450 workers that are employed in the production areas, the assembly line and the warehouse. Thanks to interviews to the managers (Human Resource Manager, Production Manager and Logistics Manager) and the representatives of the trade unions, we could collect information on: the production process, the TPS project and the way it has been implemented in the company, and the results in terms of productivity improvement. After the delivery of the training program, the workers that attended the course were asked to fill a questionnaire, the managers and the trade union representatives were met again to discuss the results, moreover the data regarding productivity of the assembly lines were collected.

Literature suggests various methods and techniques for teaching and training on lean thinking, ranging from traditional lectures, to learning factories or training on the job, computer simulation, analysis of case studies, role-playing, games, observations followed by discussions (e.g. [5, 6]). The case study discussed in this paper presents the application of a training program that blends these techniques, with the aim not only to teach *ex-cathedra* but also to facilitate the sharing of knowledge among the participants and to make the whole organization to learn. Interestingly, the participants to the training are workers of a SME: training to this kind of audience is, in the existing practice and literature, less investigated.

The remainder of the paper is organized as follows: the case study is presented in Sect. 2, Sect. 3 is devoted to the discussion of the training approach and tools used, finally, Sect. 4 discusses the main results and limitations of the training program.

2 The Case Study

The company analyzed is a medium-sized manufacturing company, which operates in Northern Italy for an international group. They produce refrigeration systems and furnishing for large-scale retailers. Customers are located mainly out of Italy. The company controls product design, production and assembly processes. The company employs more than 400 workers in the assembly line. The company's revenues grew significantly the first decade of 2000s, mainly thanks to its ability to provide customized products. To sustain such strategy, the company has experienced an increase in the variety of parts and finished products that stressed the production and logistics system. Indeed, the number of components, colors and options increased, with significant impacts on the stocks of raw materials and semi-finished products. While, the increase in the product range, thus the variability of orders, associated to an increase in the production volumes, increased the variability of production workload. To face this challenge, the company has resorted to the so-called "numerical flexibility", i.e. the use of temporary workers at peak times to saturate the production lines. However, these actions have only partially solved the problem: the use of temporary workers, if on the one hand it allowed, although partially, to face the increasing volumes of demand, on the other hand, has generated further problems of product quality.

To face the challenges posed by the market, there was the need to improve the materials management system, to reduce stocks but at the same time assuring there is no lack of materials, and movements of materials, as well as to redesign the production system, to make it more flexible and adaptable to customer demand.

The company decided to redesign the production cycle by investing in dedicated technologies and in changing the production system embracing TPS approach. The innovation project is entrusted to the Production Manager, who has already managed lean projects in other industrial plants.

The objectives of the project were improvement in delivery lead time and cost reduction. The main actions aimed at increasing flexibility to the mix of assembly lines and the versatility of the employees, and to reduce the stock of semi-finished products and materials. As for suppliers, just in time deliveries were put in place. As for the warehouse management, the system moved to deliveries synchronized with assembly lines needs, to provide the "right" material at the "right" time. As for the assembly lines, these were differentiated for product types/families, and workers were asked to develop the ability to work on all the models of each product family. Each assembly line works one piece at a time, and each piece is associated to a specific customer order.

2.1 Phase 1: The Top-Down Approach

The Production Manager managed the project with a "top down" approach and carried it out with the support of an external design team and an internal technical team. The innovation project is articulated in some interventions of re-engineering of the logistic flow, production lines and workstations of traditional production lines.

First of all the materials warehouse was reconfigured with the aim of concentrating the materials in a dedicated area, to optimize the stocks and eliminate the intermediate warehouses, and to synchronize the supply flow with the production flow. The supply of materials to the lines was ensured by a system of "trolleys" that served the production lines. The trolleys were loaded, in line with the bill of materials, with the materials necessary for the processing of the single order. The components are located with radiofrequency systems, identified with barcodes and arranged on the trolleys. The trolleys were used to transport the materials at the beginning of the production line and, once the assembly activities are over, were returned—empty—to the warehouse.

Secondly, the layout of the factory has been redesigned to overcome the organization by departments and optimize physical flow. The assembly was distributed on small lines dedicated to specific product families. However, the implementation of the new layout concerned only a limited part of the plant. Therefore, two assembly line types, with the associated different methods, co-existed in the plant: (i) the traditional assembly line and (ii) the new assembly line.

The traditional assembly line was divided into two parts: (i) the first part, where standard activities are performed, and (ii) the second part, where customization of the products takes place, in this part "artisanal" work is carried out. The semi-finished products of the first part were stored between the two parts of the line waiting to be customized. Each operator operated on a single fixed location. The processing time for each station was pre-defined only for upstream part. The processing time of the station in the customization part were longer, and there was not a standard for them. Components and materials were placed aside the assembly line and each worker picked the ones he needs to carry out his job. Due to this organization of the assembly line, the customization part was a bottleneck. Moreover, it made the assembly lines unable to efficiently product the variety required by the market, as well as it generated a lot of waste due to the high incidence of different non-value added activities that the operators performed, such as the search for materials in the line-side warehouses.

The new assembly lines were dedicated to specific product families to produce each single order. Cycle times were relatively standard. They also required polyvalent operators able to rotate between stations of the same line and between different lines. Finally, they are fed with the trolley system by a single flow of materials synchronized with the work in progress. Since only a few lines worked according to the new model, the Production Manager has also decided to start a project to improve the traditional lines. In particular, he decided to apply the principles of the 5S to the stations along the assembly line. The aim was to improve ergonomics and eliminate some not value added activities related to the search for tools, small parts and materials. This project was entirely developed by the technical team. At the end of the re-engineering interventions, a training program was developed aimed exclusively at middle managers to make them understand the need to evolve towards leading roles in line with the new organizational model. The workers were provided with minimal information on the characteristics of lean production and TPS. Phase 1 of the lean project has been developed over two years and resulted in some limited improvements especially in terms of reduction of stock and lead times. However, no significant benefits have been recorded in terms of increasing productivity and improving quality. In fact, trade union representatives reported that many workers did not understand the innovations introduced and the company's demand for versatility. The property was alarmed and asked for an explanation of the "non-return" of the investment. The rumor of a possible relocation of the factory to Eastern European countries was also circulating and nobody denied it....

2.2 *Phase 2: Involvement and Participation*

Given the situation, the Human Resource (HR) manager carried out an analysis on the workers, from which it emerges that the workers felt little involvement in the new organizational model. After two years, issues such as versatility, team work and performance bonus are still unresolved. In addition, HR manager perceived the difficulties in the dialogue between workers and the managers responsible for the continuous improvement: the innovation project seemed to be bogged down. For the workers, especially those who grew up in "Fordist" environments, it was difficult to understand the meaning of innovations and give "meaning" to the new way of working. For many it was difficult to link together the set of innovations and relate them to the expected job role. There was also a difficulty in accepting change and becoming active: custom, often the result of consolidated habits, and cultural difficulties often make it difficult to react positively to changes in work. Initiatives aimed at increasing not only the understanding of innovations but also the active involvement in the process of change of the entire population of production and logistics (over

400 employees) were therefore necessary. Company management and trade union representatives agree to work together in two directions: (i) to increase the involvement of workers in the innovation project and (ii) to define performance bonuses for the new commitment of workers, also with recognition of new professional abilities.

The idea that came forward was to develop a training cycle for all workers with two main goals: (i) involve the workers in the project, increasing their knowledge of the new lean model and its critical understanding, and (ii) in each specific work area, start a discussion with the workers, to fine-tune the new organization, especially on points of versatility, knowledge exchange, cooperation and the team. In the following section, the training approach, organization and tools used were detailed.

2.3 The Approach

The educational objectives of the training were essentially three: (i) understanding the reason for the new production model, (ii) learning the new way of working, and (iii) experimenting and finding improvements. In this context, the training was organized so to create a free and welcoming "discussion space" for everyone in the classroom, with the possibility of deepening in order to understand innovation, weigh the pros and cons, develop ideas and discuss them with the others. Training was also intended to be an opportunity to experiment together, as in a laboratory, both the new ways of working with lean management applied to the new production line, as well as the unresolved problems, such as versatility and teamwork. Finally it had to be designed jointly (company and trade union) as a solution to the impasse of the innovation project.

3 The Training Program

3.1 The Organization of the Training Program

The participants, together with their leaders and coordinators, are divided into homogeneous classes for production areas/lines and follow a three-day course. The topics addressed are: the market and the company, the Just In Time and the new production model, the streamlined flow and teamwork, the workplace and its improvement, professionalism and continuous improvement. Each training day is organized so to provide for an alternation between activities performed in the classroom and activities performed in the shop floor, in order to combine reflection with action, discussion with personal experience. The learning path started from the comparison between old and new way of working, continues with experiences of mutual help to learn how to work, and closed with the simulation of work in team and evaluation of professional repercussions. The structure of the training is presented in Table 1.

Table 1 The structure of the training

Timing	Location	Overall topic	Content
Day 1—morning	Classroom	The old and the new way of working	Introduction to the course The company, the market and the effects on the Production System Analysis of "my" actual workplace
Day 1—afternoon	Production line		Visit to the new production line and discussion of the new production model Examples of other companies: from Fordism to Toyotism
Day 2—morning	Classroom/production line	Helping and learning while working in the production line	Kit assembly classroom games and reflections on working in a production line Mutual training on other assembly positions with rotation of roles
Day 2—afternoon	Classroom/production line		Exchange of stations between "teacher" and "student" Reflections: effects on production (quality, quantity), on tasks and behavior Logistics in lean production
Day 3—morning	Classroom	Collaboration in the team and professional roles	Collaborative behaviors: classroom games and reflections Simulation of group work in line or department The workplace to collaborate: the 5S and the new production model
Day 3—afternoon	Classroom		Continuous improvement in lean production The change of professionalism, versatility and behaviors Conclusions

3.2 The Tools Used in the Training Program

All the didactic techniques focused on activation and involvement. In particular, a plethora of different training tools have been used. As far as the theoretical aspects, such as Just in Time and logistics in the TPS, these have been presented through *photos and figures*. Practical aspects, connected to the new production model, were tackled with *site visit and group reflection to summarize* together what the people have seen, as well as with practical experience in the production line. Personal experiences were discussed using *group discussion* to make issues to emerge, e.g. the effect of the explosion of product range. *Group work* was used to tackle common problem, e.g. non value added movement were discussed building together a spaghetti chart. With *didactic games*, e.g. assembly of a modular toy cars, the new and the old way of working, as well as group decision methods, are compared. Finally, *practical experiences* were done directly *in the production line* in the form of mutual training on the job: each participant played the role of teacher, when he taught his job, and that of the apprentice when he learnt the job from the others.

As for the didactic game of the assembly of the modular toy cars, this aimed to understand the benefits of collaboration and teamwork between workers in different stations in the production line. It was organized as follows:

- There are 2 Groups of 3 people each (Group A and B) that play the assembly of the small modular toy car (Yellow and Blue) in two successive hands. Each person is assigned to 1 of 3 workstations and must follow the assembly instructions in sequence.
- There are two rounds: In the first one: the 3 people assemble the 2 cars in sequence without helping each other and in silence. In the second one: the same 3 people assemble the 2 cars in sequence but helping each other, holding out the right pieces, talking or pre-assembling the wheels.
- Observers measure Group A and Group B times and assess the wastes.
- The group discusses the differences in terms of performance between the first and the second round.

3.3 The Results

To assess the quality of the training from the point of view of the participants, the responses to a questionnaire were gathered. The participants were asked to rate on a likert scale 1–5: how much the course and each topic was of interest for them, how much they liked the didactic methods used, and the teachers, and how much the course was useful for their work. Out of 349 respondents, on average, they rate 3.97 the appreciation of the discussion in class, and 3.18 the usefulness of the course for their work. Additionally, the authors had discussion with the workers, the mangers and the trade union representatives. These agreed that the objectives of the training

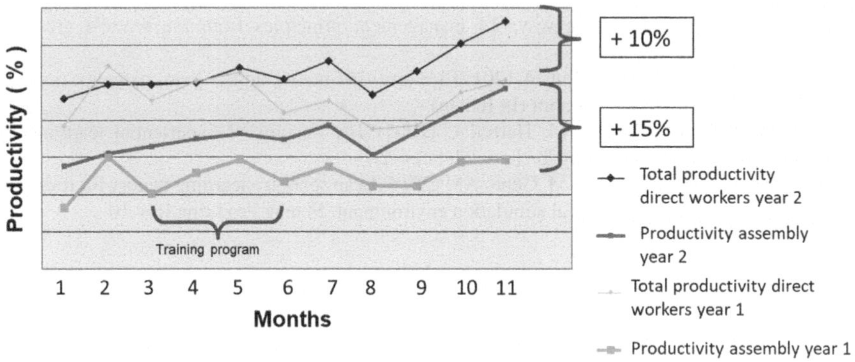

Fig. 1 The productivity trend

program have been reached thanks to the involvement of all employees and the bottom up approach.

From a performance point of view, the productivity of the assembly lines, in the year of the training (year 1) and in the year after the training program (year 2), increased as depicted in Fig. 1.

4 Conclusions

The course helped in building a new culture, because it made everyone understand how the work, the business and the market were evolving and how the factory can be improved "here and now". Moreover, through the training, people became aware of the goals and objectives of innovation and could appreciate the advantages for the company and for the workers of the new way of working. The training helped the workers to express their ideas and proposals and to be heard by the company managers. Finally, the process of change has been unblocked: the company climate has become constructive, the managers have placed themselves in a listening position, and the process of continuous improvement has begun. The approach centered on sharing and on the team, experimented with the training activity, has also been maintained in ordinary management, allowing even on traditional lines productivity gains attributable to participation.

References

1. Womack J, Jones D (1996) Lean thinking. Banish waste and create wealth in your corporation. Free Press, Simon & Schuster, Inc., New York, NY
2. Gupta S, Jain SK (2013) A literature review of lean manufacturing. Int J Manag Sci Eng Manag 8(4):241–249

3. Liker JK (2004) The Toyota way: 14 management principles from the world's greatest manufacturer. McGraw-Hill
4. Campagna L, Pero L, Ponzellini A (2017) Le leve dell'innovazione - lean, partecipazione e smartworking nell'era 4.0, Guerini (In Italian)
5. Miles M, Melton D, Ridges M, Harrell C (2005) The benefits of experiential learning in manufacturing education. J Eng Technol 22(1):24–28
6. Ruano, JP, Hoyuelos I, Mateos M, Gento AM (2019) Lean school: a learning factory for training lean manufacturing in a physical simulation environment. Manag Prod Eng Rev 10

A Migration Methodology for Factories Digital Transformation

Roberto Rocca, Filippo Boschi, Ambra Calà, Paola Fantini and Marco Taisch

Abstract Over the last years, several technologies and control systems have been developed towards the decentralization of automation control architectures for cyber-physical production systems. Nevertheless, only few of these technologies are already in use. To support their adoption in brownfield production sites migration strategies and business case evaluations are necessary. This paper presents a migration approach and a business case evaluation methodology tailored to the migration of legacy automation systems towards the Industry 4.0 paradigm. The proposed approach aims to evaluate opportunities and mitigate the risks of migration from technical, operational, human and business perspectives. The methodology follows an iterative approach starting from the definition of the current situation of the factory and the identification of business goals aiming at evaluating a set of possible migration paths and selecting the optimal one according to a cost-benefit analysis. The paper concludes with an exemplary application of the methodology in a real industrial environment, developed within the FAR-EDGE European Project.

Keywords Industry 4.0 · Digital factories · Digital transformation · Migration strategy · Business case evaluation

R. Rocca (✉) · F. Boschi · P. Fantini · M. Taisch
Department of Management, Economics and Industrial Engineering, Politecnico di Milano, Via R. Lambruschini 4/b, 20156 Milan, Italy
e-mail: roberto.rocca@polimi.it

F. Boschi
e-mail: filippo.boschi@polimi.it

P. Fantini
e-mail: paola.fantini@polimi.it

M. Taisch
e-mail: marco.taisch@polimi.it

A. Calà
Siemens AG, Corporate Technology, Günther-Scharowsky-Str. 1, 91058 Erlangen, Germany
e-mail: ambra.cala@siemens.com

M. Rossi et al. (eds.), *Proceedings of the 6th European Lean Educator Conference*, Lecture Notes in Networks and Systems 122,
https://doi.org/10.1007/978-3-030-41429-0_31

1 Introduction

"Flexible manufacturing" means manufacturing, which "allows changing individual operations, processes, parts routing and production schedules" [1]. Digital transformation represents one of the preconditions for production flexibility and to fast react to current changing conditions such as volatile market demands, increasing the frequency of new product introduction and new technological developments [2]. Digitalization represents one of the biggest challenges a company has to face nowadays [3]. In fact, it is not only a technical issue, but it requires also a cultural change that involves a slow modification of a deep and eradicated mindset in all the company structure and, therefore, requires an incremental change process. Companies not only need to be able to quickly reconfigure their production systems, allowing higher capacity-flexibility and smaller lot sizes. They, also, should finalize managerial capabilities, production management approaches, models and tools able to be compatible with rapid reconfigurations of the production system [4].

For these reasons, new automation architectures are required to enhance flexibility and scalability, enabling the integration of modern IT technologies and, consequently, increasing efficiency and production performance [5]. Nevertheless, reality today shows the dominance of production systems based on a traditional approach, characterized by a hierarchical and centralized control structure.

One of the main reasons is that manufacturers are traditionally conservative against the adoption of digital technologies. They need to make sure that any new technology is introduced with minimal disruption to their existing production operations. At the same time, they need to be confident that the new systems will improve production time, quality and cost. Manufacturers will never accept production systems that compromise the achievement of the above mentioned performances [6]. Indeed, the intent of this paper is to present a new business case evaluation methodology that gives support to strategy implementation for a smooth migration, considering a holistic vision from the technical, operational, human and business perspective.

The paper is structured as follows: Sect. 2 describes a new migration methodology for factories digital transformation; Sect. 3 describes a methodology application case for a durable goods company developed within the FAR-EDGE European Project context; Sect. 4 draws some final conclusions.

2 A New Migration Methodology for Factories Digital Transformation

To increase the adoption of future manufacturing paradigm, industries should be supported with migration strategies [7]. The term "migration" refers to the transformation of a system from its legacy condition to the desired one. In particular, this paper refers to migration as a progressive transformation, moving an existing production system towards digitalization. Migration strategies are essential to support the

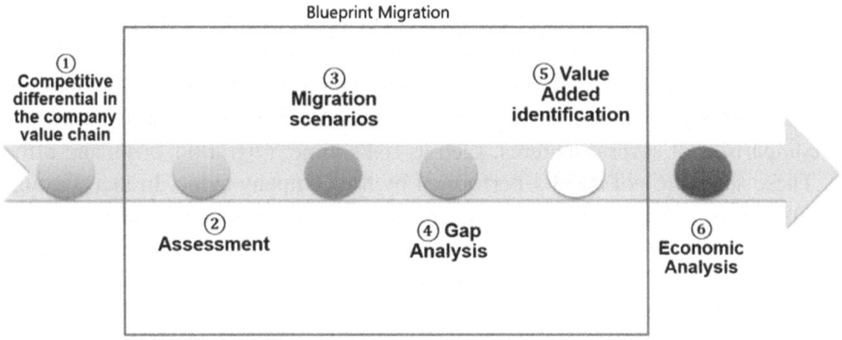

Fig. 1 Proposed migration process

implementation of digital technologies in the manufacturing sector aiming at achieving a flexible manufacturing environment based on rapid and seamless processes as a response to new operational and business demands [5].

Figure 1 shows the framework of the migration process and business case evaluation methodology proposed in this paper. It is a framework articulated in six steps. The procedure is described in the next sections, following a stepwise order.

2.1 Competitive Differentials in the Company Value Chain

The purpose of the first step is to assess from a business model perspective the opportunity for a factory to migrate towards digitalization by means of cloud and edge-based technologies. In this context, a SWOT analysis [8, 9] has been selected as one of the possible tools to carry out the identification of business changing factors. Due to its simplicity and well-known diffusion in the strategic context, the SWOT analysis can be applied to any project. The straightforward consequence of this analysis is the identification of the variables related to the migration. For this reason, a KPIs definition is a fundamental activity in order to clarify the desired impact of the project implementation, accordingly to the SWOT analysis outcome related to internal and external business changing factors. Then, Porter's value chain model is applied in the operations and management description [10, 11], in order to identify the relevant variable previously identified within the different business units of the enterprise.

2.2 Blueprint Migration Strategy

The blueprint migration strategy [6] is represented by steps 2–3–4–5 of the methodology. The term "blueprint" refers to an early plan that explains how something might be achieved. The strategy has been developed within the FAR-EDGE EU project. As described in [2], it aims at identifying the optimal migration path for a factory towards the decentralization of its current automation architecture by means of cloud- and edge-based technologies. The FAR-EDGE migration strategy implies the comparison of several features, such as risks, time, migration, costs and effort [7]. These steps are not usually performed by the company alone. In fact, through questionnaires and workshops with people involved in the manufacturing process, the methodology can support the company to define the migration goal and starting point, as well as the possible impact and the typical difficulties that the new digital solution can have.

The blueprint migration strategy starts with the "Technical, Operational and Human assessment" step. The main goal of this assessment is to obtain a clear picture of what should be changed in the company's business, by investigating the implications and preconditions of this change at technical, operational and human dimensions. To assess the various issues related to the three dimensions, a digital maturity model has been exploîted [2, 5].

The blueprint migration strategy continues with the "Migration Matrix and scenario definition" step. The migration strategy plays a crucial role in the target KPI values definition. There could be more than one way to reach the desired result and the TOBE scenario could not be unique. In fact, different migration scenarios according to the possible technology options are investigated [7] in order to identify the migration alternatives to go from the identified AS-IS situation to the TO-BE one. To this end, a tool called Migration Matrix (Fig. 2) has been developed within the FAR-EDGE project [5] to identify all the necessary improvements in the direction of the Industry 4.0 vision of smart factory based on technical, operational and human dimensions and 5 levels scale of digital maturity. The matrix provides a representation of the current state of the company suggesting how the three dimensions can be combined in order to reach the final objective (envisioned by FAR-EDGE) in a smooth and stepwise migration process.

For this reason, a collaboration with OEMs and solution providers is required at this point in order to assess the feasibility of the scenarios and provide solutions able to improve the KPIs defined in the first step. The final result implies one Migration Matrix for each scenario where it is depicted only one AS-IS status and one of the different TO-BE status.

Migration matrixes provide a clear map of the current and desired conditions of a factory, revealing different alternatives to achieve the first short-term goal in the direction of the long-term vision. These alternatives are then evaluated according to the business strategy, considering also strengths and weaknesses points outlined in the first step of the methodology.

The "Gap analysis" between the AS-IS and the TO-BE scenarios, derived by the Migration Matrix, has to be analyzed taking into account three main aspects: (i) required components, (ii) possible integrations and (iii) steps for application. Required components represent a detailed list of the additional components, operations, and new professional skills required for each scenario migration. Further possible extensions of the architecture, e.g. the integration of components, must be

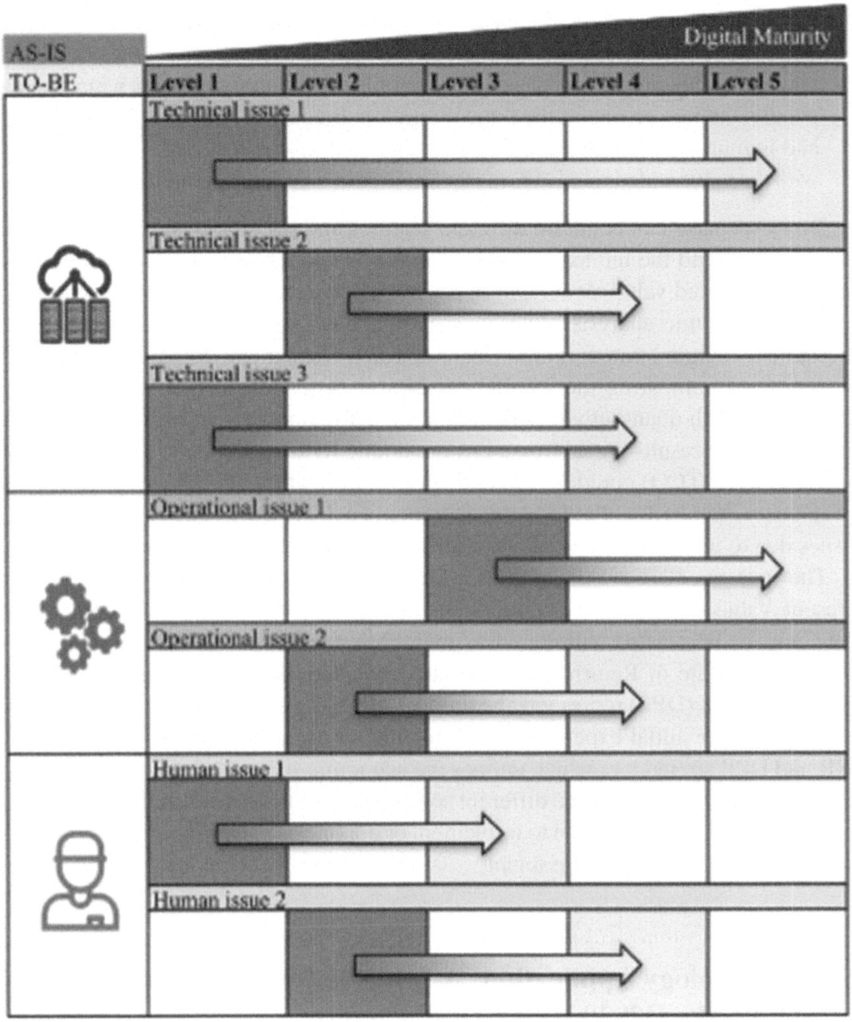

Fig. 2 Migration Matrix: AS-IS (in red) and TO-BE (in green) scenarios definition

considered to enable system flexibility and easier future updates. Then, the steps required for the implementation of each scenario are defined. They consist in general instructions for the implementation of the technology starting from the AS-IS status. The gap analysis provides also a cost and time evaluation for the overall migration completion, which is very relevant for the economic analysis in the last step.

The "Value added identification" step aims to justify the efforts spent by the company on the use case implementation. The complete definition of the migration scenarios allows to evaluate the quantitative and qualitative improvement of the system. Value added has a double meaning:

1. KPI improvement: a quantitative and measurable increase of system performance with respect to the AS-IS situation.
2. Unmeasurable advantage: a benefit derived from the new system which has a positive impact on one of the three migration dimensions (technical, operational and human). The unmeasurable advantages of the scenario have to be coherent with the value added relative to the technological enablers selected in the scenario.

Every scenario can count on different features and peculiarities, hence the KPIs improvement and the unmeasurable advantages could differ from one another. For this reason, added value identification has to be performed for each scenario.

The "Economic analysis" step consists of a cost-benefit analysis to justify the investment in digital transformation is the last step of the methodology. This analysis is performed comparing the TO-BE situation with respect to the AS-IS situation and it deals with quantitative results. First, the KPIs improvement is translated into money and the results are addressed as Economic KPIs or Benefits. The Total Cost of Ownership (TCO) could be adopted for cost evaluation [12]. It is recognized as an industry-standard method for the economic analysis for IT and other enterprise issues due to its holistic view of costs across enterprise boundaries over time.

The investment is evaluated with a Discounted Cash Flow approach. The Net Present Value (NPV) is an economic index that measures the value acquired by the company associated with the investment. Moreover, the Profitability Index (PI) and Internal Rate of Return (IRR) are also considered in Step 6. The Discounted Payback Period (DPP) represents the number of years necessary to break even from undertaking the initial expenditure recognizing the time value of money. NPV, PI, IRR and DPP are indexes which defines the economic success of a project, and they are also a measure to compare different solutions, as in the case of multiple scenario evaluations. The final decision to implement or not one or more of the defined project is the arrival point of the assessment.

3 Methodology Application: A Durable Goods Company Within the H2020 FAR-EDGE Context

The application use case is provided by a world's top white appliances manufacturer company. The company needs to overcome the rigidity of the sorter that distribute the finished goods at the end of the assembly lines to different bays in order to be stocked in the warehouse. The static logic implemented by the sorter in the AS-IS situation does not allow the required flexibility in the use of the bay resources, and to adapt to the actual mix of goods produced.

Specifically, the FAR-EDGE based solution is composed of smart objects implementation that will be managed by an independent Edge Node, which is responsible to process the information from the field in order to modify the dispatching policy accordingly. All these elements communicate within a Distributed Ledger smart contract called Collaborative Sorting, which sets the sorting method depending upon the state of each bay. In fact, based on the operators' decision or based on the current workload each bay can be on-line, off-line, filled, empty or partially filled.

Furthermore, a smart object implementation is used for installing a circular smart conveyor that will be surrounded by many smart bays that, using a Distributed Ledger, can collaborate as peers towards the common goal of a new flexible streaming system. In this case, an algorithm is needed to provide indications to manage the physical flow of products flowing between the conveyor, the sorter and the bays, and the information flow, namely the different communications that the Edge nodes exchange with each other.

In this methodology application case, the TO-BE scenario is clearly defined by the innovative architecture proposed by the project. This way, pointed out the above described TO-BE scenario, a SWOT analysis has been applied in order to comprehend the strengths, weaknesses, opportunities and threats of the FAR-EDGE Architecture implementation within the company plant. The analysis underlined how the sorting system of the plant represents an issue which could be solved and improved by FAR-EDGE. The support activities involved are the firm infrastructure which should lead the change. In fact, the connectivity between the sorter and the bays will be implemented, installing information access points and instantiating security mechanisms in order to fully integrate the information exchange between production and logistic processes.

The technology development which should implement and maintain the new architecture enabling improved data management. Also, the human resource management will be impacted as it is responsible for employees training and hiring digital skilled people. From the operational perspective, a good ability for processes management will be requested, supported by a particular attention to utilize automation and analytics activities as enabling procedures to decrease not value-added activities and to gain more competitiveness. At this point, the KPIs were identified after to having get a clear idea about the potential application field of the project: (i) sorter OEE and (ii) sorter reconfigurability.

The gap analysis identifies two aspects of the FAR-EDGE functionalities could be instantiated for the TO-BE scenario realization: (i) Cyber Physical System, as all the smart objects are represented and configured on the Distributed Ledger; and (ii) Simulation Services needed to suggest an optimized sorting policy exploiting real-time data from the field.

The final consultation with the experts estimates a possible growth of the KPIs identified in the first step. The Sorter OEE will increase by 5% and the sorter reconfigurability will improve significantly, coherently also with the company lean strategy. Regarding unmeasurable advantages, it is possible to list: (i) a synchronization between field and simulation that could be further used for other purposes; (ii) an innovative machine autonomy which creates machine-to-machine collaboration in the operational dimension. For the human dimension, the number of stressful emergency breakdown situations are reduced and the reconfiguration of the system is not manual any more. On the other hand, the IT system becomes more complex and requires advanced widespread technical skills from the employees. The economic appraisal is then performed. First, the upfront costs and the recurring costs have to be computed. As a second phase, the improvement KPIs are transformed into economic benefits:

- An OEE improvement has a direct effect on the throughput of the system. It has been verified that an additional 5% of the sorter OEE implies an additional 5% of the system throughput.
- The improvement of the system reconfigurability has been considered exploiting a situation analysis: the flexibility acquired will impact on the Total Cost of Change (TCC) of the system in three different ways.

Finally, the Discounted Cash Flow appraisal is calculated. Alongside a consistent initial investment, the NPV is significantly positive after 10 year. The DPP occurs slightly after the third year. In the end, the IRR has been computed, with a value of 34.9%. Since IRR is greater than the discount rate has been utilized (10%), also this index confirms the profitability of the investment in the TO-BE scenario. The results cannot be evaluated without considering the FAR-EDGE context and all the specific hypothesis has been considered.

4 Conclusions

In spite of the benefits, the important impact of the potential technological and organizational changes has prevented many companies from investing in Industry 4.0 capabilities, especially for two relevant common barriers: (i) high investment costs due to a lack of Industry 4.0 suitability of the existing production infrastructure and (ii) missing transparency or quantification of benefits [13]. For these reasons, this paper presents a holistic methodology for business case evaluation related to digital transformation, that has been developed and applied to FAR-EDGE project use cases. Overall, the application case has confirmed the company's perception of the benefits that the adoption of the FAR-EDGE solution can bring for the use case. However, the advantage that the company has acknowledged is not only in the calculation of the economic indicators, but also in an increased awareness about the migration activities, with the associated efforts, costs, uncertainties and expected benefits.

In general, the revolution that the new automation paradigms are introducing involve as the first actor the employee and its way to adapt to this kind of changes in operations.

Acknowledgements This project has received funding from the European Union's Horizon 2020 research and innovation programme under grant agreement No. 723094.

References

1. ElMaraghy HA, Wiendahl H-P (2009) Changeable and reconfigurable manufacturing systems
2. Calà A, Luder A, Boschi F, Tavola G, Taisch M (2018) Migration towards digital manufacturing automation—an assessment approach. In: Proceedings—2018 IEEE industrial cyber-physical systems, ICPS 2018, pp 714–719
3. Aboud C, Buvat J, Crummenerl C, El Aoufi H, Kar K, Sengupta A, Solis B (2017) The digital culture challenge: closing the employee-leadership gap
4. Boschi F, Zanetti C, Tavola G, Taisch M (2018) From key business factors to KPIs within a reconfigurable and flexible cyber-physical system. In: 23rd ICE/ITMC conference- international conference on engineering, technology, and innovation ICE/IEEE ITMC international technology management conference, January 2018
5. Calà A, Boschi F, Fantini P, Lüder A, Taisch M (2019) Migration strategies towards the digital manufacturing automation. In: Soldatos J, Lazaro O, Cavadini F (eds) The digital shopfloor: industrial automation in the Industry 4.0 era. River Publisher
6. Calà A, Soldatos J, Boschi F (2018) Migration strategies to digital automation in the Industry 4.0 era—a guide for smooth transition to Industry 4.0
7. Cala A, Luder A, Cachada A, Pires F, Barbosa J, Leitao P, Gepp M (2017) Migration from traditional towards cyber-physical production systems. In: Proceedings—2017 IEEE 15th international conference on industrial informatics, INDIN 2017, pp 1147–1152
8. Gürel E (2017) Swot analysis: a theoretical review. J Int Soc Res 10(51):6–11
9. Helms M, Nixon J (2010) Exploring swot analysis—where are we now?: a review of academic research from the last decade. J Strateg Manag 3(3):215–251
10. Azzone G, Bertelè U (2017) L´impresa - sistemi di governo, valutazione e controllo
11. Hergert M, Morris D (1989) Accounting data for value chain analysis. Strateg Manag J 10(2):175–188
12. Grobelny MS (2017) Evaluating the total cost of ownership for an on-premise application system
13. Pessl E, Sorko SR, Mayer B (2018) Roadmap Industry 4.0—implementation guideline for enterprises. Int J Sci Technol Soc 5(6):193

Lean Healthcare: How to Start the Lean Journey

Fabiana Dafne Cifone, Alberto Portioli-Staudacher and Alberto Silla

Abstract In this paper we present how ASST Cremona, an Italian hospital belonging to the National healthcare system, started its lean journey. In order to spread out the lean culture within the company and to engage people, the first initiative launched was a 4 months improvement project within Internal Medicine department. This was carried out following A3 framework and was aimed at improving the service level provided to the patients, in terms of the critical KPI represented by the length of stay. The improvement project was successful leading both to a strong reduction in the length of stay and to the spread of improvement culture within the hospital, that will continue its lean journey in different departments.

Keywords Lean healthcare · Italy · Discharge process

1 Introduction and Company's Profile

Healthcare systems—especially hospitals—face serious challenges, from a rapid growth in patients demand to higher quality expectations, in terms of both application of new expensive treatments and service level as shorter waiting and processing times [1]. Furthermore, hospitals have limited budgets, which force than companies in utilizing scare resources effectively and efficiently [2]. There is extensive literature about hospitals that search for solutions to increase productivity and apply lean management concepts to operations (e.g. [3, 4]). As a matter of fact, in the US, lean

F. D. Cifone (✉) · A. Portioli-Staudacher
Department of Management, Economics and Industrial Engineering, Politecnico di Milano, Via Lambruschini 8/B, 20156 Milan, Italy
e-mail: fabianadafne.cifone@polimi.it

A. Portioli-Staudacher
e-mail: alberto.portioli@polimi.it

A. Silla
Operations Management Department, ASST Cremona, Cremona, Italy
e-mail: a.silla@asst-cremona.it

M. Rossi et al. (eds.), *Proceedings of the 6th European Lean Educator Conference*, Lecture Notes in Networks and Systems 122,
https://doi.org/10.1007/978-3-030-41429-0_32

experiences within hospitals seem to have significantly increased over time, and in the UK, the government chose lean management as a mean to reform its public sector.

Despite the interesting results achieved by those organizations, in Italy only few hospitals attempted to implement a system-wide approach including lean thinking in their operations. Italian healthcare is still in its early stages of searching for operational excellence [5]. Hence, there is a wide room for improvement in Italian healthcare in order to decrease costs and increase service level (e.g. [6]). According to 2018 annual report of Observatory on Healthcare Organizations and Policies in Italy [7], hospitals are now more opened to new managerial tools and methods aimed not only at redesigning programs and control systems, but also at strengthening operation management, intended as patient logistics and asset management.

The example we present hereby is from ASST-Cremona, an Italian hospital belonging to ATS Valpadana: its board saw in the lean culture the proper way to cope with industry's challenges. But how to start a lean journey within a hospital? For ASST-Cremona the preliminary step was to create a new department in charge of improving the service level, studying and planning all activities related to surgery and bed management. Then, the first proper lean initiative launched was within the critical department of Internal Medicine, where patients coming from the Emergency room are nursed. According to the literature, not only around 27% of patients in Emergency room are then hospitalized in the Internal Medicine department, but also 60% of them is elder than 75 years old. All these factors led to significant criticalities in the discharging process of patients, with a related increase in the average hospital stay, causing problems in both management and clinical activities. This context easily justifies the need of the improvement project carried out as first lean activity within the hospital.

2 The Project

The improvement project lasted 5 months and was carried out by Lean Excellence Centre of School of Management of Politecnico di Milano in collaboration with and supervision of the Operation Manager, Bed Manager, Ward Chief, Head nurse and Medical direction of the hospital. The total workload assigned to the project considering all involved people is around 700 h. Moreover, the project was developed according to A3 framework, as a useful structured problem solving tool.

2.1 Problem Background

The medical ward of the department under analysis has a capacity of 58 beds splitted into two blocks of 30 and 28 each. Personnel is the same for the two blocks and rotates according to work shifts. The ward hospitalizes multi-pathological elderly patients, with an age average of 76.8 years, who often require assistance even after being discharged by the hospital. According to 2018 data, the ward is characterized by high saturation of beds (99.9%) and around 30% of patients has a length of stay (LoS) higher than the maximum permitted by law (11 days). Furthermore, the LoS variance is about 52 days. All these aspects lead to face a high managerial complexity in the ward, that affects not only patients turnover (availability of beds), but also the service level provided to patients and the overall hospital cost structure. Managing high/variable LoS will enable the company to better manage beds, thus to satisfy the real demand of the ward, to reduce waiting times for patients, to provide higher service level. However, even the high expectations related to this project, internal staff was not convinced about its success due to resistance to change, lack of knowledge about lean approach, and low commitment of the department.

2.2 Breakdown the Problem

Thanks to historical data and to interviews, the overall LoS has been studied and split into six different main activities: (1) Ward entrance, (2) Clinical activity, (3) Blood tests, (4) Instrumental tests, (5) Medical Advice, (6) Discharge process. These activities and related times are analysed through Pareto in order to understand the ones most responsible for the long LoS. Clinical and discharge times contribute to 80% of the total time. However, due to technical reasons, the focus of the project is only on discharge time (which accounts for 37% of the total LoS).

Moreover, discharge process is further clustered by the hospital into 9 groups[1] depending on the different typologies. Related number of patients and average time were analysed and the results are reported in Table 1.

Thanks to a crossed ABC analysis, we focused on code 8, that showed the highest number of occurrences as well as an average time to be processed significantly higher than the overall ward average.

After several Gemba walks and interviews, we used a Makigami chart to study and map the discharging process of code 8. It highlighted that around 99% of discharging time was waiting time. Discharging is indeed a really fragmented process, where several actors are involved that have only partial information. The only structured moment when all the information regarding the patients are shared among all actors was the briefing held by the Multidisciplinary equip (i.e. Head Nurse, Bed Manager, Case Manager and Social Worker). However, physicians, who have a critical role in the discharging process, did not take part to this moment. It is worthy to stress that communication among actors inside and outside the ward was mainly paper-based and the FAX was the primary tool used. Many delays were caused for example by

[1]Discharge classification: 1. Ordinaria domicilio del paziente; 2. Volontaria; 3. Trasferimento ad un altro istituto di cura per acuti; 4. Deceduto; 5. Dimissione ordinaria presso Residenza Sanitaria Assistenziale; 6. Dimissione al domicilio del paziente con attivazione di ospedalizzazione domiciliare; 7. Trasferimento ad altro regime o tipologia di ricovero all'interno dello stesso istituto; 8. Trasferimento ad un altro istituto pubblico o privato non per acuti; 9. Dimissione ordinaria con attivazione di assistenza domiciliare integrata.

Table 1 Discharge codes

Discharge code	Number of patients	Average required time
1	1480	8.27
2	15	3.6
3	10	18.8
4	98	7.56
5	160	7.9
6	1	22
7	131	13.17
8	146	17.26
9	2	14
Total	2043	9.19

not readable forms that needed to be re-filled. This situation was even emphasized by the lack of a responsible of the process, who monitor and check the progress of discharging process.

2.3 *Target*

Must have goal regarded the reduction of the average LoS for patients discharged with code 8 from 17 to 14 days (−17.6% of the total time). A *Nice to Have* target was also set regarding a further reduction of average LoS to 11 days (−35% of the total time).

2.4 *Root-Causes Analysis*

We interviewed different actors involved in the process, checked emails and medical records to grasp the causes of long duration of discharging process for code 8. At the end, Ishikawa diagram allowed to give structure to discovered causes and to define root-causes (Fig. 1).

Thanks to a FMEA, causes were linked to the relative failure mode and failure effect. Assigning a score to the severity of the effect, occurrence of the failure mode, and to detectability enabled to compute Risk Priority Number (RPN) for each cause. According to RPNs, an ABC analysis was performed to prioritize causes to address our effort coherently. Table 2 summarizing the results of the ABC analysis.

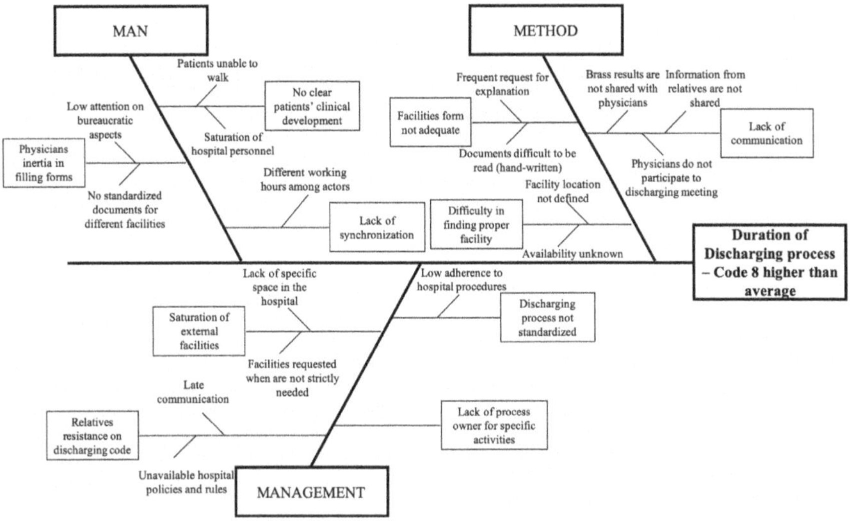

Fig. 1 Ishikawa diagram

Table 2 ABC analysis on causes

Causes	RPN	Cum. %	Class
Saturation external resources	1000	15.2	A
Lack of long-term care facilities	1000	30.3	A
Lack of monitoring system for availability of external facilities	1000	45.5	A
Patient is not more able to walk	810	57.8	A
Patient needs subacute centers	500	65.4	A
No check before sending forms	450	72.2	A
Talks with relatives not defined in terms of time and methods	450	79.0	A
Uneffective information coordination and sharing	300	83.6	B
Lack of control on the overall process	300	88.1	B
Hospital policy on discharge not defined	280	92.4	B
3 different forms to fill for different facilities	160	94.8	B
Usage of paper forms	100	96.3	C
No structured tools to share information	96	97.8	C
Resistance from family members on the discharge mode selection	80	99.0	C
Responsibilities not defined on forms sending	30	99.4	C
Synchronization among activities	14	99.6	C
Use of Fax as communication system	12	99.8	C
Facilities' form not correctly filled	12	100.0	C

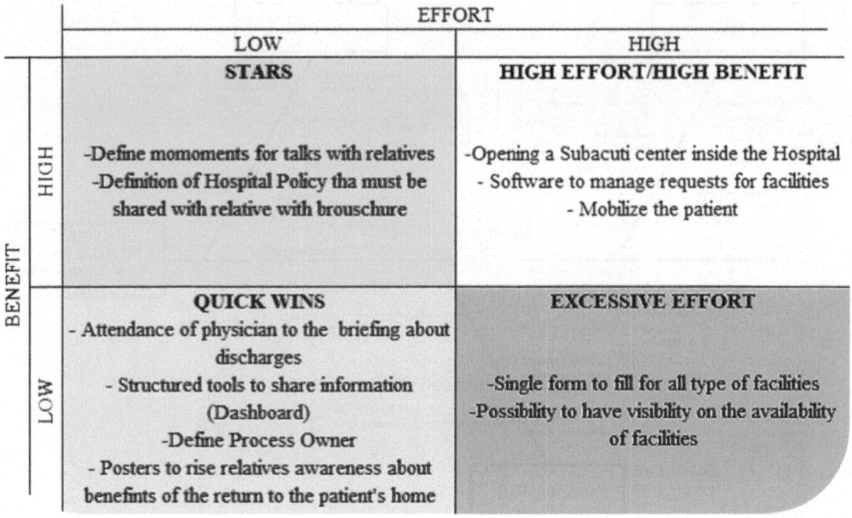

Fig. 2 Effort-benefit matrix

2.5 *Countermeasures*

A list of countermeasures was developed after several brainstorming sessions with the involved actors. In order to pick the most profitable countermeasures and to address there our effort, an effort-benefit analysis was carried out. Here following the results of the analysis (Fig. 2).

2.6 *Implement Countermeasures*

According to the results presented and due to time constraints, we focused on quick-wins solutions. A Gantt chart was developed to guide the implementation.

We defined rules and structure for the briefing about discharges. As briefly explained above, physicians usually did not attend the briefing meeting about the discharge, causing longer coordination and waiting time. The natural consequence was to set a meeting in which physicians must participate in order to share the required information on their patients and to receive feedbacks from the multidisciplinary equip. Moreover, we relied on some visual management tools to tackle two different countermeasures. Firstly we designed posters to raise awareness about the risks connected to an improper stay in the Hospital: they are attractive, easy understandable by everyone and placed in locations frequently attended by relatives. Secondly, we created a visual dashboard not only to share information in the ward about critical discharges in order to avoid information's fragmentation among stakeholders, but also to monitoring results of LoS overtime. Lastly, we identified a Process Owner of discharge activity named as "Ward Discharge Manager". The Head Nurse was selected for this role, that implies continuous monitoring of the discharge process, summarizing all the patients' information updating the dashboard, identifying problems and stops in the process, soliciting activities to be done in case of delays.

2.7 Monitor Results and Process

After a trial period of two weeks during which we performed several Gemba walks, we were able to summarize criticalities and further improvements of our countermeasures in a Snag list. Based on this, some adjustments to our countermeasures have been done. For example, a final version of the dashboard was defined in agreement with Chief ward and Head nurse, so to make operators directly involved in the project. Indeed, this was mainly done to increase both the commitment and the knowledge about the lean approach of department.

The final layout allows to keep under control the main steps of the patients' care process adding important information, like exams, and eliminating the useless one, with respect to the first draft of the dashboard.

In order to check if the implemented countermeasures had a positive impact on the organization and allowed to achieve the *must to have* target set in the initial phase of the process, a monitoring phase was needed. In the period immediately antecedent to the implementation, the average LoS was of 18 days (green line) while after the implementation the average decreased to 13 days. The *must have* target has been successfully achieved. There are just 4 points (22%) out of control over 18 observations: these are due to the intrinsic nature of hospital processes. Following the achieved results (Fig. 3).

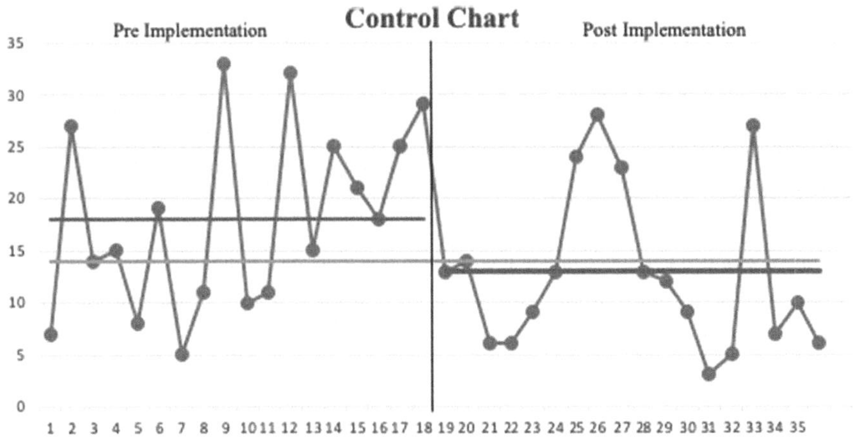

Fig. 3 Control chart

2.8 *Standardize and Share Success*

In the closing phase some future steps for the company have been identified for further improvements: (1) due to the positive effect obtained in Medical ward, countermeasures can be shared and standardized also in other wards of the Hospital; (2) the dashboard can be replaced by a digital and interactive screen directly interconnected to the digital medical record. Moreover, we also suggested a training session in order to easier the change management and to make operators more involved.

3 Discussion and Conclusion

Even considering the strong effort required for the initial phases of the analysis, the lean approach revealed to be fundamental to address and tackle the root-causes of the long patients' stay in the Medical ward. Thanks to this successful first lean initiative, ASST-Cremona started the lean journey, aimed at improving the quality of service provided to patients. Through standardization, visual management tools and monitoring system Medical ward is now able to deliver a better quality to patients. Thank to this initiative, ASST-Cremona started to spread out the lean culture. Indeed, the success of the project within the Medical ward allow people to understand that a deep training on the topic is not needed to make the difference when guided by an expert. This deeply increases people commitment, affecting their resistance to change, because they can speak their voce loudly being heard by the management. According to this, ASST-Cremona is willing not only to start new projects, but also to define professional positions as lean experts, and train or hire people able to support the management in this transition.

Acknowledgements We would like to thank *Dr. Rosario Canino*, Medical Director of ASST Cremona, *Dr. Matteo Giorgi Pierfranceschi*, Director of Operations of Internal Medicine, *Serena De Marco* and *Marco Spadari*, brilliant students of Politecnico di Milano, for their collaboration and support.

References

1. Graban M (2016) Lean hospitals: improving quality, patient safety, and employee engagement, 3rd edn. Productivity Press
2. Bhasin S (2015) Lean management beyond manufacturing: a holistic approach. Springer
3. de Souza LB (2009) Trends and approaches in lean healthcare. Leadersh Health Serv 22(2):121–139
4. Waring J, Bishop S (2010) Lean healthcare: rhetoric, ritual and resistance. Soc Sci Med 71(7):1332–1340

5. Portioli-Staudacher A (2008) Lean healthcare. An experience in Italy. In: Koch T (eds) Lean business systems and beyond. IFIP—The international federation for information processing, vol 257. Springer, Boston, MA
6. Zidel TG (2006) A lean guide to transforming healthcare. ASQ Quality Press
7. Cergas (2018) Annual report of observatory on Healthcare Organizations and Policies in Italy

Towards a Data-Based Circular Economy: Exploring Opportunities from Digital Knowledge Management

Federica Acerbi, Claudio Sassanelli, Sergio Terzi and Marco Taisch

Abstract Manufacturers are called to cope with always more complex challenges. Among all, the reduced amount of resources available in our planet forecasts a dramatic future scenario. Circular Economy (CE) is one of the most promising strategies. In view of the increasing interest in both Lean Thinking and CE, a first systematic literature review (SLR) has been performed to investigate the interlink between these two concepts. It unveiled common grounds and gaps, as the limited number of studies dealing with knowledge management (KM). Considering the new value obtained by knowledge, a second SLR has been conducted to study the application of KM as support for CE. Researchers explored KM application on all the three levels of CE paradigm (micro, meso and macro), almost ever highlighting the strategic role of technological systems to exploit data management under a circular perspective. Directions for future researches are proposed as main contribution of this paper.

Keywords Lean thinking · Circular economy · Data management · Knowledge management · Digitization

1 Introduction

Nowadays, the reduced amount of resources available on our planet forecasts a dramatic future scenario. Primary material consumption is expected to double, reaching a top value of 160 gigatonnes in 2060 [1]. To limit this problem, countermeasures

---

F. Acerbi (✉) · C. Sassanelli · S. Terzi · M. Taisch
Department of Management, Economics and Industrial Engineering, Politecnico di Milano, Via R. Lambruschini 4/b, 20156 Milan, Italy
e-mail: federica.acerbi@polimi.it

C. Sassanelli
e-mail: claudio.sassanelli@polimi.it

S. Terzi
e-mail: sergio.terzi@polimi.it

M. Taisch
e-mail: marco.taisch@polimi.it

M. Rossi et al. (eds.), *Proceedings of the 6th European Lean Educator Conference*, Lecture Notes in Networks and Systems 122,
https://doi.org/10.1007/978-3-030-41429-0_33

have been studied and the most promising among the possible strategies is the Circular Economy (CE) application, shifting from a make-use-dispose paradigm to a make-use-reuse one. The extant literature suggests Product-Service Systems (PSS) business models as the most suitable to address CE and its benefits [2–4], suggesting to integrate CE and Industry 4.0 to better pursue the ReSOLVE's "Exchange" archetype, leading to the adoption of new technologies [5]. It must not be neglected that, information technology, strictly connected with advanced manufacturing practices, is impelling the shift towards CE, aiming at de-materialization and closing traditional linear lifecycles of both products and industrial systems [6]. Manufacturing companies are not completely ready to cope with this complex scenario. They have to deliver complex solutions, managing huge quantities of data along the extended and circular lifecycle of the solution. Aimed at the design and delivery of the traditional product, several tools and methodologies deriving from Lean Thinking (LT) [7–9] have been proposed in the extant literature to manage, exploit and share lifecycle knowledge in a valuable way. On the heels of [10], the goal of this paper is to understand how the application of different CE strategies can be supported by implementing LT approach. A special focus is dedicated to understand the role of digitization phenomenon, data and Knowledge Management (KM)-driven systems, to propose future research directions.

The paper is structured as follows: (1) introduction, (2) research context, (3) research methodology, (4) literature review results, (5) conclusions and further researches.

2 Research Context

In this section, the research context about CE, Lean Thinking and KM and digital technologies is provided.

2.1 *Circular Economy (CE)*

CE is defined as "*an industrial economy that is restorative by intention and design*" [11] and it aims at slowing, narrowing and closing the resources loop. This paradigm is studied under the wider umbrella of sustainability [12] which is based on three pillars: social environmental and economic, also called Triple Bottom Line (TRL). However, since "*sustainable development is development that meets the needs of the present without compromising the ability of future generations to meet their own needs*" [13], CE differs from sustainability adding to this concept the need to reduce resources consumption by closing the loop. The CE paradigm is based on the following three main principles: (i) preserve and enhance natural capital, (ii) optimise resource yields and (iii) foster systems effectiveness. These can be addressed by removing wastes and retaining as much value as possible from the products and materials involved [14] also using renewable energies and avoiding toxic chemicals. Moreover, combining different building blocks as product design, new business models and reverse logistics [15], this paradigm can be applied at three main levels: micro (firm or product level), meso (interfirms' initiatives) and macro (regional and national initiatives and policies) [16].

2.2 Lean Thinking and Knowledge Management

Lean Thinking was introduced by Womack et al. [9], who applied its principles mainly to manufacturing and operations management according to the main Lean Manufacturing concept, "Doing more with less". Throughout time, Lean Thinking has been defined as a "*dynamic, knowledge-driven, and customer-focused process through which all people in a defined enterprise continuously eliminate waste with the goal of creating value*" [17]. In this conception, customer satisfaction is of paramount importance and wastes, the so-called *muda*, represent everything not aligned to this. As a result, Lean gathers all the practices able to remove and elude wastes and add value to the solutions delivered [18]. Wrapping up, Lean Thinking can be defined as based on waste identification and value focus [19]: in this process, knowledge acquires a strategic role, increasing its value related to both the solutions delivered and the entire organization involved.

2.3 Knowledge Management: Systems and Technologies

KM is becoming a fundamental business function for organizations who realize that competitiveness depends on effective management of intellectual resources [20]. KM "*is complex and multi-faceted concept supporting the creation, transfer, and application of knowledge in organizations*" [21]. Under this perspective, integration of Process, People and Technology has to be taken in consideration [22]. While process is aimed at creating value and avoid wastes, people are selected and trained to foster collaboration and exchange of knowledge in a cross-functional way. Last, technologies incorporate design and manufacturing supporting tools (computer-aided design systems, machine technology and digital manufacturing), and also soft tools supporting the efforts of people during the development process. In this new era, technologies are covering a prominent role. There are different solutions for archiving and accessing the increasing amount and set of information types [23], both internally and externally to companies' boundaries: traditional information systems, digital technologies, IT collaborative systems (e.g. PDM, PLM, databases, web platforms) and authoring software (e.g. CAD, simulation tools, etc.).

3 Methodology

In order to achieve the goal mentioned above, a literature review has been conducted. It was first investigated CE and Lean Thinking, analyzing how Lean Thinking has supported the implementation of the CE paradigm in manufacturing companies. A first search by keywords was conducted on Scopus applied on title, keywords and abstract without time and document type limitations. Using "Circular economy" and "lean", 19 results were obtained: after a title and abstract analysis, 1 was discarded since not in line with the research scope. After the entire manuscripts analysis, 15 have been selected. However, based on the results obtained through this first analysis, a second one was needed to explore more deeply the role of data and KM systems in CE. To do this, a second search by keywords has been conducted on Scopus. Following the same criteria of the first one, the string ("Knowledge management" OR "data management") AND "circular economy") led to 23 articles: 11 were excluded through title-abstract-keywords analysis as considered out-of-scope. Then, the entire manuscript evaluation led to 7 articles selected. To be more precise, papers discarded, along the entire process, dealt with policies issues not compliant with the topic of this research. The remaining papers were analyzed and in the following section the results obtained are reported.

4 Literature Review Results

This section wants to clarify the link between Lean Thinking and CE, analyzing the extant literature in this twofold context. Section 4.1 provides a glimpse of the results coming from the first analysis on the two concepts, leading to Sect. 4.2, which reports a major focus on the role of data and KM in the context of CE.

4.1 Lean Thinking Supporting Circular Economy

Even though the two paradigms support each other, few researches have been conducted to explore their interrelations, mostly focusing on either business models or operations management. In particular, Peter Nadeem et al. [24] investigated from a general viewpoint what the two pillars, value and waste, mean, from the two perspectives, and their potential coalescences. Waste, under a Lean lens, is any activity which does not add value to the customer, while represents "food" from a CE perspective. Value, from a Lean perspective, depends on customers' requirements ("everything they are willing to pay"), while for CE is waste reduction. Regarding the link between the two concepts in operations management, Kurilova-Palisaitiene et al. [25] studied how the lean philosophy supports the improvement of remanufacturing process, which is one of the possible CE strategies at micro level. Moreover, Rieckhof and Guenther [26] match the value stream mapping together with the material flow analysis in order to support the CE under the TBL umbrella. Bocken et al. [27] transferred the concept that Lean Start-up has to bigger entities in order to perform the experimentation of circularity in lager organizations. Wrapping up, one of the main challenges unveiled with this review is the incapacity to manage in a structured way information and knowledge to allow their usage on the long-term horizon with a CE aim [25].

4.2 Knowledge and Data Management in Circular Economy

The analysis of the contributions selected investigating how KM supports CE has been structured on a threefold level (micro, meso and macro), based on [16].

Concerning the micro level, Jensen and Remmen [28] analysed how KM supports different employees or divisions in the application of CE in manufacturing companies. Indeed, starting from an analysis of three different sectors (Automotive, Aerospace and Ship), with a focus on the Product Stewardship, the authors stressed the challenges and opportunities faced by Original Equipment Manufacturers in applying CE paradigm. In particular, the common problem emerged among the three sectors concerned the information management along the entire life cycle of the product. This led authors to propose and implement information systems such as the Enterprise Information System (EIS) or the Product Life Cycle Management System (PLM) to store all data concerning each single product and share them among different actors. Focused on the entire products life cycle, with an attention to the maintenance activities, Olivier et al. [29] proposed a framework to boost sustainable performance management by covering the TBL. It is underlined the need of digital platforms to store data under the light of KM for the monitoring of the product life cycle. Process-Aware Information Systems (PAIS) is the technology used, accompanied by the development of an ontology to grasp also the semantic meaning capability.

Indeed, at micro level, the necessity to define standard data models to structure the information collected is highlighted as one of the main challenges for firms. Some technologies and systems, as digital platforms, have been studied as potential support to achieve this goal aiming at both collect and distribute knowledge. However, dealing with the sharing and distribution of knowledge, when different industrial actors are involved, information security and confidentiality issues arise as problems to consider.

At meso level, in most of the papers analysed, scholars studied how companies, within or across sectors, might exchange waste and by-products as resources with the ultimate goal of reducing waste and resources consumption. Under this perspective, Mulrow et al. [30] have studied how to implement Industrial Symbiosis (IS) focusing on the sharing of services more than the exchange of by-products. Moreover, different authors investigated the existing solutions in terms of digital tools developed to pursue the objective of exchanging waste converting them into resources. These are usually input-output (I-O) matching tools or collaborative platforms [31, 32]. In particular, Raabe et al. [32] proposes a "By-products exchange network model" that is a system architecture to support the decision-making process and to enable companies to exchange waste or by-products as resources. This model verifies and stress the economic feasibility through the simulation of different scenarios. Halstenberg et al. [31] explores possible I-O matching tools describing the different data sources used while identifying the criticalities encountered by companies in implementing IS. Even though promising results could be achieved, few industrial implementations are detected. As for the micro level, digital technologies and information systems could support in overcoming this problem, although: "*A challenge regarding IT support for IS lies in the identification, extraction, and provision of relevant data to sufficiently describe a by-product and allowing it to be evaluated for its usefulness in other industries*" [31]. Indeed, in inter-firms' initiatives, the focus is placed on the "exchange" and "share" of by-products, services and waste and these strategies are reinforced by usage of I-O tools and technologies or collaborative systems.

Finally, the KM arose as an essential aspect also at macro level. Indeed, in order to implement CE in the economy system, adequate sustainable policies are necessary to be studied. Indeed, to achieve this goal, Marra et al. [33] propose to combine different disciplines and knowledge domains to ease the decision-making process and to take the best decisions for all the sustainable aspects at macro level. Here digital technologies are not discussed and it is not proposed a method to structure the tacit knowledge in order to enable its reusability with a long-term view. On the other hand, digital technologies are reconsidered at macro level for the management of complex systems including different factors and their relations [34]. Luo et al. [34] proposes the structural design of a three-layer web-technology based network management information system platform, where the server is coupled into web-server and database server to ensure flexibility. This platform includes both enterprises and industrial production information management systems at regional level in order to use and share all the information needed to pursue the objective of regional cleaner production management and to support the decision process.

At macro level is evident a misleading interpretation of the CE concept, which is confused with the more general "sustainable" notion. Further researches should be undertaken in this field to explore more CE oriented policies being supported by a structured KM system which could be reinforced by using technological systems.

5 Conclusions and Further Researches

This paper has proposed a two-step literature review, analyzing the link between (i) Lean Thinking and CE and (ii) CE and KM. Despite of the great involvement of academics in the contexts of Lean Thinking, CE and KM, taken singularly, contributions dealing with CE taking in consideration also the potential role of KM methods and tools, guided by Lean principles, are limited. This suggests that there is room for further studies. Results of this study unveil that KM has been considered crucial to pursue the CE application in the manufacturing sector and thus, to reduce waste generation and resources consumption. Under the twofold light of KM and Industry 4.0, the digitization of such an approach has been studied. Indeed, for a structured collection of data that ease both their interpretability and their transformation in usable knowledge, some systems and technologies have been proposed in literature. Nevertheless, some gaps arise and future research directions are proposed:

- To investigate what information and thus, what data are necessary to be collected according to the phase of the life cycle of the product/solution delivered.
- To explore how to structure data collection. It has been stressed, in the papers analysed, that to enable the sharing of information, standard languages are necessary and also a standard data model is required.
- To explore which technologies and models can be used by SMEs, often not ready to use PLM systems from both an organization and financial perspective, in order to pursue the CE implementation and use an appropriate KM.
- To explore which technologies and models are most used according to the level at which CE is tackled (micro, meso and macro).
- Considering as a further analysis dimension the ReSOLVE framework proposed by The Ellen MacArthur Foundation [11], to detect which are the dimensions involved in this context.
- To clarify the concepts of sustainability and CE, that become blurred when applied in the manufacturing sector, being declined both in the TBL.
- To deeper investigate the concept of knowledge under the twofold light of Lean Thinking and CE. Indeed, focusing on the value represented by knowledge, one challenge is to understand how to share the right information to the right actors in the right moment, supported by different technologies.

Finally, to answer to some of the above open points, this research has to be improved through the adoption of a more structured, extensive and replicable method as a systematic literature review. Other scientific databases and keywords will be employed: in this analysis will be considered and tangled different concepts related to the digitization area, product lifecycle management and circular business models.

Acknowledgements This work has received funding from the European Union's Horizon 2020 research and innovation programme under grant agreement No. 760792.

References

1. OECD (2019) Global Material Resources Outlook to 2060. OECD
2. Rosa P, Sassanelli C, Terzi S (2019) Circular business models versus circular benefits: an assessment in the waste from electrical and electronic equipments sector. J Clean Prod 231:940–952
3. Tukker A (2015) Product services for a resource-efficient and circular economy—a review. J Clean Prod 97:76–91

4. Rosa P, Sassanelli C, Terzi S (2019) Towards circular business models: a systematic literature review on classification frameworks and archetypes. J Clean Prod
5. Bocken NMP, Short SW, Rana P, Evans S (2014) A literature and practice review to develop sustainable business model archetypes. J Clean Prod 65:42–56
6. Porter ME, Heppelmann JE (2014) How smart, connected product are transforming competition. Harv Bus Rev, 64–89
7. Ohno T (1988) Toyota production system: beyond large-scale production, vol 15, no 2
8. Womack JP, Jones DT (1996) Lean thinking, vol 55, no 1
9. Womack JP, Jones DT, Roos D (1990) The machine that changed the world: the story of lean production
10. Hall RW (2010) Compression: meeting the challenges of sustainability through vigorous learning enterprises. CRC Press
11. The Ellen MacArthur Foundation (2015) Towards a circular economy: business rationale for an accelerated transition
12. Merli R, Preziosi M, Acampora A (2018) How do scholars approach the circular economy? A systematic literature review. J Clean Prod 178:703–722
13. WCED (1987) Report of the world commission on environment and development: our common future
14. Kirchherr J, Reike D, Hekkert M (2017) Conceptualizing the circular economy: an analysis of 114 definitions. Resour Conserv Recycl 127:221–232
15. Planing P (2015) Business model innovation in a circular economy reasons for non-acceptance of circular business models. Open J Bus Model Innov, 1–11
16. Ghisellini P, Cialani C, Ulgiati S (2016) A review on circular economy: the expected transition to a balanced interplay of environmental and economic systems. J Clean Prod 114:11–32
17. Murman EM, Allen T, Bozdogan K, Cutcher-Gershenfeld J, McManus H, Nightingale D (2002) Lean enterprise value: insights from MIT's Lean Aerospace Initiative. Palgrave, New York
18. Soderborg N (2008) Lean product development. In: WCBF design for Six Sigma conference. Las Vegas, Nevada
19. Morgan JM, Liker JK (2006) The Toyota product development system: integrating people, process and technology, vol 24
20. Grover V, Davenport TH (2001) General perspectives on knowledge management: fostering a research agenda. J Manag Inf Syst 18(1):5–21
21. Alavi DE, Leidner M (2001) Review: Knowledge management and knowledge management systems: conceptual foundations and research issues. MIS Q Manag Inf Syst
22. Rossi M, Morgan J, Shook J (2017) Lean product and process development. In: Netland T, Powell D (eds) The Routledge companion to lean management. Taylo & Fr, New York, NY, Abingdon, Oxon, UK, pp 1–26
23. Sassanelli C, Pezzotta G, Rossi M, Terzi S, Cavalieri S (2015) Towards a Lean Product Service Systems (PSS) design: state of the art, opportunities and challenges. Procedia CIRP 30:191–196
24. Peter Nadeem S, Arturo Garza-Reyes J, Anosike AI, Kumar V (2019) Coalescing the lean and circular economy
25. Kurilova-Palisaitiene J, Sundin E, Poksinska B (2018) Remanufacturing challenges and possible lean improvements. J Clean Prod 172:3225–3236
26. Rieckhof R, Guenther E (2018) Integrating life cycle assessment and material flow cost accounting to account for resource productivity and economic-environmental performance. Int J Life Cycle Assess 23(7):1491–1506
27. Bocken NMP, Miller K, Weissbrod I, Holgado M, Evans S (2017) Business model experimentation for circularity: driving sustainability in a large international clothing retailer. Econ Policy Energy Environ 1:85–122
28. Jensen JP, Remmen A (2017) Enabling circular economy through product stewardship. Procedia Manuf 8:377–384
29. Olivier S, Pires SP, Loures ERF, Santos EAP, Cestari JMPA (2015) Knowledge management for sustainable performance in industrial maintenance. IIE Annual Conference & Expo 2015, June 2015

30. Mulrow JS, Derrible S, Ashton WS, Chopra SS (2017) Industrial symbiosis at the facility scale. J Ind Ecol 21(3):559–571
31. Halstenberg FA, Lindow K, Stark R (2017) Utilization of product lifecycle data from PLM systems in platforms for industrial symbiosis. Procedia Manuf 8:369–376
32. Raabe B et al (2017) Collaboration platform for enabling industrial symbiosis: application of the by-product exchange network model. Procedia CIRP 61:263–268
33. Marra A, Mazzocchitti M, Sarra A (2018) Knowledge sharing and scientific cooperation in the design of research-based policies: the case of the circular economy. J Clean Prod 194:800–812
34. Luo D, Zhang H, Chen D, Zhang F (2011) Design of information management system for regional cleaner production based on B/S structure. In: 2011 second international conference on mechanic automation and control engineering, pp 2066–2069

Using Process Automation for Optimizing Engineering Practice

Angelo Corallo, Anna Maria Crespino, Mariangela Lazoi, Antonio Margarito and Claudio Rocco

Abstract By an operational point of view, the process automation is today one of the most important technique to optimize flows of documents/information/activities passing through process actors according to a set of rules and it is often the basis of the implementation of a Business Process Management Systems (BPMS) in order to improve efficiency and efficacy. Their use, in fact, is growing up in many different industries as a combination of ICT and BPM methodologies, offering a better and scalable process modelling, gaining automation, simplifying workflow design and supporting collaborative working. According to the previous considerations, the present paper tries to show BPM benefits thanks to process automation. In particular, the paper is focused on the improvement of the Design Rectification Request (DRR) and the Testing Management processes in an aerospace company. The results provide a description of the activities carried out and a discussion about their impacts.

Keywords BPM · Industrial process · Process modelling · Workflow management systems · Aerospace industry

A. Corallo · A. M. Crespino (✉) · M. Lazoi · A. Margarito
Dipartimento di Ingegneria dell'Innovazione, Università del Salento, Lecce, Italy
e-mail: annamaria.crespino@unisalento.it

A. Corallo
e-mail: angelo.corallo@unisalento.it

M. Lazoi
e-mail: mariangela.lazoi@unisalento.it

A. Margarito
e-mail: Antonio.margarito@unisalento.it

C. Rocco
Rende, Italy

M. Rossi et al. (eds.), *Proceedings of the 6th European Lean Educator Conference*, Lecture Notes in Networks and Systems 122,
https://doi.org/10.1007/978-3-030-41429-0_34

1 Introduction

By scientific literature, it is possible to observe a significant growth of BPM and BPMS suites to support quality, efficacy and efficiency of a lot of processes in the creation of products and services. Consequently, enterprises have the possibility to be supported in terms of methods and also by several IT solutions that try to enhance business activities, especially in context characterized by a high degree of technology and knowledge such as: mechanics, IT services, healthcare, management and many branches of engineering.

Among these last areas of interest for industrial researches, aerospace industry provides good examples of processes that may be simplified and streamlined. It is often a complex context, where errors made during the design phases not only generate problems for future engineering and manufacturing processes, but also requires time and efforts to restore the correctness of a wrong technical choice. These issues, then, have to be solved quickly and with right tools.

This challenge is addressed by the present paper with the aim to show benefits of web based BPMS to support processes in the engineering department of an aerospace company, considering that BPM is still a complex management practice sometimes not easy to implement and to develop towards higher stages of maturity [1].

Through the description of two applications for Design Rectification Request and Testing Management processes realized with a BPMS, the paper wants to highlight the relevance to automate business processes in an engineering department. Awareness about the performed processes increases and also, errors due to human and manual operations are reduced and in some case avoided.

The paper is structured as follows: BPM basic concepts and examples of application for processes automation are outlined in Sect. 2; the Sect. 3 is dedicated to the description of the research approach; Sect. 4 describes the development and the implementation of two processes; Sect. 5 about discussions and Sect. 6 about conclusions end the paper.

2 Background

According to some definitions used in the scientific literature the Business Process Management includes concepts, methods and techniques to support design, administration, configuration, enactment and analysis of business processes [2]. Together with a methodological approach, BPM is often implemented though Business Process Management Systems (BPMS) as technological suite, which is designed to help process re-engineering at an operative level. A BMPS, in fact, is composed of a set of software tools to support all stages (analysis, design, execution, monitoring and optimization) of implementation of a business process, as part of a BPM project.

Benefits of BPM introduction are recognizable in organizations and industries belonging to different sectors. Some of them can be cited as examples of success. A research carried out in food industry shows the impacts of BPM approach combined with the introduction of SAP ERP system [3]. One of the main findings of the study is the reduction of the high degree of reworks during organization processes (excessive activities, greater complexity, etc.), a higher automation, the better reliability of the information generated and the decreasing of large amount of forms, parallel systems, and spreadsheets used.

Another interesting research is a case study performed in Danfoss Power Electronics, a Danish manufacturing company. Here, BPM is employed to enhance the process execution (managed through SAP ERP in the early 90s) by means of data mining activities in the company. The findings of BPM project introduction show the success of the new process thinking: simplified business rules (especially in customer service a pricing area); 30% of time saving in shipping processes; better automation of invoice activities.

A third example of success can be detected at Audi Japan KK [4]. In this enterprise, the BPM implementation present as result faster decision making, process control and visibility. Focusing on the Request for Approval (RFA) as process case study, other specific benefits are highlighted: (a) cost savings; (b) time reduction; (c) increased revenues; and (d) higher productivity. Also in this case the enhancement is obtained by a rigorous BPM implementation cycle and a correct use of ERP suite.

In the scientific literature there are also some interesting hints of the BPMS applications in many industrial/service sectors and firms.

Chen et al. [5] and Pagani [6] give an in-depth analysis of BPM and web services in the context of C-Commerce, which is defined as "an application of an interorganization information system for electronic collaboration between business partners and organizational employees". Another example of BPMS is found in the research of Tomaz et al. [7] with the evaluation of the BPCE (Business Process Cooperative Editor) tool. The results present evidences that such a tool can minimize the time to develop new models, reduces the differences among similar business processes conducted in distinct organization units, enhances the quality of process design and promotes reuse. Another case study is represented by Jalali et al. [8]. In their paper, a BPM solution is proposed using an aspect oriented business process model based formally on Coloured Petri Nets. Through this case study, they also demonstrate that the adoption of aspect oriented modularization increases the reusability, while the complexity of business process models is reduced. Veldhuizen et al. [9] instead, discuss about the implementation of a BPM systems in small and medium enterprises (SMEs).

BPMS can be also considered as a sort of middleware that collects and provides inputs to others companies systems such as ERP and PLM. The role of a BPMS is different from an ERP or a PLM workflow. BPMS are not designed to handle large databases like customer master data, nor transactional data such as production or accounting documents or sales orders. As a consequence, they need to be integrated with ERP systems that store this kind of data and transactions. Indeed, some companies are using web supported BPMS to integrate intercompany business processes without disrupting the transactional systems of each of the organizations involved

[10]. The implementation of a SOA based PLM system architecture may request services from legacy systems as BPMS technology for business process management for a greater flexibility on the 'process definition', 'process execution', 'process measurement and analysis' and 'process improvement' [11].

Besides previous papers, it is interesting to underline also that some researches discussed about comparisons between BPMS and other traditional methods. A contribute is provided by Van der Aalst et al. [12], who considers it as an extension of classical Workflow Management (WFM) systems. BPM approach, in fact, supports the diagnosis phase, allowing the support of operational processes. Rudden [13], instead, compares three different approaches to drive process improvement: to buy an application, to extend an existent one or to use a traditional one developed internally. This last alternative is affected solely by a poor performance in the attaining of requirements and the time to market, that are factors of success normally enhanced by BPMS [14]. This last kind of context, where knowledge represents a relevant asset but also a criticism in design and manufacturing of complex and multi-parts products, is the main area of interest of this paper.

3 Research Method

This paper is focused on the optimization of technicians/engineers tasks for the design and manufacturing of aircrafts that are critical processes and where the efforts related to time and resources are very huge and with an high impact. Due to the product complexity, in fact, each design error in the engineering units (e.g. a mistake in a join of two or more parts) affects manufacturing sub-processes and safety even more gravely.

The research question that this paper tries to answer is: How the processes automation using BPMS can foster processes awareness and optimization in an aerospace domain?

To address the research question, a case study approach is carried out in an Italian aerospace company. To better support the findings two applications are developed for processes automation: the DRR (Design Rectification Request) process and the Testing Management one. The required information to analyze the case study was collected by means of interviews (semi-structured and open questions) carried out on key company employees working in the engineering design unit for identifying roles, activities and responsibilities and to develop the current As-Is models. Final feedback are collected validating the results with the final users. For the applications development, Bonita BPM software is used.

4 Results

4.1 Application 1: The DRR Process

A document of Design Request Rectification (DRR) is emitted in relation to a non-compliance detected during the preparation and verification of a cycle or in the course of manufacturing activities. It is solely due to design problems, whose solution involves a change to the drawings and schemas, and the module consists of a single sheet. When it is required a modification of an existing design, a business unit edits a form. If the object is related to a change that already exists, the operator indicates the number of PM (Proposal of Modification) and the MOD (document of MODification). In case of a new modification, the operator selects the corresponding box. The form is submitted to the competent technical service, which ensures the numbering of the document, its evaluation and the forwarding to the specialist unit of the engineering design and quality control. The quality control unit, furthermore, ticks the form ensuring its traceability.

In this context, the DRR workflow for the new solution is summarized according to the following five steps:

1. DRR process characterization;
2. Modelling in BPMN the DRR processes through the: definition of involved actors; specification of the flow of process activities; and determination of rules to follow in the decision points.
3. Collection of the necessary information to make the model executable, such as: allocation of roles; definition of data consumed in the course of the process; configuration of possible connections to others software.
4. Graphic design of the user interface (GUI);
5. Execution of the process through the web application automatically generated.

A set of interfaces is proposed to support DRR management. It is useful to consider a situation where an employee belonging to the Production Engineering Unit sends a proposal of DRR to his/her Production Engineer. In this particular case study, the employee accesses to the system by means of a web interface on the basis of the assigned profile and privileges. Then, a login procedure is set to enter the username and password and operates on Bonita. At this point he/she can open a new DRR, sets its priority and edits the form and can compile mandatory and optional fields with data necessary to describe the request such as: title, plant, priority, drawing number, etc.

The system generates the document and assigns a code number that identifies the DRR, so it can be saved and managed by the actors involved in the process. In this case, a DRR can be opened, edited and saved on the system. The Production Engineer follows the state of the DRR through a monitoring of a task manager dashboard provided by the software. Successively, Production Engineer evaluates the document and rejects or accepts the request using a form created "ad hoc", communicating the decision by an additional form. At the end of the process, the employee receives the DRR with the motivation of the rejection or acceptance.

4.2 Application 2: Testing Management Process

Some activities related to the testing process of physical prototypes (test feasibility studies, planning, design, implementation, and reporting) are supported by different IT tools. These tools are not well integrated and this does not allow an effective control of the overall testing process, its efficient execution, and a proper information exchange. This is true, in particular, for the management and the consistency between test data (in Tec.Manager software), test design data (in Teamcenter PLM software) and test documents (in several document management systems).

In each step of the process, the activities need to manage a high volume of information from different resources and tools, as shown Fig. 1.

This process facilitates cooperation between all the subjects involved in the testing process and improves some key activities about structural tests and prototypes manufacturing. The application helps to improve the document management process in order to speed up significantly the approval procedures of them. The data and information of the test product are stored on the server and it is possible to access them using the authentication procedures. The information available in the process is very high, so that the number of documents submitted is, therefore, high.

In this view, the main purpose of the proposed approach has been to optimize the testing process with an advanced automation and document management. To address this purpose, a technological layer responsible for the management of testing activities and the information flow is underlined. Several connectors are developed to

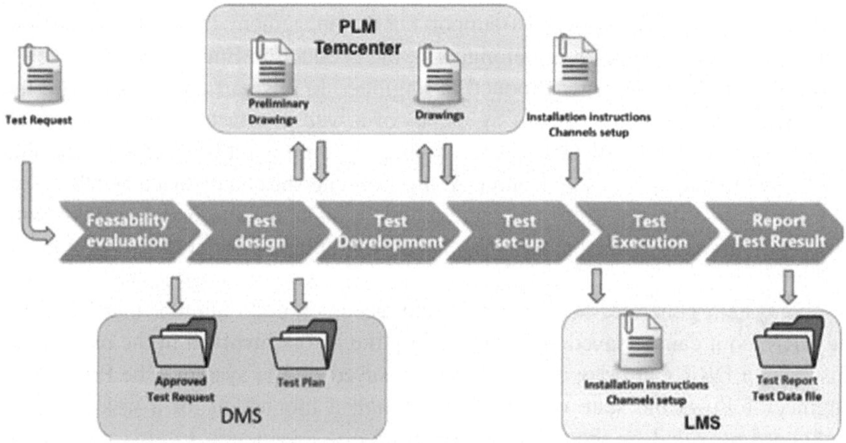

Fig. 1 Structural test lab—processes and main outputs

extract information and data from various software tools. For example, the connector to TeamCenter has been developed in JAVA using the SOA services who is an API that openly exposes Teamcenter's Business Logic Server capabilities to Web Services, as well as to language-specific programs.

As previous introduced, the BPMS is Bonita but in this application the form designer of Bonita is insufficient to create the user interfaces appropriate for the process instantiation and for human tasks. To compensate this lack, the Java Technologies for Web Applications is used to develop the appropriate user interfaces.

Therefore, after overcoming the technological issues, a general process composed of six processes is designed and is composed by: Feasibility Evaluation; Test Design; Test Development; Test set-up; Test Execution; Report Test Result. For each of the six process, specific sub-processes are implemented.

Interfaces for each tasks of the sub-processes are developed to allow the complete execution of the process. These interfaces let users interact with the functionalities like login access, management documents, monitoring process status and sending a reminder.

A relevant role is played by the Administrator profile available in the portal that is responsible for managing the processes, assigning tasks, managing the actors and documents. The User profile is, instead, responsible for performing the tasks for which they are a candidate and for starting new cases of the Processes to which they have access.

5 Discussion

The users expressed a set of positive feedback on the implementation and the improvements about some important issues:

- Accessibility to information, through a more structured management process that enables greater control on the activities flow;
- Classification and distribution of information, through a decrease of time-consuming for extraction and processing;
- Data inconsistency control relating to particular documents (e.g. drawings);
- Integration with internal enterprise documentary/information systems;
- Faster development of the first application prototype;
- A better communication between engineering design unit and the IT unit;
- Easier collection of users feedback during the development;
- A dynamic list of tasks to manage for the users;
- Easier modification of the process and of the application;
- Preserving knowledge generated by the users, through guidelines and good practice for the users about future problems.

On the other side, some downside are also observed and discussed:

(1) The resistance at the introduction of a new system in the company;

Table 1 Qualitative indicators

Indicator	Value	Description	Motivation
Level of engineering and formalization of the process	Increased	It measures the level of process optimization for data management	BPM allows the modelling and standardization workflows, the map of users involved, and the management of complex data
Quality of the process	Increased	It provides an estimate of consistency of the workflow	The process of managing data set includes different human operations, with a high level of repetition and errors problems that was overcame
Process logging	Increased	It refers to the estimation and quality of the storage data process	The processes management allows storing all tasks and operations process
User-friendly interface	Increased	It evaluates the interface usability	The user operability increases for the standardization and optimization of user interface, which includes only useful information and allowing to faster create new knowledge

(2) The licence costs of a new system;
(3) The training of almost one company's employee to use the BPMS as administrator;
(4) The training of the users;
(5) In some cases, a change management strategy is required;
(6) Define governance rules and assign responsible organization for maintaining services throughout their lifecycle.

Nevertheless during the validation phase, impacts were collected using qualitative performance indicators in following table and are listed in Table 1.

6 Conclusion

The paper discusses the introduction of BPMS applications in an aerospace industry. The research activities are focused on the design and implementation of new solutions supported by software in order to optimize their performance.

By results of the business process automation (by Bonita), it can be observed that the implementation fits to the industrial case study considered in the paper and also to related contexts characterized by analogous activities. Thanks to its scalability, in fact, the software seems to be suitable for other design and manufacturing processes into the company. This results also allows to better explain and support the usability of BPMS in combination with PLM and ERP. Using BPMS, it is possible to manage processes in a structured way, in contrast with point-to-point integration via tool chains and custom code development that can be developed in others cases. In a BPMS using development tool kit to create connectors may be interact and access an external SOA architecture like that PLM that supports the use of web services to guarantee interoperability between different systems.

The paper contributes to enlarge the scientific state of art about BPM. Through a case study, in fact, applications, benefits and indicators of BPM introduction in a company are underlined and can lead to new studies and explorations in this field. They can be considered as a starting point to design future research and to collect evidences reinforcing the described results.

Considering future outlooks, future researches may discuss the criticisms of BPMS web applications implemented and used for particular classes of devices or their adoption in other industries different from the aerospace context. Together with these studies, new perspectives for these tools can be investigated, for example their use in the design, manufacturing and, more in general, the development of sustainable product and new businesses more careful to the environment.

References

1. Rosemann M, De Bruin T (2005) Application of a holistic model for determining BPM maturity. In: BPTrends. McGraw-Hill, pp 1–21
2. Weske M (2012) Business process management: concepts, languages, architectures. Springer
3. Borges GHA (2013) Business Process Management (BPM) using ERP systems in the food industry. Glob Res Soc 3(6):18–27
4. Fischer L (2012) Delivering competence advantage through BPM: real-world business process management. Future Strategies Inc., USA
5. Chen M, Zhang D, Zhou L (2007) Empowering collaborative commerce with Web services enabled business process management systems. Decis Support Syst 43:530–546
6. Pagani M (2005) Encyclopedia of multimedia technology and networking. Idea Group Inc (IGI)
7. Nt JA, Xexéo GB, Souza JM, Tomaz LFC (2009) Collaborative process modelling and reuse evaluation. In: 5th International conference on collaborative computing: networking, applications and worksharing
8. Wohed P, Ouyang C, Johannesson P, Jalali A (2013) Dynamic weaving in aspect oriented business process management. In: On the move to meaningful internet systems: OTM conferences
9. Veldhuizen R, van Ravesteijn P, Versendaal J (2012) BPMS implementations in SMEs: exploring the creation of a situational method. In: BLED 2012
10. Carreño-Vargas JE, Vega-Mejía CA, Castellanos-Arias JS, Hernández-Martínez YP, Aguirre-Mayorga HS (2012) Evaluation of integration approaches between ERP and BPM systems. Ingeniería y Universidad

11. Lee T, Lim J, Shin J, Myung S, Choi M, Baek S, Kim J, Oh J, Lee D, Han Y (2007) An implementation methodology of SOA based PLM system. In: Proceedings of international conference on product lifecycle management, pp 303–310
12. Ter Hofstede AH, Weske M, Van der Aalst W (2003) Business process management: a survey. In: Proceedings of the international conference on business process management. Eindhoven: No. 2678 in Lecture Notes in Computer Science. Springer
13. Rudden J (2007) Making the case for BPM: a benefits checklist. In: BPTrends
14. Forrester R (2003) Business technographics study. Survey of Corporate Executives
15. bonitasoft, "bonitasoft, " [Online]. Available https://www.bonitasoft.com/business-process-management-bpm. Accessed settembre 2019

A Processes Engineering Initiative for Lean Performing Arts Organizations

Manuela Marra, Lorenzo Quarta and Aurora Rimini

Abstract Cultural and Creative industry (CCI) is a wide sector with different specializations and application areas. It refers to the creation and distribution of services or product of a cultural or artistic nature. The study investigates the lean thinking methodology application to the management of the production processes in the cultural and creative industries (CCIs) sector for realizing live performance. Applying a lean methodology to this processes costs and production times can be reduced and the overall performance increase. To address this need, a theoretical processes' framework is proposed for process re-engineering and better allocation of human and physical resources. This result is reached by the interaction between researchers with many years of experience in business process and the expertise of four main cultural operators of the Apulia region.

Keywords Cultural and creative industries · Performing arts · Processes framework performance optimization · Arts process handbook

1 Introduction

Cultural and Creative Industries (CCIs) are industries whose principal purpose is production or reproduction, promotion, distribution or commercialization of goods, services and activities of a cultural, artistic or heritage-related nature [1]. In recent years, CCIs have been characterized by great income (US$2250 billion of global revenues in 2013) [1] and are the target of many public and private initiatives (regional and European) to drive economic growth and create jobs.

M. Marra · A. Rimini (✉)
Department of Innovation Engineering, University of Salento, 73100 Lecce, Italy
e-mail: aurora.rimini@unisalento.it

M. Marra
e-mail: manuela.marra@unisalento.it

L. Quarta
Naica Società Cooperativa, 73100 Lecce, Italy
e-mail: lorenzoquarta@naicasc.com

M. Rossi et al. (eds.), *Proceedings of the 6th European Lean Educator Conference*, Lecture Notes in Networks and Systems 122,
https://doi.org/10.1007/978-3-030-41429-0_35

Despite their importance and potential value, there are few definitions about what Industries are part of the Cultural and Creative domain. One of these has been provided by UNESCO and includes the following eleven sectors: Advertising, Architecture, Books, Gaming, Music, Movie, Newspaper and Magazines, Performing Arts, Radio, TV and Visual Arts [1].

Whereas events represent one of the main source of income and a perfect means to publicize a cultural and creative product, there is a need to organize them more and more appealing and efficiently in order to attract the attention of media and as more paying spectators as possible.

The cultural and creative products like festivals, concerts, events, tours, exhibitions, theater performances, etc. have their own life cycle. They can be considered a "complex product" and could be managed in a more efficient way introducing Product Life Cycle Management (PLM) and Business Process Management initiatives.

The principal problems in the Performing Arts Sector are: low processes standardization; long information transmission times; inadequate distribution of work to employees; and difficulty in increasing the customer satisfaction.

Therefore, the CCIs share with the manufacturing firms same weaknesses that can be overcome by lean management methodologies and approaches. This point of view is new in the literature.

As in general all the organizations also the art and cultural ones have to deal with their business processes which are subject to constant changes and influenced by internal and external factors in their environment.

The dynamism and complexity of the business processes of this domain is proportional, however, to a lack of design, organization and awareness.

In fact, the socio-economic process and most of the government assets up to the last decades have underestimated the economic potential and reduced this category of enterprise activities to merely leisure activities. This scenario has prevented them from finding the necessary funding and cultural incentives to improve and seek organizational innovation, differently from other sectors (such as aerospace and manufacturing) characterized by a solid and stable culture of enterprise and engineering. There are no Enterprise process Framework, such as APQC's PCF [2], focalizing on the specific features of the sector either.

The paper aims to fill this gap, to methodize organizational activities but above all to transmit a culture oriented to business process management, highlighting processes and related processes owners in order to invite companies to a better systematization of their activities and to sustain organizational improvements.

For leading this goal, a comparative analysis was carried out in the Distretto Puglia Creativa. It is a productive cluster of companies operating in the CCI. The research involved companies in the Performing arts sector (one of the areas of the cluster) and four companies have been analyzed using interviews and focus groups.

The final result is a processes' framework leading a lean thinking. It is based on the APQC Classification structure but customized for the specific peculiarities of Performing Arts. The framework suggests at the performing arts companies a guideline in the execution of their activities and the awareness of the net of processes activated for realizing their products/services. For consulting and IT companies the framework suggests a complete picture of the sector and can lead to the definition and development of new techno-organizational solutions based on the execution of specific processes or on integrated systems for managing and track data and information among processes and activities.

2 State of the Art

Historically, the lean approach to production management was created by Taiichi Ohno, who defined an innovative methodology for reducing costs and eliminating waste: the "Toyota Production System (TPS)". It was developed during the second half of the '900. The methodology considers the production time line from the moment with which the order is received until the moment with which it is withdrawn; the objective is to reduce the time line eliminating waste that does not have added value. The methodology identifies seven types of waste in the organization: transport, inventory, movement, waiting, overproduction, processing and defects. The interesting aspect of the methodology is its inclination to identifying unproductive efforts that do not create value for the end customer. In the production world, this methodology is called by "lean", and describes systems that provide superior results with less use of resources. The diffusion of the methodology in the different sectors and the related studies have identified the Toyota Production System as the basis for the "lean production" movement that has influenced production trends (together with Six Sigma). The application of the Toyota production system in an organization is identified as a lean enterprise [3]. Process modelling is a way of estimating wastes and improve the process, in order to understand more about the information transfer, process movement, the responsibilities of acting and steering. The process modelling focuses on recording as well as mapping processes, roles and information flows between business areas to increase the transparency of value-added operations. Waste-relevant requirements were deduced from the study and general Lean Management aspects to provide the basis for the modelling regulations [4]. In addition, to reduce waste, it is important to identify the process owners and the related competences that can be used to increase business productivity. Competences management becomes, therefore, relevant. The identification and formalization of competencies and roles involved in a process are key factors for success [5].

The initiatives of engineering of processes in literature are present but explain how to improve the process of performance execution by the use of software that through new audio/video technologies increase the performance of artists, such as the recent work of Mokhov [6]. On the other side, there are studies that discuss the problems related to the performing arts sector and its difficulty in optimizing costs compared to the effort incurred for the production of live events.

The main publications that have inspired this study are the following: starting from the theory of Baumol and Bowen [7] that define the performing arts such as a "stable productivity sector", that is without any prospect of improvement per hour-man.

Instead, the study of Cuadrado-Perez [8] analyses the operational process of a theaters selection and come to conclusion that there are factors that, with a suitable process management, can produce positively effect on the overall event performance.

Finally, the recent article by Sacco [9] states that in reality the cultural and creative industries, if properly managed, can release an economic potential, kept unexpressed null now.

Lean initiatives of this kind are present in various sectors, manufacturing [4], pharmaceutical [10], healthcare [3] and so on. In the performing arts sector, a lean approach has never been applied, but the authors of this paper believe that the lean methodology can lead to increase the value for the customer, reduce waste and optimize usage and competencies of resources for the processes activated by the organizations of the performing arts sector. This is on the assumption that the life cycle of the live event has the same complexity as a manufacturing product [11].

3 Industrial Context

The industrial case is represented by Distretto Puglia Creativa. This is a cluster recognized in 2012 by Apulia Region, in the south of Italy, according to Regional law 23\2007, representing more than 100 Apulian cultural and creative enterprises bringing together some heterogeneous Apulian (and beyond) organizations which work on different areas and competences in the cultural sector both on a regional and national level.

In particular, the study and the results have been carried out with the involvement of four companies of the cluster, whose businesses are an important example of the sector:

- Cantieri Teatrali Koreja: a 30-year history innovation theatre.
- Coolclub Cooperative Society: it has been involved for over 10 years in the conception, planning, organization and promotion of cultural events, press office and communications, especially in the music sector.
- Bass Culture: which is mainly concerned, together with CoolClub, with the music area. It was founded in 2001 and within six years has become a member of Assomusica, an agency associating the biggest producers of live events in Italy.
- Officine Cantelmo Cooperative: known in the territorial area for the organization of divulgatory and business-oriented live events.

All these organizations have well-established businesses and great potential for improvement, but all suffer from a lack of structuring of their internal processes. In fact, there are different activities that are repeated in a redundant manner, and that could be automated, drastically reducing the effort. Furthermore, mainly the biggest problem is the lack of communication between the various actors in the process for each information flow. Each business unit is a black box compared to the others, and this multiplies the efforts as well as generating results qualitatively worse.

4 Research Method

The research activities have been realized during an action research involving University researchers and industrial representatives of the involved companies. A comparative analysis [12] is realized exploring the four companies (Fig. 1).

The study started from the analysis of the available processes framework, above all the APQC'PCF Cross Industry [2]. It is a relevant reference and provides insights for the processes analysis in different context. Based on the structure of this framework and on the related areas, the developed industrial case study has aimed to propose also for the CCI a framework, similar in the structure, for analysing and discovering their processes.

As previous explained the four analysed companies are part of four different subsectors of Performing Arts, therefore, the analysis carried out with as first task a set of group interviews in each company in order to understand their organization, managed issues and roles.

From these group interviews and for each company, it has emerged that all the activities can be classified in four areas: event design, production, administration and promotion & communication.

Based on this classification, single interviews and focus groups have been administered. The single interviews were managed in each company with the persons in charge of the different activities in each area. At the interviewees were asked to describe their activities, internal and external relationships, the used software and the data in input and output. The results of interviews were modelled using a high level value chain and also BPMN 2.0 diagrams. These last ones were used to go in deep in the activities and to validate together with the interviewees the flow of activities and how they are organized.

These first evidences were summarized for areas and 4 focus groups were organized, one per area. The focus group was used to confirm the contents available, for generalizing the findings and for defining in an unambiguous way the processes

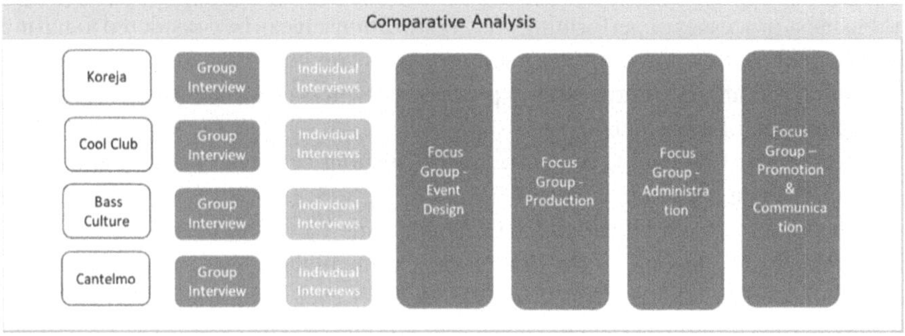

Fig. 1 Research method phases

owners. The focus groups were managed cross-sub-sectors involving almost one representative for each company of the analysis. To support the validity of the findings also participants from others companies of the cluster were invited.

All the collected information was structured in a processes' framework for the Creative and Cultural Industries and is described in the next section.

5 Results: A Process Framework and a Tool for Process Exploration

To compare the processes activated to manage a cultural product, more specifically a live event, the similar information among the companies was studied. During the study it was observed that employees are unaware that even if they produce different kinds of cultural products they put into practice the same processes.

In general, the analyzed companies do not have a unified view of their own business, each function has its own understanding of the activities. The study showed that there are not significant differences compared to other companies on the Distretto Produttivo Puglia Creativa about the fundamental processes macro-area, such as "Event Design", "Executive Production", "Administration" and "Promotion and Communication". It was observed that the development of a business strategy is a process carried out informally. Business processes such as event coordination and planning, ticketing, budgeting, analytics, assets management, promotion and communication and safety management are typically for small performing arts organizations in the cultural and creative industries sector.

Based on the results of this analysis and the APQC's PCF (Process Classification Frameworks) it was possible to present a primary version of a reference model for business processes of companies in the performing arts subsector. It aims to provide an overview of the classification of processes, so that it can help companies to clearly define best practices.

Figure 2 shows the high-level structure of the processes' framework for improving the business processes of performing arts organizations. It can be considered a starting point to analyze how the organizational processes can be classified. It is composed by a structured library of processes, a processes' owners classification and a tool for the exploration of its contents.

It is composed by twelve categories distinguishing between operating processes and management and support ones.

In this processes classification for CCIs there are 3 levels of detail. First Level is represented by "Category" of the Processes: it is the highest level within the model, being indicated by integer (from 1 to 12); Second Level is represented by "Process Groups" (from 1.1 to 12.3) and the third level is composed by "Processes" expressed in one decimal digit (from 1.1.1 to 12.3.11).

The main result of this industrial case is a library of processes applied in the performing arts sector useful to provide awareness, order and innovative elements

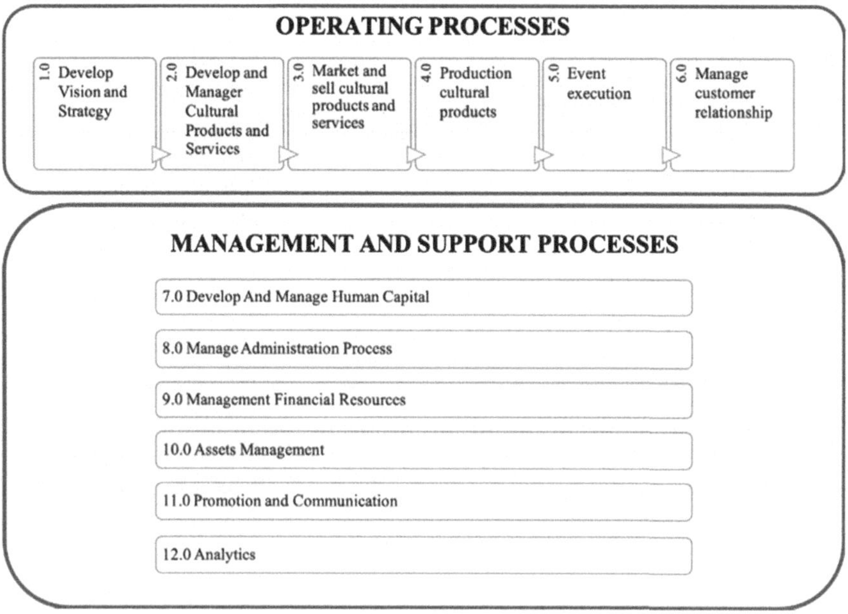

Fig. 2 High level business processes' framework for cultural and creative industries operating in the Performing Arts Sector (*Source* APQC's PCF re-elaborated by the authors)

in an unstructured context and leading a lean thinking. The selected "Processes" for each "Process Groups" follow existence criteria: to be inserted in the framework, a process has to be applied and available in almost one of the analyzed companies.

In addition to the processes' library, the study also suggests the Processes Owners involved for each "Processes Group" as emerging from the discussion in the interviews and focus groups. In this way, the processes framework is enriched by a view also on the organizational aspects that can be used for re-engineering the internal processes and also to optimize the allocation of human resources. The detailed distribution of "Processes Group" and related Processes Owner is available in Table 1.

To provide evidence of the applied criteria, to reinforce the validity and prove the feasibility of the study, a highly usable web platform has been designed, so as to offer the possibility to explore the methodology, and to share and exploit, in the Performing Arts sector, a business process-oriented culture. In other words, it will be easily accessible to all CCIs that want to improve their business processes.

As is evident in the figure, in an intuitive way, it is possible to navigate all the process areas, the related sub-processes and all the details related to them. In addition, each phase is characterized by the definition of metrics capable of quantifying performance (Fig. 3).

Table 1 Processes owners and process categories

ID	Processes categories	Process owner
1	Develop vision and strategy	Art director
2	Develop and manage cultural products and services	Production manager Art director Administrative manager
3	Market and sell cultural products	Administrative manager
4	Produce cultural product	Production manager
5	Execute the event	Administrative manager Production manager Technical manager
6	Manage supplier relationship	Technical manager
7	Develop and manage human capital	Administrative manager
8	Manage administration process	Administrative manager
9	Manage financial resources	Administrative manager
10	Manage assets	Technical manager
11	Promote and communicate	Communication manager
12	Analyze data	Communication manager

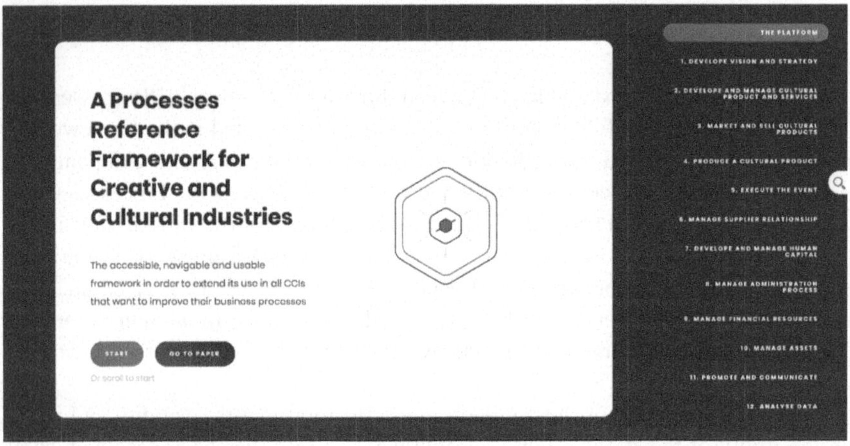

Fig. 3 Web Platform user interface

6 Discussion

Traditionally CCIs organizations do not rely on processes. Considering CMM (Capability Maturity Model) levels [13], the companies involved in the case study, but we could assume those of this sector, are all at level 1 and do not have organized processes. We could refer to these immature organizations as companies based on heroes;

managers are usually focused on solving immediate crises and schedules and budgets are not based on realistic estimates. If someone asks how long something will take, or what resources will be needed, those answering the question are just making a guess (based on experience) but they do not have a structure of their internal processes or the data needed to provide accurate answers.

This is the situation that we find in the design and production of a cultural event, whether it be a concert, a tour, a play or a conference. The artistic director and the technical manager but also the administrative manager carry out their activities following their own experience and a way of working that, although consolidated, is only in the heads of the most experienced people.

The main roadblock was to convince the companies' actors to be careful to processes and especially, to their hidden need to better execute their work. They were invited to think about their way of work and on the different modalities to organize their daily activities. Comparison has been used as the methodological way of building the proposed process architecture. Specifically, we started the analysis by comparing the processes, activities and professional figures of the four companies involved, then generalizing the results by categories of performing arts and finally build a standard view for CCIs. The use of interviews and focus groups allow to grasp details and to invite the participants for a better reflection on actors involved and activities performed and allow them to understand similarities and best practice that can be applied inter-sub-sectors.

The final result is the awareness acquired by the companies about their own processes and related processes owners and by the creative Apulia district about the similarities existing between its members; even if they produce different kinds of cultural products they put into practice the same processes.

During the interviews, the technological topic emerged different times and a big roadblock was to highlight and explain the role played by specific ICT tools that can be used to make the difference on the processes and on the overall organization. The low value and the lack of use of centralized ICT tools has not allowed, for example, the application of process mining techniques, useful to extract from the information systems direct information on the real flow of activities and the rules that manage their executions [14].

As a visual support to share and exploit, in the Performing Arts sector, a business process-oriented culture the web platform has been designed.

The results enable the adoption of lean principles in CCI, allowing companies to investigate their processes and identify the value-added and non-value-added activities. The awareness acquired on the processes, and the use of the web platform for their exploration, represent the first steps of a process mapping activity within a company. This supports CCI lean transformation by identifying opportunities for waste elimination.

7 Conclusion

The described results represent a real input for innovation and efficiency of CCIs. The results are original because starting from the APQC's PCF Cross Industry a new reference framework has been created for the Performing Arts organizations. Furthermore, the analysis contributes to previous studies proposing the processes that are the result of the merging of the companies' best practices working in the same sector but specialized in the production of different cultural products.

Based on these evidences, these "Processes" best practices can be applied with the appropriate evaluation for each specific case by companies producing different types of cultural products. In fact, using the proposed framework, it's possible to apply a unique vision of the processes among different organization and to lead sectors innovation both operational and technological ones applicable in the Performing Arts sector of cultural and creative industries.

This study has strongly confirmed the authors' intuition of how it is possible to innovate the sector of Cultural and Creative Industries through Business Process Management starting from a clear and general view on the used processes in the daily activities.

Acknowledgements The authors would like to thank all the MYA project partners, for their valuable contributions to the study. In particular, the authors would like to thank Cesare Liaci and Osvaldo Piliego of Coolclub Cooperative Society, Anna Petrachi, Salvatore Tramacere and Marco Chiffi of Cantieri Teatrali Koreja, Vincenzo Bellini and Silvia Rossini of Bass Culture and Marco Cataldo, Eleonora Leucci and Andrea Sanità of Officine Cantelmo Cooperative, for sharing their experiences, skills and processes.

Funding This research was funded by the European Union and the Regione Puglia Local Government through call Innolabs, project MYA—Manage Your Art.

References

1. EY (2016) Italia Creativa
2. APQC's Process Classification Framework (PCF)® (2019) Available at https://www.apqc.org/pcf
3. Parkhi SS (2019) Lean management practices in healthcare sector: a literature review. Int J, Benchmarking
4. Magenheimer K, Reinhart G, Schutte CS (2014) Lean management in indirect business areas: modeling, analysis, and evaluation of waste. Prod Eng Res Devel 8(1–2):143–152
5. Fortunato L, Lettera S, Lazoi M, Corallo A, Guidone GP (2011) A methodology for engineering competencies definition in the aerospace industry. In: WMSCI 2011—The 15th world multi-conference on systemics, cybernetics and informatics, Proceedings, vol 2. Orlando, Florida, USA, pp 296–301, 19–22 July. ISBN: 978-193633830-6
6. Mokhov SA, Song M, Chilkaka S, Das Z, Zhang J, Llewellyn J, Mudur SP (2016) Agile forward-reverse requirements elicitation as a creative design process: a case study of illimitable space system v2. J Integr Des Process Sci 20(3):3–37

7. Baumol WJ, Bowen WG (1965) On the performing arts: the anatomy of their economic problems. Am Econ Rev 55(1/2):495–502
8. Cuadrado-García M, Pérez-Cabañero C (2005) The process of programming in the performing arts: an empirical research in Spain. In: Colbert E (ed) CD ROM proceedings: 8th international conference on arts and cultural management. HEC Montreal, Montreal
9. Sacco PL (2012) Le industrie culturali e creative e l'Italia: una potenzialità inespressa su cui scommettere. IlSole24 Ore, Milano, Italy
10. Petrusch N, Sieckmann F, Menn JP, Kohl H (2019) Integration of active pharmaceutical ingredient production into a pharmaceutical lean learning factory. Procedia Manuf 31:245–250
11. Corallo A, Lazoi M, Marra M, Quarta L (2018) Innovating performing arts management through a product lifecycle management approach. In: IFIP international conference on product lifecycle management, July 2018. Springer, Cham, pp 420–431
12. Bryman A, Bell E (2015) Business research methods. Oxford Press
13. Paulk MC, Curtis B, Chrissis MB, Weber CV (1993) Capability maturity model, version 1.1. IEEE Softw 10(4):18–27
14. Ceravolo P, Azzini A, Damiani E, Lazoi M, Marra M, Corallo A (2016) Translating process mining results into intelligible business information. In: 11th international KMO conference, the changing face of knowledge management impacting society, 25–28 July. Hagen, Germany. https://doi.org/10.1145/2925995.2925997. ISBN: 978-1-4503-4064-9

The Development of Continuous Improvement in SMEs and the Supportive Role of the A3 Tool

Jannes Slomp, Gerlinde Oversluizen and Wilfred Knol

Abstract This paper focuses on the supportive role of the A3 tool in the development of continuous improvement in SMEs. We performed two case studies at companies where already several students used the A3 tool in improvement projects. Both companies embraced the concept of the A3-method and extended its' use to all improvement projects in the company. Furthermore, both companies developed hoshin kanri to align the various improvement projects. In this paper, we describe what the key elements were of the continuous improvement journey of the two companies. Next, we indicate how principles of the A3 tool supported these elements.

Keywords Continuous improvement · SME · Lean manufacturing · A3 tool · Hoshin kanri

1 Introduction

Many managers embrace the concept of lean management but are struggling with the development of continuous improvement practices in their companies. Extensive research on the development of lean and continuous improvement in SMEs has been performed by Knol et al. [1–3]. Based on multiple-respondent self-assessments from 33 manufacturing SMEs, they conclude that in the initial stages of the lean journey, companies could improve their lean practices in a bottom up manner with only having

J. Slomp (✉) · G. Oversluizen · W. Knol
Faculty of Engineering, HAN Lean QRM Centre, HAN University of Applied Sciences, P.O. Box 2217, 6802 CE Arnhem, The Netherlands
e-mail: jannes.slomp@han.nl

G. Oversluizen
e-mail: gerlinde.oversluizen@han.nl

W. Knol
e-mail: wilfred.knol@han.nl

W. Knol
Institute for Management Research, Radboud University, P.O. Box 9108, 6500 HK Nijmegen, The Netherlands

M. Rossi et al. (eds.), *Proceedings of the 6th European Lean Educator Conference*, Lecture Notes in Networks and Systems 122,
https://doi.org/10.1007/978-3-030-41429-0_36

additional attention on local success factors such as communication and a learning focus. When lean practices are more advanced, company-wide factors play an important role: top management support, a shared improvement vision, and a supplier link [1]. In another study on the development of improvement routines in the lean journey of companies, they conclude that in the beginning no specific improvement routines are needed to realize some lean practice implementation. More attention for various improvement routines (getting the improvement habit, understanding improvement, leading the way and improvement of improvement) becomes important for more advanced lean practitioners [2]. In a detailed case study of three carefully selected industrial cases, Knol et al. [3] show that a careful balance between lean practice implementation and the development of continuous improvement routines is needed during the development of a company. The study presented in this paper, can be seen as an illustration and an extension of the findings of Knol et al. [1–3]. Through two case studies we illustrate the logic behind the development of continuous improvement in the companies. We present a five-stage development process which is in line with the findings of Knol et al. [1–3] but further specifies the key activities in the company. We also show how the A3 tool supports the various stages.

In his book 'Managing to Learn' Shook [4] presents and illustrates the A3 management process to solve problems, gain agreement, mentor and lead, as it is used at Toyota for more than 50 years. The A3 problem solving tool communicates the background, current state, targets, analysis, future state, implementation plan and follow up. Several authors further explain the elements and the logic of the A3 problem solving tool [4–7]. Oversluizen and Slomp [8] study how students use the A3 problem solving tool in industrial projects and indicate the strength of the tool to structure problem solving and to support communication. Although the importance of the A3 tool is stressed in several books and papers (see e.g. [9]), its relevance for developing continuous improvement in companies is not gaining attention in the scientific literature nor in practice. This paper fills this gap.

In Sect. 2, we will dive deeper in the role of the A3 tool at Toyota. Therefore, we interviewed Isao Yoshino, one of the first managers of John Shook at Toyota, also mentioned in the acknowledgments of Shook's book. Yoshino was highly involved in the education of managers at Toyota and taught them to use the A3 tool. Through this interview, we broaden the meaning of the A3 tool. At Toyota, it is more than a logical sequence of PDCA activities. Section 3 presents the continuous improvement journey of two industrial cases. Students used the A3 method in industrial projects at these companies (see Oversluizen and Slomp [8]). Section 4 discusses the logic behind the development of continuous improvement in the companies and, furthermore, indicate the role of the A3 tool. Section 5 presents major conclusions of our study.

2 The Use of the A3 Tool at Toyota

The A3 tool has been used at Toyota for many years. It was a means to communicate in a time without computers. Deming, the promotor of Total Quality Control (TQC) at many Japanese companies, introduced PDCA-thinking at Toyota which also has been incorporated in the A3 communication at Toyota. Masao Nemoto, a strong promotor of TQC at Toyota and mentor of Isao Yoshino, was a strong supporter of creating the link between A3 and PDCA. He recognizes the elements of the A3 problem solving tool 'as the order in which presentation of QC improvement projects is made'. But, as he indicates, it is more than that. 'It is also the order through which improvement activities are conducted. In other words, you can follow this order to engage in your improvement activities and then follow the same order when you make your presentation' [10, pp. 155]. An A3 is, according Nemoto, the ideal format for giving a presentation of 10 min [10, pp. 16].

In order to gain a deeper understanding of the use of A3 at Toyota, we interviewed Isao Yoshino, who was visiting our university several times in 2017–2019. This interview illustrates how at Toyota it is linked to organizational development. Here, we summarize this interview by clustering Yoshino's sayings in several topics. We will refer to this clustering in our discussion in Sect. 5. The text has been approved by Yoshino.

Isao Yoshino is a 40-year Toyota leader (14 years of which was in the U.S.) and a former NUMMI training manager based at Toyota HQ in Japan. In 1979 and 1980, he was in charge of the Manager Development Program ("Kan-Pro" in Japanese) for all the managers at Toyota headquarters. In 1983, he was assigned to the manager of a newly established training section for NUMMI's shop floor leaders. During this time, he hired John Shook and other staff members from within Japan to develop a hands-on training program for American shop floor group and team leaders who travelled from Fremont, California to Toyota City to learn the Toyota Production System. In June 1984, Yoshino and his team started training American shop floor leaders for three weeks at Toyota's plant. Yoshino has held management roles at Toyota in research, human resources, corporate planning and production control, and manufacturing. He retired from Toyota in 2006. Yoshino is currently a lecturer at the Nagoya Gakuin University and travels internationally to teach and speak about Toyota leadership, hoshin kanri and coaching.

2.1 The Use of A3 at Toyota

A3 is already used for many years at Toyota. It started as a problem-solving tool. Dr. Deming introduced PDCA at Toyota which gave more structure to the A3. People liked it so much. They extended it further for other purposes. It is now very common in Toyota.

A3 is also used for proposals and for status-update reports. A3 is a very useful tool to convey your idea or plan and to share it with others, so that everybody of the group is at the same page. It is a communication tool. It serves organizational learning in Toyota. It is an organization tool. I don't think that PowerPoint can have the same impact.

A3 is not an order of the management. Sometimes it comes from a subordinate, they know everything. Managers do not know everything. So, subordinates may start with an A3. They may expect that the manager gives a good advice. A perfect A3 is not needed. It may just be memo to review.

2.2 Targets

Sometimes targets should come after the analysis. But in general, I agree with your sequence in the introduction. A target is what you want to attain. You must set the target regardless of the analysis. It is a serious problem. If you set the target after the analysis, then you may get scared, and be too careful. The target depends on what you need, for instance zero defects, or what the current result is. The target is your challenge. Not realizing the target means that the problem is not completely solved yet. It may be the start of a new project.

I always search for the target in an A3. Without a target, your document is nothing. 'Where is your target' is often my first comment. You must put it somewhere. Sometimes it is difficult to specify an indicator for the target. But that is no reason for not being clear about the target. You must think about the target and an indicator. An indicator may not be perfect, but you must select one. As long as the challenge is clear and measurable, it's fine.

2.3 Development of People and the Organization

A3 is a very powerful tool to develop people. Without A3, only verbally, it is very difficult. Misunderstanding can happen. A boss can better give you advise, when you have an A3 in front of you. For groups it is even more important. If you have ten people and they discuss the same A3, then they are focused to the same goal. I have been in meetings without A3, and the discussion went everywhere. With an A3 you can share your target. It strengthens the understanding of the own goal. It also helps to make faster decisions. All key points are on the A3—everybody can see it.

2.4 Training and Coaching

A3 is just a tool. Training is needed to use this tool in an effective way. It is about 5S. If you are good in 5S, then you can sort out unnecessary information and arrange important things in your A3. You must understand that you cannot handle everything. The limited space of an A3 is a great opportunity to learn how to distinguish necessary and unnecessary things. The more experience, the better people apply A3.

I don't care if I am not perfect when I write A3. Everybody has its own opinion. It reflects your thinking. It is a good tool to show to your boss or subordinate. My advice, if you write something, put what is on your mind on paper. So simple as possible. Then you can improve. Don't try to be perfect.

The elements of A3 are important for problem solving, but I like an easy way of thinking about A3. If coaches take the elements of A3 too serious, then employees will not use A3 too openly. It is good that employees try. Coaches must give them trust. Then the A3 will gradually improve.

2.5 Hoshin and A3

There is a clear link between Hoshin and A3. Hoshin is the way to come to shared goals, action plans and timing. A3 is a tool to show hoshin, to communicate about it. At Toyota, we periodically review A3s. A3 is a very important tool to check if Hoshin works well.

3 The Continuous Improvement Journey of Two Industrial Cases

This section presents two industrial cases of SMEs where the continuous improvement journey developed for several years. The companies are both partner of the HAN Lean QRM Centre (HLQC) and regularly get students from the minor World Class Performance (WCP) performing improvement projects in the company. Both companies currently participate in some (applied) research projects of the HLQC.

3.1 Case 1

The first company is a manufacturer of instruments for soil and water research. It employs around 150 people. The company has a component manufacturing department as well as an assembly department. The manager of both departments became enthusiastic about lean management several years ago and has initiated various lean projects. The company started their lean journey around 2012 and was initially focused on performing waste reduction projects. Students of the minor WCP participated in several of these projects and introduced the A3-method. The company adopted the A3 method for their own improvement projects. Along with the use of the A3-method, performance management has been developed in the company. There is now a better understanding of what to measure and how to measure. There was also a need to coordinate the various lean initiatives. In one of the first years, the manager therefore made an 'objective-tree' to create linkages between the local student projects and the more strategic objectives of the company.

However, during the lean journey the manager recognized that using the A3-method and applying the 'objective-tree' was not enough. He felt that more common focus was needed to gain more synergy in the company. It was not always clear for everybody whether an A3-project really contributed to the company goals. Therefore, the company started to experiment with Hoshin Kanri. The manager and the team leaders made an overall Hoshin document, which they called the A3-compass. This overall document described the ideal state (true north) and the medium-term goals and actions per team. This supported the communication about improvement in the company substantially.

The company also invested in gaining a clear improvement structure. Team leaders are trained to support the employees in improvement activities. Employees are getting more and more involved. By setting up a daily stand-up meting and structurally using the Hoshin document and the A3's for communication, the company moves in the direction of a learning organization. About two years ago, the company hired a recently graduated student to become the lean facilitator.

Currently, the operations manager, who is responsible for two departments, experiences the need that also other departments, such as sales and product development, start with a lean and continuous improvement journey. Key problems in his department comes from these other departments.

3.2 *Case 2*

The second company is a manufacturer of paving tiles. It employs around 500 people spread over 4 production sites. The production process is quite simple but asks for a detailed configuration for producing tiles of high quality. For four years the company is actively using lean methods to improve their processes. This was especially initiated by a new, lean minded, director operations. He stimulated the setup of lean improvement projects. He also attended, together with the operations managers of the production sites, several lean workshops organized by the HLQC.

The company became an enthusiast user of the A3-method. A new project now only starts if there is an approved A3 proposal, a champion and properly selected team members. If appropriate, a student from the minor WCP is also assigned to a project. Although the company is happy with all lean projects, they also recognized difficulties. It is hard to close the PDCA-circle. The Check of the PDCA is often not done, which subsequently leads to fuzzy insights in the results and contributions of the project to the goals.

Another important problem was the lack of overview: there were (too) many A3-projects running in the company without a clear focus. To handle this issue, the management of the company decided to start with Hoshin Kanri and constructed, with the help of the HLQC, an X-matrix which aligns strategic objectives, departmental goals and improvement projects. The X-matrix, now in use for two years, is helpful for selecting the improvement projects to be done in the next year.

The company experiences the change towards a lean culture as challenging. The company hired several minor WCP students after they finished their study. They perform as lean facilitator and help the employees in their lean journey. In some cases, middle managers not able to adapt to the lean philosophy had to leave the company. The company recently started with a leadership-training for middle managers.

The director operations is part of the management team. He was able to convince his colleagues that lean management and continuous improvement are a strategic choice and need to be adopted in the whole organization. Following this strategy in the other departments, however, is another issue.

3.3 *Stages in the Continuous Improvement Journey of the Two Companies*

We interviewed managers, and some other employees, in the two companies to understand how lean thinking got deeper rooted to the core of the companies. In the analysis of the interviews, we used the qualitative case analysis method of context mapping [11]. This resulted in a mapping of related topics concerning the integration of continuous improvement in the companies.

Figure 1 summarizes our findings. We first looked for situations, events and measures that determined the continuous improvement journey of both companies and put them in time order, as far as possible (middle part of the figure). Next, we defined the logic of continuous improvement development in the companies, and probably in more SME companies (left part of the figure). Finally, we identified the role of the A3 tool in the continuous improvement journey of the companies (right part of the figure). The left and the right part of the figure are explained in Sect. 4.

The lean and continuous improvement journey in both companies started with a manager who embraces the lean philosophy. The initial focus of both managers was on the reduction of waste and the realization of flow in processes. They involved employees in improvement projects and experienced enthusiasm and performance results. However, they also experienced that improvement projects do not automatically lead to new projects. Without other measures, the lean and improvement drive would weaken. The managers of both companies recognized the need to develop improvement routines. They stimulated the development of performance management and, furthermore, initiated hoshin kanri in the company in order to streamline

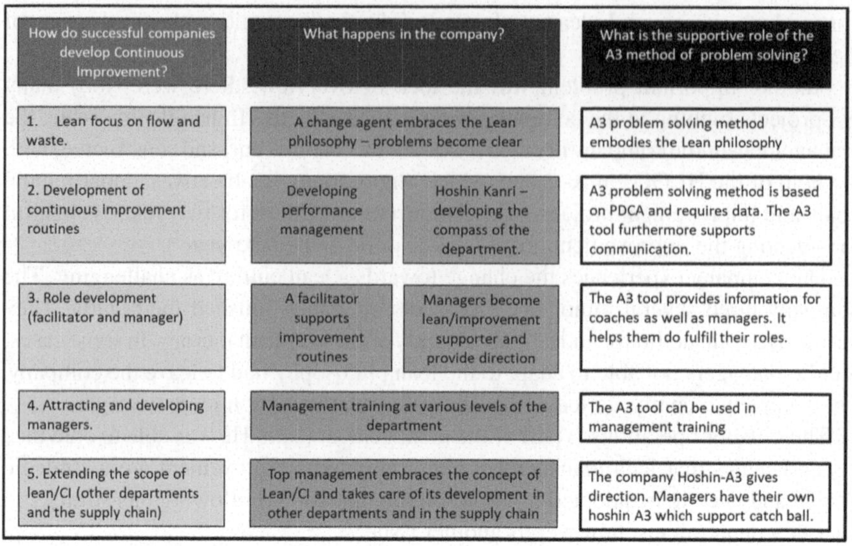

Fig. 1 The development of continuous improvement in the companies

and link the various improvement projects within their departments. Subsequently, both companies hired a lean facilitator to support employees in the development of improvement projects. The facilitators can be seen as the carriers of improvement routines. The managers of the companies also experienced the need to develop their own role, and those of other managers, in the company. Instead of just being the boss, managers recognized their role as lean coaches, stimulating their employees/managers to search for improvement opportunities which supports the company performance. Subsequently, they experienced the need for training of their team leaders and other managers working under their responsibility. Finally, leadership training and attracting new managers with lean knowledge was seen in the company as being essential for the creation of a sustainable continuous improvement culture.

4 Discussion

Looking carefully to the development of continuous improvement in both companies, we recognize a pull mechanism. There is a key driver: the wish to develop lean and continuous improvement practices in order to gain performance advantages. This driver is, initially, only embraced by one, or a limited number of change agents. Their first attempt to implement lean and continuous improvement in one department has led to an implementation process that can be broken down into five stages, where each stage asked for the next one.

In the *first stage*, flow and waste reduction receives attention within the company and lead to some performance results. The enthusiasm and the goodwill factor of the manager, the initiator of the lean movement in the company, is essential. The focus on flow and waste creates understanding of the problems in the company. The A3 tool, as used by the students in the companies, is helpful to structure the various improvement projects. Furthermore, lean knowledge is integrated in the various elements of the A3 method, helping to solve the problems (see Oversluizen and Slomp [8]). In both companies, students supported the spread of the required lean knowledge in the company.

Although projects were successful, managers of the company experience difficulties to hold the improvement drive. It is not obvious to start new projects after the end of other projects. Also, the enthusiasm goes down in the course of time and the goodwill factor probably deteriorates. This is a risky situation. In our experience, several SME companies get lost in this situation and then decide to stop their lean journey. The A3 tool may help to overcome this stage: PDCA is the basis of the elements of the A3 tool and defining a follow-up project or activities is key. Students doing A3 projects in companies, however, find it difficult to specify clear follow-up projects (see Oversluizen and Slomp [8]). This difficulty may be because of a lack of experience. Another reason may be the absence of clear direction for improvements in the company.

The *second stage*, developing continuous improvement routines, can be seen as an answer to the difficulty to decide for follow-up improvement projects. In this stage, management takes care of the development of performance management and hoshin kanri. These are essential elements for the development of meaningful improvement routines. Performance management gives the information of the current performance and the discrepancy from the norm. It gives an important trigger for improvement. Hoshin kanri provides the required direction of improvement. Communication (catch-ball) is essential for making hoshin kanri work. Using the A3 tool asks for performance data and is furthermore an important support in the communication about improvement projects. So, the use of the A3 tool by students in both companies, asked for and supported this stage.

Performance data and hoshin kanri support required actions for continuous improvement but do not offer the knowledge and ability to perform the improvement activities. This triggered managers to develop the role of lean or continuous improvement facilitation and to rethink their own role. This can be seen as the *third stage* of the continuous improvement journey. The managers in both companies attracted a lean facilitator, responsible for supporting continuous improvement, bringing lean knowledge and supporting employees to do the required analysis and development work. In the first stage, this role was fulfilled by the managers themselves. Consequently, they also had to rethink their own role. Based upon the needs of the continuous improvement process, they developed their roles towards being the supporter of improvement projects, creating the require conditions, and the leader which initiates or supports the development of new improvement projects which are linked to the needs of the company. The A3 tool is a good support in this third stage: it becomes easy to understand for a manager to see what is happening in the improvement projects, how they are linked to the needs of the company (background element of the A3) and what follow up is needed (follow up element of the A3). It is interesting that students experience, in their industrial project, these elements of the A3 as most difficult. Managers have a task in this. The A3 also helps the facilitator in his coaching task. It is easy to recognize were support is needed, which element of the A3 required more effort.

Although the roles of facilitator and manager are important, the real work within continuous improvement must be performed by the workers themselves and organized by team leaders and middle managers. The facilitator and manager in both companies recognized the need for training of the team leaders and middle managers. Without some training it becomes difficult to coach and to lead. The *fourth stage* in the continuous improvement journey is the lean and continuous improvement development of team leaders and middle managers. They must embody the culture of continuous improvement on the shop floor. They also should feel ownership and the ability to realize improvement projects. Both companies initiated training for their managers. The problem solving A3 can be used in the training, since it inhabits lean as well as continuous improvement elements.

Currently, lean and continuous improvement routines are well integrated in the operations function of both companies. The operations manager in both companies now experience the need to extent the scope of lean and continuous improvement to other departments, such as sales and product development, in order to further improve the performance of the operations department as well as the whole company. This is not easy without having a champion in these other departments. In our experience, this is a major problem in many SMEs, a reason why a successful lean operations manager may leave his company. The company wide development of lean and continuous improvement is the *fifth stage* in our model. Companywide hoshin kanri is important to support this state. Hoshin A3s for all managers may further support the alignment of improvement activities between the various department.

The five-stage model (left part of Fig. 1), is derived from the two cases and is in line with the findings of Knol et al. [1], who show that initially attention is needed for local success factors and later for the more company-wide factors. The model also fits the findings of Knol et al. [2] with respect to the development of improvement routines.

We explicitly mentioned the role of the A3 tool to support the various stages in the development of continuous improvement in a company (right part of Fig. 1). As Yoshino mentioned, see Sect. 2 of this paper, it has served organizational learning in Toyota and, therefore, can be seen as an organization tool. The key elements of the A3 tool which require attention at companies where it is used [i.e. mentioned in Sect. 2 (clear targets, development of people and the organization, training and coaching, the link with hoshin kanri)], all relate to the improvement routine of 'getting the improvement habit', which is, according Knol et al. [2], the most important routine to focus on while developing an organization.

5 Conclusion

Based upon two case studies, this paper presents a five-stage model for the development of continuous improvement in a company. The model is consistent with the findings of Knol et al. [1, 2] which were based on data from cross-functional self-assessments among managers of more than 30 companies. The pull character of the five-stage model is probably the key for the successful implementation of continuous improvement: a strong focus ('getting the improvement habit') and a pull of activities/stages starting at the work floor, the place where value is created.

It is hard to overestimate the possible role of the A3 tool in the five stages of continuous improvement development. In our experience, managers of partner companies who embrace the concept of lean and continuous improvement, are enthusiastic about the use of the A3 tool and have adopted the tool for all their improvement projects. In this paper, we indicated what the strengths are of the A3 tool in each of the five stages. It is interesting to note that value of the A3 tool in the five stages show a balance between the focus on lean knowledge and continuous improvement principles. This balance is a key element in the development of lean company (see Knol et al. [3]).

References

1. Knol WH, Slomp J, Schouteten RLJ, Lauche K (2018) Implementing lean practices in manufacturing SMEs: testing 'critical success factors' using necessary condition analysis. Int J Prod Res 56(11):3955–3973
2. Knol WH, Slomp J, Schouteten RLJ, Lauche K (2019) The relative importance of improvement routines for implementing lean practices. Int J Oper Prod Manag 39(2):214–237
3. Knol WH, Lauche K, Schouteten RLJ, Slomp J (2019) The duality of lean: organizational learning for sustained development. Academy of management proceedings, no 1
4. Shook J (2008) Managing to learn. Lean Enterprise Institute, Cambridge
5. Sobek D, Smalley A (2008) Understanding A3 thinking: a critical element of Toyota's PDCA management system. CRC Press, Boca Raton
6. Liker J (2004) The Toyota way: 14 management principles from the world's greatest manufacturer. McGraw-Hill, New York
7. Liker J, Meier D (2006) The Toyota way fieldbook: a practical guide for implementing Toyota's 4Ps. McGraw-Hill, New York
8. Oversluizen G, Slomp J (2019) Use of A3-method in industry projects. Paper presented at ELEC2018, improved and submitted to production planning and control
9. Shook J (2009) Toyota's secret: the A3 report. MIT Sloan Manag Rev 50(4):30–33
10. Nemoto M (1987) Total quality control management—strategies and techniques from Toyota and Toyota Gosei. Prentice Hall, Inc., New Jersey
11. Stappers PJ (2012) Teaching principles of qualitative analysis to industrial design engineers. In: Proceedings of the 14th international conference on engineering and product design education, Antwerp, Belgium, pp 1–6

5 Conclusion

Based upon three case studies, this paper presents a five-stage model for the development of continuous improvement in companies. The model is consistent with the findings of Kim et al. [1, 2] which were based on data from cross-functional self-assessment among managers of more than 100 companies. [illegible]

[illegible] who embrace the concept of lean and continuous improvement [illegible] about the use of the A3 tool and have adopted the tool for all their improvement projects. In this paper, we indicated what [illegible] of the A3 tool in each of the five stages. [illegible]

References

[illegible]

Investigating Maturity of Lean Culture in Product Development Teams

Torgeir Welo, Geir Ringen and Monica Rossi

Abstract Nowadays, many companies attempt to expand the lean concept from manufacturing to Product Development (PD). However, there are very few practical examples reporting how lean can be successfully applied outside the manufacturing floor. The benefits of lean can only be realized in PD once the concept is scaled to the business environment where it is employed. In this paper, we use a maturity framework for identifying gaps between current lean culture capabilities and those deemed necessary for competitive PD practices in the future. The general objective is to identify differences between companies as to how they assess capability gaps relating to lean within their operational context. A lean assessment framework using a continuous descriptive five-level maturity grid method has been used in the PD environments of nine companies. Focus has been placed on assessing characteristics associated with a lean culture in PD. The results show significant differences between the companies as to how they assess the gap between their current lean cultural capabilities and those deemed necessary in the future. Overall, the use of *simple and visual communication* means in PD activities is identified to provide the larger capability gap among the companies studied. In the other end of the scale, showing the smaller gap, we identified the perceived *role of digital tools and technology* in PD.

Keywords Product development · Lean · Culture · Assessment

T. Welo (✉) · G. Ringen
Department of Mechanical and Industrial Engineering, NTNU, Rich Birkelandsv. 2B, 7491 Trondheim, Norway
e-mail: torgeir.welo@ntnu.no

G. Ringen
e-mail: geir.ringen@ntnu.no

M. Rossi
School of Management, Politecnio di Milano, Milano, Italy
e-mail: monica.rossi@polimi.it

M. Rossi et al. (eds.), *Proceedings of the 6th European Lean Educator Conference*, Lecture Notes in Networks and Systems 122,
https://doi.org/10.1007/978-3-030-41429-0_37

1 Introduction

Over the years, manufacturing companies have implemented various measures in response to increasingly competitive markets. Lean [1] is perhaps the most significant concept that has emerged in the manufacturing community in modern times. However, lean production has shifted from its early days as a competitive frontier to today's role as the industry standard. Many companies have thus established strategies for moving the lean concept beyond the factory floor and into knowledge functions such as Product Development (PD) [2–5]. However, these efforts over the two last decades have yielded relatively little progress in arriving at a unified understanding of lean when this concept is applied in PD.

Rather than dealing with physical objects that go through a linear process, PD concerns the creation and transformation of information into retrievable and (re)useable knowledge [6–8]. This transformation process takes place through mostly interrelated activities with the overall outcome to mitigate risk(s) to acceptable level [9]. Thus, due to the inherent nature of PD, it would be naïve to believe that the implementation practices of lean manufacturing can be used in their original form in PD. Furthermore, due to the important role of people in knowledge transformation processes, such as PD, it is questionable if practices established in the corporate culture of Toyota Motor Company will work equally well in other contexts. We hypothesize that people and, thus, culture are particularly important aspects for applying and implementing Lean in PD environments.

With very few exceptions, such as Browning and Heath [10], the scientific evidence of how contextual factors influence lean practices in knowledge environments is mostly undependable. Furthermore, there is lack of attention given to gaining insights into how interdisciplinary PD teams evaluate their current lean capabilities and practices relative to those deemed necessary to sustain competitive in the business context that they operate. This is particularly important since the application of lean in any organizational environment builds heavily on a culture for continuous improvement and change [11, 12]. This is particularly demanding in PD environments due to the intangible nature of the 'process' and the lack of instant outcome metrics [13]. Overall, there is a strong need to gain additional insights into how the lean concept can be successfully adapted to the context of different knowledge organizations—the way it already has revolutionized production environments all over the world during the past three decades.

This research seeks to determine lean cultural capabilities in PD environments, using a hierarchical capability maturity framework for assessing PD teams. A lean assessment tool is used for gathering data for Lean PD maturity levels [14]. The tool is based on an interpretation of lean that goes beyond manufacturing [1] and into PD [2]. Its structure serves the purpose of facilitating discussions on key PD capabilities and practices within the assessment team. The main objective of this research is to answer the following research question: *How do PD teams in knowledge-intensive manufacturing companies rate their current Lean cultural capabilities relative to the need for increased value creation in the future?*

The remainder of the paper is organized as follows: Sect. 2 describes the basic of the assessment tools along with its structure and composition. Section 3 presents the assessment process and an overview of the companies. Section 4 gives the results and a brief discussion and Sect. 5 summarizes the paper.

2 Assessment Tool

2.1 Basic Strategy

Implementation of Lean in PD calls for a methodology for assessing current capabilities, defining future goals and establishing metrics to measure progress towards those. Maturity models in various forms have been applied to assess improvement efforts within different functional areas of an organization, including PD [15]. The basic methodology includes describing in a few statements the typical behavior of an organization at a number of levels of maturity for an assortment of key characteristics of the process area to be assessed. The different levels provide the opportunity to codify what might be regarded as practice in accordance with a specific performance characteristic, along with some transitional stages. The benefit of the maturity grid is the descriptive text tied to a scale of each characteristic of performance for each level. The drawback is that the descriptive text becomes increasingly difficult and complex as the number of levels increases.

In this research, the maturity grid approach was chosen since it provides a means for process improvement and management based on longitudinal process data [16]. It can also be used as an interactive research survey tool for collection of cross-sectional data related to where different companies identify their current capabilities and capability gaps. Furthermore, according to Fraser et al. [15], "the typical maturity model used in PD tends to be structured according to existing notions of good practice, and generated using experience-based principles, and tested by qualitative approaches such as interview and beta testing, but with little quantitative or statistical analysis". This was an important consideration for choosing this strategy because of the sample size considered, which was largely dictated by the time frame needed for completion of the audits in each company. The main inspiration for the format and structure of the assessment forms came from The Lean Enterprise Self-Assessment Tool (LESAT) due to its prowess and documented capabilities, Nightingale and Mize [17]. However, our focus was directed toward lean PD practices, rather than lean business practices which was done in [17]. It should also be noted that due to the large contextual differences between the companies and the aim to identify Lean improvement initiatives in this study, focus was primarily placed on identifying and analyzing capability gaps between current and future practices in each company, rather than focusing on capability maturity levels.

2.2 Structure

The dimension *Lean culture* was divided into a set of five main characteristics, see Table 1. Each of the characteristics was decomposed further into a subset of three capabilities. Each capability was given by situational descriptions at different maturity levels, used to assess the actual practice in the company [18], see Fig. 1. For each capability, the description was tied to a maturity scale, where the auditee could identify the current situation and required future situation. To reduce amount of text, statements for only three different maturity levels were codified and linked to a Likert-scale (1, 3 and 5). For each characteristic, the result (current and future levels) were obtained by averaging over the three capabilities.

The main purpose of this research is to identify capability *gaps* (G) between current and future foreseen practices, and the second one is obtaining maturity levels

Table 1 Characteristics and capabilities covered in assessment of lean culture

Characteristic	Overall question	Capability
1. Trust, respect, and responsibility	To what extent are trust, respect, and responsibility core values?	• Openness and respect of opinions/ideas • Coaching and delegating management • Involvement and commitment to decisions
2. Fact-based decision making	Is fact-based decisions part of culture at all levels in the organization?	• Openness and respect of opinions/ideas • Coaching and delegating management • Involvement and commitment to decisions
3. Creativity and entrepreneurship	Is creativity encouraged, valued and part of product and technology strategy?	• Strategic role of NPD • System/process for capturing ideas • Separated research from development
4. Digital tools in product D&E	What is the perceived role of digital tools (relative to others) in achieving goals?	• People and process over tool and technology • Stabilization before automation • Implementation, rather than tool, as a competitive factor
5. Simple and visual communication	To what extent is visual communication anchored in the culture?	• A3s used for visual communication • Visual management practices/environment • Visual communication for learning and problem-solving

6:3	Culture for Creativity and Entrepreneurship				
	Level 1	**Level 2**	**Level 3**	**Level 4**	**Level 5**
Is creativity and entrepreneurship encouraged and valued in the company, and does it take a central role in the product and technology strategy of the company?	• New products and technology development have a minor role in meeting future business goals. • No system exists for generating, handling and evaluating new ideas. Quality is secured through minimizing innovation. Conservatism and protectionism holds development of new ideas at a low level. • There are no separate resources appointed for research, innovation, or internal entrepreneurship.	Interpolate between Level 1 and Level 3	• New products and technology development have a part in the company's long-term strategic plan. However, the strategy for new products is not very aggressive. • When ideas come up they are met by interest and encouragement. However, no good system exists for handling or capitalizing on new ideas; decisions and actions are to some extent arbitrary. • Resources for research, technology development, innovation and entrepreneurship belong to the product development organisation. Long term resource needs are forced to compete with (and loose to) short term project needs.	Interpolate between Level 3 and Level 5	• New product and technology development have a very central role in meeting long term business goals. • An effective system for creating, handling and evaluating ideas does exist. Innovative and creative thinking is encouraged and there are mechanisms to capture and capitalize on out-side-the-box-ideas. Curiosity and open-minded interest are always the response to a new idea. • Separate resources are appointed for research, innovation and internal entrepreneurship. Exploration of new business models, markets or products is encouraged and colors the mindset of the entire company.
	[ ] Current [ ] Desired	[ ] Current [ ] Desired	[ ] Current [ ] Desired	[ ] Current [ ] Desired	[ ] Current [ ] Desired
Background					

Fig. 1 Example of assessment sheet, main characteristic: *culture for creativity and entrepreneurship*

(C). The dimension Lean culture then is divided into totally fifteen capabilities, see Table 1.

3 Method

3.1 Implementation

The case study was implemented in different steps, starting with an initial scoping event to create buy-in in the company. This also served to educate the research team on the challenges that the company was facing as important input to the main assessment. The result from the scoping event was summarized in a so-called A3 report, or Knowledge brief, which is a commonly used Lean PD too.

After the initial scoping event, the main assessment was scheduled. A typical assessment team was multi-disciplinary, consisting of a representative group of individuals (4–12 people) involved in PD projects. In total, 57 project team members from nine different companies provided their input through face-to-face conversations, discussions and dialogues while completing the assessment.

The main assessment started with reviewing the agenda and objectives for the assessment. Then, the company representative(s) gave a briefing on the latest developments since the initial scoping documented in the A3 report. The remaining part of the event was dedicated to completing the assessment. The structure of the form, audit process guidance, and the aim of focusing on capability maturity gaps ($\boldsymbol{G}^k$) rather than maturity levels ($\boldsymbol{C}^k$) were communicated, see Fig. 1. The assessment included the research team giving a 30 min presentation on the underlying theory and the practices associated with Lean culture in PD [19]. Then, the capability maturity gaps were assessed for each of the fifteen different capabilities on an individual basis. The research team then gathered and compiled the results, before presenting the individual scores to the audit team. If there were notable differences in the way individuals rated a given capability, a discussion was facilitated to reach consensus before the team collectively decided on a final score.

After completing the assessment, the research team analyzed, compiled and synthesized the results into a format to fit within an A3 report, serving as a part of the research protocol.

3.2 Case Companies

All nine case companies in this study, named by the letters A-I, are multinational and operate in various industrial sectors, including oil and subsea, automotive, consumer goods and defense and aerospace with PD as a central part of their business strategy. Only one company (I) operates in the B2C segment, whilst the rest is B2B firms. Unlike the other companies, company H does not have own production or product ownership, but are providing product development and production integration services as their main business. The companies all had different operational modes, while the common denominator is development of advanced products with relatively high value-added. Their size ranges from just above 100–4,700 employees.

Companies G and F (I) are suppliers of mechanical systems, structural chassis components, etc. to automotive OEMs and system integrators. The products are mainly produced at high annual volumes. Their PD process is relatively repetitive and predictable in nature, and mostly driven by new model and platform development of the OEMs.

Companies A, B and C operate within the industry sector of defense, missiles, space, rocket engines and aviation. The customers are mainly state committees, governments in different countries, NATO (contractual obligations), etc. Unlike G and F, where R&D costs are typically included in the sales price of the product, R&D costs are usually a part of the contract paid up-front by the customer.

Companies D and E are suppliers of equipment for oil and gas, drilling, subsea and maintenance to oil and drilling companies. These two companies have a very strong project-oriented organizational structure, where its functional organization is organized according to products and product categories.

4 Results and Discussion

4.1 Capability Maturity Levels

The assessments were made by input from up to twelve people from each company, altogether 57 respondents. The research methodology included face-to-face interactions and discussions before arriving at a collective score. Note that Company E decided to round off the average of the individual ratings to the nearest integer. Companies B, F and I rounded off to the nearest multiple of 0.5 and the remaining rounded off to the nearest multiple of 0.1.

Table 2 summarizes the results of the assessment of current capability maturity and capability gaps. Although the sample size is limited, it should be kept in mind the number of people involved for each data point ($\boldsymbol{C}^k$) and the fact that each data point is based on evaluation of three individual capabilities (see Table 1). However, note that the current ratings (C^k) are not suitable for identifying which Lean cultural characteristics of a company are more mature. The reason is that the situational descriptions in the assessment sheet were tied to ordinal scales, which generally fail to capture information that will be present in other scales due to unintended differences between the levels of the various ordinal scales. Therefore, the rationale of use is comparing capability maturity levels *between* different companies. This is interesting from an academic view, but less useful for using the assessment as a starting point for implementation of Lean; in other words, a capability maturity level of, say, 2.0 could be sufficient for one company but not for another company, depending on their business and PD strategies.

The collective average score across all companies was 3.07, ($\widehat{\boldsymbol{C}}$). Some variations are observed between each company and between the three capabilities (see Table 1) that belong to the same characteristics within a single company. Overall, company G (an automotive supplier) gives the highest average rating ($\widehat{\boldsymbol{C}}^{(G)} = 3.45$), mainly due to the high rating of 'Simple and visual communication'. This again may be related to the fact that the company is very manufacturing-oriented with strong background in Lean production. The majority of the other companies rates this capability low. Company B (a defence and aerospace company) gives the lowest rating of overall Lean culture capability ($\widehat{\boldsymbol{C}}^{(B)} = 2.60$), mainly due to low score on visual communication tools.

4.2 Capability Gaps

Table 2 lists the Lean culture capability maturity gaps identified when using the assessment framework as an interactive research tool. The maturity gaps (G^k) have a somewhat broader applicability than the maturity levels (C^k) listed above, since G^k represents a variation from a given reference. Therefore, these are less sensitive to any

Table 2 Current maturity scores (C) and gaps (G) for companies A-I obtained for lean culture

Firm	A	B	C	D	E	F	G	H	I	$\widehat{S}^k_{avg}$
C^1	3.3	2.5	3.3	3.2	5.0	1.5	3.0	2.8	3.0	**3.07**
C^2	3.0	3.5	3.6	3.0	1.0	3.0	3.5	3.4	4.0	**3.11**
C^3	2.7	3.0	3.7	3.7	4.0	3.0	3.0	3.2	3.0	**3.26**
C^4	3.7	3.0	3.2	3.4	5.0	3.0	3.5	2.6	2.0	**3.27**
C^5	2.3	1.0	2.5	3.2	2.0	3.0	4.3	2.8	2.5	**2.62**
$\widehat{C}$	**3.00**	**2.60**	**3.28**	**3.30**	**3.40**	**2.70**	**3.45**	**2.96**	**2.90**	**3.07**
G^1	1.0	1.5	1.3	1.5	0.0	3.0	1.5	1.4	1.5	**1.41**
G^2	1.0	1.5	0.9	2.0	4.0	2.0	1.0	0.8	0.5	**1.52**
G^3	1.6	1.0	1.0	1.3	0.0	2.0	2.0	0.8	2.0	**1.30**
G^4	0.0	1.0	1.4	1.4	0.0	0.0	1.3	1.2	2.0	**0.92**
G^5	2.0	3.5	1.5	1.8	3.0	2.0	0.8	1.6	1.5	**1.97**
$\widehat{G}$	**1.12**	**1.70**	**1.22**	**1.60**	**1.40**	**1.80**	**1.30**	**1.16**	**1.50**	**1.42**

unintended bias between the scale levels of the individual characteristics. The gaps are useful to identify and select capabilities as a starting point for implementation of Lean (PD) in a given company based on the following: the magnitude of gap identified; the resources required; and the importance of the capability—i.e., whether the capability is a bottleneck for the output from one of the 'value streams' within the PD operation, or not.

The Lean culture capabilities depict relatively high maturity gaps. In particular, the capability gap of 'Simple and visual communication' shows a very high gap ($\widehat{S}^{5}_{\mathbf{avg}}$ = 1.97) across the different companies, with companies B (defence and aerospace, $G^{5(B)} = 3.5$) and E (subsea supplier, $G^{5(E)} = 3.0$) as the major contributors to this average large capability gap. Also, the Lean cultural capabilities 'Trust, respect, and responsibility' (C^1) and 'Fact-based decision' making (C^2) show gaps of $\widehat{S}^{1}_{\mathbf{avg}} = 1.41$ and $\widehat{S}^{2}_{\mathbf{avg}} = 1.52$, respectively, with companies F ($G^{1(\mathrm{F})} = 3.0$) and E ($G^{2(\mathrm{E})} = 4.0$) providing the highest gap scores. Here, company F is an automotive supplier and company E is a supplier of subsea installations. The capability 'Role of digital tools' $\widehat{S}^{4}_{\mathbf{avg}} = 0.92$ provides the lower capability gap among the five cultural Lean culture capabilities (Table 1). Several companies (A, E and F) do not assess a gap at all, as to how they view the 'Role of digital tools' in meeting business goals. In other words, they rate *people and process* over *tools and technology*, and have focus on process and automation but first after stabilizing the process. Companies C, D, G, and particularly I, have a somewhat different perception of the role that digital tools play in their daily operation.

5 Conclusion

This study presents a tool for assessing the maturity of Lean culture in PD organizations along with its implementation in nine different companies. The companies operate in different businesses and vary largely in size. The results show that the companies rate their Lean cultural capabilities differently, depending on business context. However, the capability '*Simple and visual communication tools*' is the one showing the largest gap across the sample of nine companies investigated herein. Also, the capabilities '*Trust, respect and responsibilities*' and '*Fact-based decision-making*' show significant gaps between current and desired Lean culture capabilities. In the other end of the scale, the '*Role of digital tools in design and engineering*' provides the smallest capability gap, in which three of the companies not assessing a gap at all. From this, it can be concluded that the assessment has demonstrated its capability as a tool for identifying and initiating Lean PD improvement initiatives.

Further work includes identifying and studying practical implementation strategies. In this connection, it is of interest to study how such implementation strategies could be tied to the context of the company; for example, does a certain company context call for a specific implementation strategy, or could the same approach be used for different companies with different PD contexts? Moreover, which ones, if any, are the main dimensions defining a PD context (important to Lean improvement initiatives)? It is necessary to understand if the initiatives should be piloted and tied to functions or to individual projects/teams.

Acknowledgements The authors want to acknowledge the Norwegian Research Council for supporting this study through the project KPN VALUE, grant 267768.

References

1. Womack JP, Jones DT (1996) Lean thinking: banish waste and create wealth in your corporation. Free Press, New York
2. Ward AC (2007) Lean product and process development. Lean Enterprise Institute, Cambridge
3. Sobek DK, Smalley A (2008) Understanding A3 thinking, a critical component of Toyota's PDCA management system. Productivity Press, New York
4. Karlsson C, Åhlström P (1996) The difficult path to lean product development. J Prod Innov Manag 13:283–294
5. Haque B, James-Moore M (2004) Applying lean thinking to new product development. J Eng Des 15(1):1–31
6. Kennedy MN, Harmon K, Minnock E (2008) Ready, set, dominate: implement Toyota's set based learning for product development. The Oaklea Press, Richmond
7. Drucker PF (2008) Management. Harper Collins, New York
8. Browning TR (2006) On customer value and improvements in product development processes. Syst Eng 6:49–61
9. Reinertsen DG (1997) Managing the design factory. The Free Press, New York
10. Browning TR, Heath RD (2009) Reconceptualizing the effects of lean on production costs with evidence from the F-22 program. J Oper Manag 27:23–44
11. Rother M (2010) Toyota Kata—managing people for improvement, adaptiveness and superior performance. McGraw-Hill, New York
12. Kotter JP (2012) Leading change. Harvard Business Review
13. Driva H, Pawar KS, Menon U (2000) Measuring product development performance in manufacturing organisations. Int J Prod Econ 63(2):147–159
14. Welo T, Ringen G (2016) Beyond waste elimination: assessing lean practices in product development. Procedia CIRP 50:179–185
15. Fraser P, Moultrie J, Gregory M (2002) The use of maturity models/grids as a tool in assessing product development capability. In: IEEE international engineering management conference, Cambridge, UK, pp 244–249
16. Chiesa V, Coughlan P, Voss CA (1996) Development of a technical innovation audit. J Prod Innovation Manag 13(2):105–136
17. Nightingale DJ, Mize JH (2002) Development of a lean enterprise transformation maturity model. Inf Knowl Syst Manag 3(1):15–30
18. Welo T, Ringen G (2018) Investigating organizational knowledge transformation capabilities in Integrated manufacturing and product development companies. Procedia CIRP 70:150–155
19. Ringen G, Welo T (2015) Knowledge based development practices in systems engineering companies: a comparative study. In: 2015 annual IEEE systems conference (SysCon) proceedings

Frictionless Innovative Slick® Process to Keep Track of Companies' Knowledge

Alessandro Carpentari, Davide Cattaneo, Marco Milani and Monica Rossi

Abstract Processing and sharing knowledge is a common problem for any kind of company. Software companies, and more in general companies whose main business is based on continuous and innovative evolution in knowledge, this problem can be critical. Developing the Slick® methodology IdeaTech decided to face this problem using the A3 approach (Saad et al. in A3 thinking approach to support problem solving in lean product and process development, Cranfield University, Cranfield, Bedfordshire, Shrivenham, Oxfordshire, England [1]) to problem solving, with the aim of define a management process to all levels of knowledge and the different levels of knowledge awareness. We started analyzing the Unknown Unknown level—the situation in which people do not know what they not know, then we investigated the Known Unknown level—the situation in which awareness come to mind, even though the knowledge is still missing, than we studied the Known Known level—the situation in which one is aware of knowing. We finally describe a process to guide team members from a UU level to KK level and a methodology to track all the knowledge that reached the KK level using an efficient, frictionless, low-footprint process.

Keywords Unknown/Unknown · Known/Unknown · Known/Known · A3 · Agile · Lean · Retrospective · SkillTag · Slick®

A. Carpentari · D. Cattaneo · M. Milani (✉)
IdeaTech srl, Via Balicco 59, 23900 Lecco, Italy
e-mail: marco.milani@ideatechsrl.com

A. Carpentari
e-mail: alessandro.carpentari@ideatechsrl.com

D. Cattaneo
e-mail: davide.cattaneo@ideatechsrl.com

M. Rossi
Politecnico di Milano, Piazza Leonardo da Vinci 32, 20133 Milan, Italy
e-mail: monica.rossi@polimi.it

M. Rossi et al. (eds.), *Proceedings of the 6th European Lean Educator Conference*, Lecture Notes in Networks and Systems 122,
https://doi.org/10.1007/978-3-030-41429-0_38

1 Introduction

There are a lot of sources of inefficiency [2] that continuously affect any kind of companies, including IT companies, and their profits. These issues can be vastly different in nature, causes and economical effect. Often there's not enough time to define common approaches to solve these problems and there is not the awareness and the ability to identify the real problems. As a result, the productive capacities and performances of the company, as well as the quality of services and products can be compromised. This paper aims to illustrate the ways in which an industrial reality has addressed the problem of knowledge management and capitalization of knowledge. We finally developed a process, part of the Slick® methodology, that successfully manages the Known Knowledge in an innovative, frictionless and cost effective fashion. The result of the application of the process described in this paper is a structured database that allows team members to know in any moment what are the topics part of the company's knowledge, who is the most expert about it, and where to find the supporting documentation useful to improve personal skill about a topic without reinventing the wheel. This database is maintained and automatically expanded by the processes already executed by the team members while developing software.

2 IdeaTech and the Problem to Solve

IdeaTech is a software development company based in Italy. Founded in 2002, it consists of 25 developers dealing with custom software development, mobile app development, and web development. One of IdeaTech's problems was sharing knowledge inside the company, inside a team and between different projects' teams of the same company [3].

In the last few years we have had evidence of problems that we have faced several times from the beginning, with consistent waste of time and resources. Similarly the introduction of new team members led to inefficient training because the existing knowledge couldn't be trained easily. This was due to a lack of communication between teams regarding existing solutions and knowledge.

We have been knowing that the company has been a strong background knowledge, unluckily this knowledge has been "stored" in the memory of individuals.

We started with this awareness in mind to find a way to fix this problem.

In the company, over the last few years, we have been using an agile approach for the management of software projects called Slick®, which allows us to move dynamically and efficiently in the entire life cycle of the development.

Thanks to experience we extended Slick® with new processes to better deal with the development activities. In this context we use IdeaTask, one of our self-developed software, able to manage the project backlogs and the various assigned tasks.

Using the A3 methodology, we described the problem of keeping track of knowledge.

3 Applying A3 Approach to Support Knowledge Management at IdeaTech

Analyzing the various problem solving techniques we came across the A3 approach [1].

The method got its name from the ISO A3 [4] size paper in which a process of achieving a goal or solving a problem was illustrated.

The A3 [5] Report is based on the Plan, Do, Check, Act (PDCA) method and it's divided in the following sections:

1. Background
2. Current Conditions
3. Target/Goal(s)
4. Causes Analysis
5. Proposed Countermeasure(s)
6. Implementation
7. Process rule(s).

3.1 Background

In IdeaTech there wasn't communication between teams and there wasn't any knowledge-sharing process. It often happened that a team solved a problem or discovered something new without sharing what learned with other teams. Furthermore, the introduction of a new resources was problematic, in fact a new employee couldn't take advantage from any kind of written did not have access to a business information library, but had to train himself orally asking other team members.

3.2 Current Conditions

There was not any knowledge sharing process. There was no written legacy, we had, indeed, the corporate wiki, but it was not used because it was out of any standard process and it was not even useful.

In order to better understand the situation, we interviewed a group of employees. We formed the group randomly, choosing three project leaders and eight developers. One of the questions asked was: "When you explore new technologies, or a new way of doing an action or new features, how do you leave a legacy in order to share with others your experience in the future?". 70% of employees replied that they did not know how to leave a comment or legacy trail.

3.3 Target Goal(s)

The objective was to be able to track the experiences, knowledge and problems experienced by the company employees without any extra work.

3.4 Causes Analysis

We organized different meetings in order to identify the causes. During these brainstorming we asked any relevant 5W (What, When, Where, Who, Why) and 2H (How, How many/How often) questions and we realized an Ishikawa's diagram. We identified the following causes:

- There was a lack of communication between teams working on different projects.
- The software (Wiki) was not used or maintained.
- There was a few time and it could not be used in the sharing of knowledge.

These first four points are referred to the Plan section of the PDCA technique and allowed us to visualize the problem as never before.

Once the causes were identified we could start to propose the countermeasures.

3.5 Proposed Countermeasures

Based on the causes identified above, we have devised a first draft of possible countermeasures. The first proposed strategy was to try to define a process to share knowledge in the following way:

- The project leader had to keep track of the technologies, problems and experiences faced during the developments by him and his team.
- Periodically the project leaders had to organize a meeting between themselves. In this meeting they had to discuss what they learned and moreover; they had to decide which of these new experiences should be officially included in the Wiki.
- The relevant experiences had to be entered by project leaders in the wiki before the next meeting.

Applying these countermeasures we would have been able to break down the causes we had identified. The meeting among the project leaders would have increased the communication between the teams and the writing on the Wiki would have given us a structured library with constantly updated contents.

Sharing the proposal with the project leaders, we realized that it would be too time consuming for them and that the developers would not have been encouraged to use the Wiki. We had to think about a new solution.

In our company we use a software tool, built internally, which allows us to develop software following the SCRUM AGILE methodology. This software, called IdeaTask, allows us to keep track of the time dedicated to each task.

In IdeaTask, all the activities of a project are collected in a container called backlog. Every backlog is divided into themes and each theme contains activities, called stories.

A story is a task on which one or more developers can work on.

When someone works on a task he has to record the time spent using the functionality in IdeaTask.

Our idea was to modify the current software, in particular the story, theme and in backlog entries, in order to permit the insertion of hashtags, which we called SkillTag.

This new field was intended to allow to keep track of new technologies and/or new themes faced during the developments through hashtags, in a social networking style.

During the SCRUM meetings and therefore during the creation of backlog and stories, the development team had to define which were the Skill Tags related to the project, the theme or the single story.

The time tracking of every single activity allows us to collect specific data in hours that indicates how much time a resource has worked on a specific technology or issue.

Once we had the list of skill tags and the hours dedicated to them from each resource, we could get a kind of leaderboard related to the knowledge of a particular topic for each.

Through this idea we were able to know who had faced what and for how long, but there was still no structured way to share knowledge.

To solve this problem, we decided to introduce a new role called Knowledge Manager. The person in this new role had to supervise the corporate wiki and to verify that the process was properly applied.

The Knowledge Manager periodically, had to extract from IdeaTask the list of the Skill Tags. For each of them he had to create a dedicated section on the wiki. For each section he had to define a curator, called a Skill Master. The Knowledge Manager choose the Skill Master from the list of resources that had worked longest on a particular subject.

However, the skill master did not have to be responsible for entering the text in the Wiki.

As we will see later when we will describe the process, it will be the resource that needs to learn a certain topic that has to write in the wiki once the new subject is learned.

Furthermore, we had to find a structured way to share knowledge, so we thought about the retrospective meeting, a meeting that already exists in Agile but that we have never used. An Agile retrospective meeting is a meeting held at the end of an iteration in the Agile software development. During the retrospective meeting, the team reflects on what happened during the iteration and identifies action steps for improvement(s) going forward.

We decided to apply and modify it according to our needs:

- The teams would be divided into groups of up to 8 people.
- A 90 min (approximate) meeting would be scheduled every two months.
- The meeting owner would be the project leader of the biggest project.
- During the meeting each resource would have 10 min to discuss both positive and negative experiences detected during the previous two months.
- In addition, the attendees had to review the Skill Tags entered and insert new ones in case in case any have been overlooked. In particular, those tags that could be considered synonyms, had to be grouped under a unique Skill Tag.
- At the end of the meeting, the meeting leader had to provide a minute with a summary of the meeting.
- The meeting leader also provided leadership with a report with any foreseeable issues or problems.

Every four months, the owners of the retrospective meetings have to organize another meeting, called retrospective meeting of meetings.

These meetings covered knowledge sharing of existing issues and the review of new technologies and innovations. Information discussed and collected was then summarized and presented to management.

We began to test the processes described above with one team consisting of eight people.

The resources involved had to insert the Skill Tags in the stories to which they were working on and to meet every month (we had reduced the frequency to a month in order to have more feedback) for the retrospective meeting.

At the end of the two months, we interviewed again the employees to which we had previously asked questions and the results were surprising.

During the interviews, we asked:

- Do you think that today there are effective methods in the company to collect the experience gained in the projects?
- When you have a problem to solve or a doubt during the development of a project, how do you try to solve it?
- When you explore new technologies, or a new way of doing an action or new features, how do you leave a legacy in order to share with others your experience in the future?

If to the question "Do you think that today there are effective methods in the company to collect the experience gained in the projects?" On the first round, 80% answered no, on the second round 80% replied yes.

The answers to the first round of interview certified the existence of the problem, while the answers to the second highlighted that the measures taken were improving the situation.

3.6 Process Rules

All the resources have to continue to insert the Skill Tags in the projects they dealt with.

The project manager is responsible for:

- Define the groups for the retrospective meetings.
- Identify the main project leader of each group.
- Organize the retrospective meeting of meetings every four months.

The main project leader of each group has to:

- Organize the retrospective meeting every two months.
- Oversee the meeting and lead and conduct it as described above.
- Draft the meeting minute.
- Share with the management any suggestions for improvement.
- Join the retrospective meeting of meetings every four months. If he was nominated as the main project leader of all the projects, he would also have had to deal with the drafting of the minute for the retrospective meeting of meetings.

The knowledge manager is responsible for:

- Updating the wiki periodically.
- He has to identify a Skill Master for each Skill Tag based on data obtained from the hours dedicated to each skill tag by the resources.
- In addition the Knowledge Manager has to verify that the contents included aligned (instead of compiled) with company standards.

Every resource that needs training has to access the IdeaTech wiki and verify that there is the Skill Tag of its interest.

In case the skill tag is present but the answers are not are not comprehensive, all-inclusive or complete for the resource, they can ask for a meeting with the Skill Master. If the skill master has knowledge and experience to support the resource then he takes care of train the resource, otherwise he has to define with the Knowledge Manager a training strategy. Once the resource (employee) is trained, they are required to write a summary explaining the new information or skill gained in the wiki.

The Skill Master is then required to review and verify that the resource's summary was correctly entered in the wiki.

In the case where a specific Skill Tag is not found, the resource (employee) and the Knowledge Manager, must identify a new training strategy for this new Skill Tag.

Once the new Skill Tag and training have been created, the Skill Master must create a new section in the wiki to covering the new Skill Tag and Training.

Figure 1 shows schematically what has been described above.

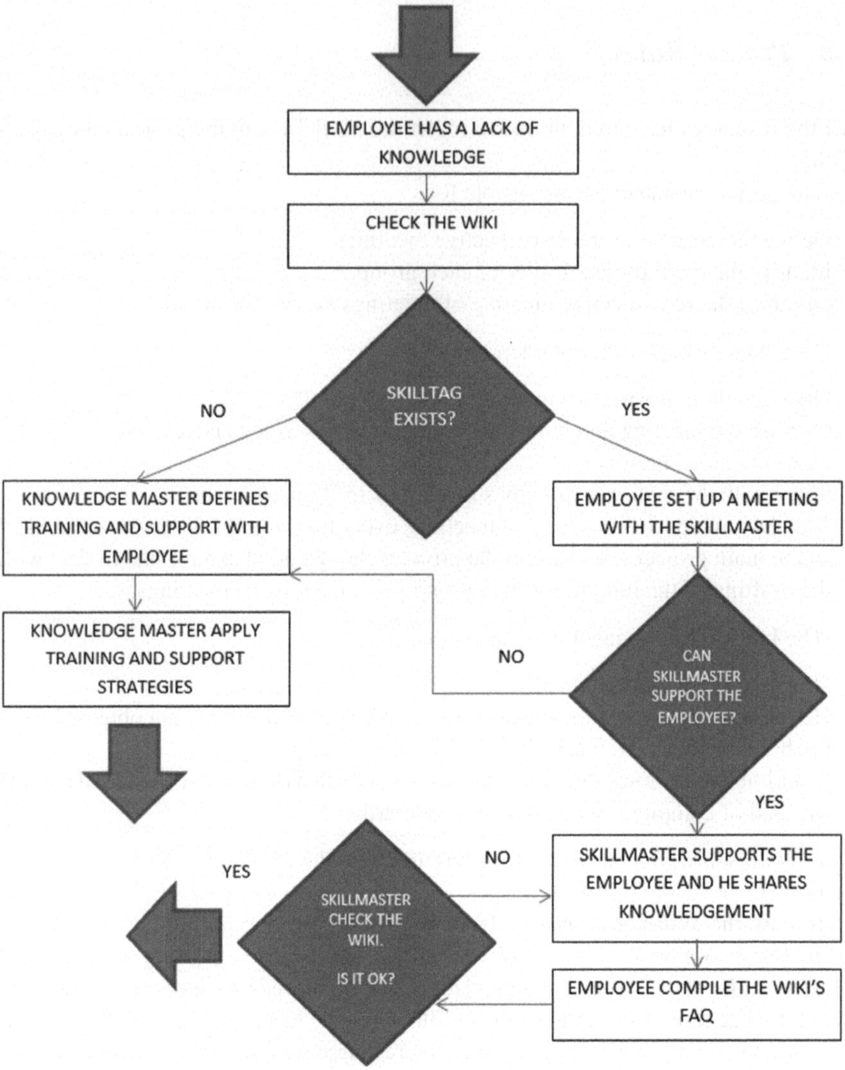

Fig. 1 The process that the employees has to follow when they have lack of knowledge

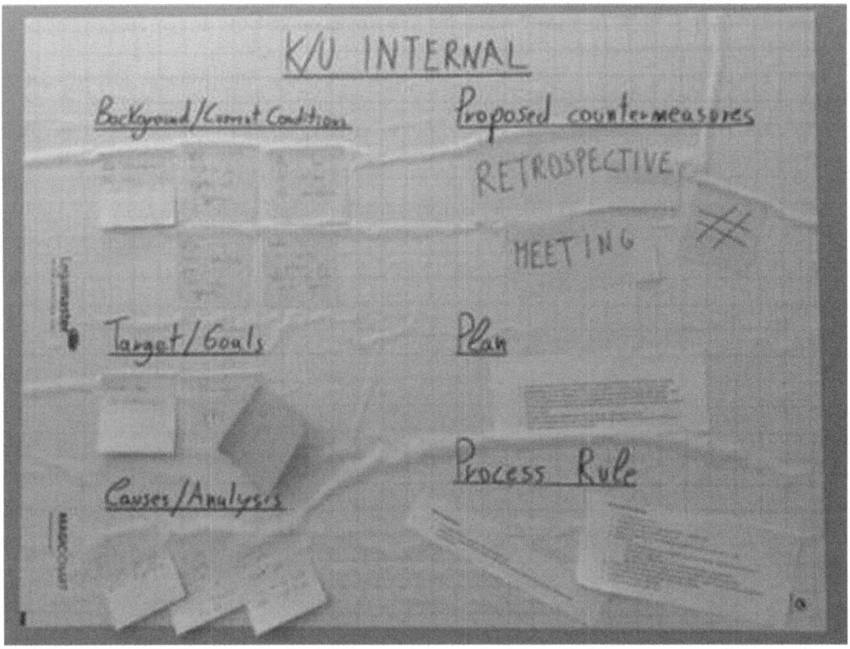

Fig. 2 The A3 used for the analysis and resolution of the problem

4 Conclusions

Using the A3 approach [6] IdeaTech has been able to solve the problem of knowledge sharing (Fig. 2).

In particular, we have defined an efficient and lean process that supports team members during the transition from the UU to the KK level.

Furthermore, we introduced two new solutions that allows us to keep track of the knowledge acquired at the company level: the Skill Tag method and the retrospective meeting. Both became part of IdeaTech business processes and are currently successfully used by all team members.

References

1. Saad NM, Al-Ashaab A, Shehab E, Maksimovic M (2013) A3 thinking approach to support problem solving in lean product and process development. Cranfield University, Cranfield, Bedfordshire, Shrivenham, Oxfordshire, England
2. Rossi M, Kerga ET, Taisch M, Terzi S (2011) Proposal of a method to systematically identify wastes in new product development process. Politecnico di Milano, Milan, Italy, Università degli studi di Bergamo, Bergamo, Italy

3. Rossi M, Terzi S (2017) CLIMB: maturity assessment model for design and engineering processes. Politecnico di Milano, Milan, Italy
4. Sobek DK II, Jimmerson C (2004) A3 reports: tool for process improvement. Department of Mechanical and Industrial Engineering, Montana State University, Bozeman, MT
5. Sobek DK II, Smalley A (2008) Understanding A3 thinking: a critical component of Toyota's PDCA management system. Productivity Press, New York, NY
6. Shook J (2009) Managing to learn: using the A3 management process to solve problems, gain agreement, mentor and lead. The Lean Enterprise Institute, Cambridge, MA, USA

Agile Management Versus Budget Control: Learn How to Win Both with Slick-Farm

Marco Milani and Monica Rossi

Abstract Slick® is a disruptive agile methodology based on Agile Scrum that offers clear and quantitative insight into any kind of innovative project. Thanks to the three main Slick® indexes, even non-technical managers can understand the progress of a project and define strategies to keep it on-track. We created the serious game Slick-farm to get through to Slick® and its benefits. We used the lean start-up methodology to create the game, we run through three different MVPs and we pivoted the wrong hypotheses based on feedback collected in the trial workshops. Applications of Slick Farm shows that the game is effective to communicate the benefit coming from Slick® as an Agile methodology used in an uncertain context and, even more, as a way of actively manage strict budget and risk control without losing in "agility".

Keywords Slick® · Agile scrum · Innovative project management · Budgeting and innovation · Risk management in agile · Slick farm · Serious games

1 Introduction

Agile methodologies are widely used in software industries and are spreading in non-software projects because of their abilities to guarantee deliverables quality even in dynamical contexts and even in situations where changes in project requirements are frequent [1–3].

Agile offers a wide range of pros, although it is affected by a set of cons that often makes it not appealing for real application in companies.

The main obstacles to the correct application of any kind of Agile methodologies are: the chaos around budgeting an Agile project, the lack of knowledge at the mid/top management level about Agile methodologies, the difficulties to understand how a

M. Milani (✉)
IdeaTech srl, via Lorenzo Balicco 59, 23900 Lecco, LC, Italy
e-mail: marco.milani@ideatechsrl.com

M. Rossi
Politecnico di Milano, Via Lambruschini 4b, 20156 Milan, Italy
e-mail: monica.rossi@polimi.it

M. Rossi et al. (eds.), *Proceedings of the 6th European Lean Educator Conference*, Lecture Notes in Networks and Systems 122,
https://doi.org/10.1007/978-3-030-41429-0_39

project is really going and the lack of formal processes capable of controlling risk and costs given the high agility promoted by the methodology itself [4, 5].

In order to overcome the budgeting problem, the agile ecosystem proposed a new way of contracting between customer and provider, based on trust and progressive step evaluation, with the aim to avoid formal quotation in advance (#NoEstimates) [6]. This is of course an ideal situation in which most of the company cannot think over; the real world asks for budgets that are certain, quotations that are fixed and confident results.

In order to address and overcome these common problems we developed Slick®, that is a low-footprint, super-set of processes and metrics added to a light version of the Scrum Agile methodology. Slick® offers strict control on projects risk and costs, without limiting the team's agility. The main benefit of this methodology is the ability to identify performance and cost problems immediately when they pop up, well before they can compromise the overall project performance.

In order to demonstrate that Slick® is effective in managing any innovative project, we also developed the Slick-Farm serious game using a lean start-up approach.

In this paper we will highlight the key values that are the base of Slick® as an evolution of Agile and the process we followed to successfully design and create the serious game Slick-Farm.

2 Slick®—Agile Management Focused on Economical Sustainability

Slick® bridges the gap between agility, promoted by the agile methodologies, and the need of strict cost controls, ensuring the company success and prosperity. Thanks to the 12 principles of its inspiring manifesto [7]. Agile methodologies are the state of the art of the approaches to develop a product in a fast changing environment. Unfortunately, this manifesto does not address economical nor financial aspects.

Consider the typical selling cycle for any kind of project: commercial quotation, development, deployment, and maintenance. Agile covers just the "development" phase, and it takes place when a budget is already set. The budget is almost untouchable in every "real-life" customer/provider relationship, even though it is usually based on limited and unclear set of requirements.

In a fast changing world, agility is a critical key successful factor, but changes do mean costs and risk. This is why Agile is loved by project and product managers and this is why CFOs, Budget managers and CEOs prefer to avoid it!

Slick® allows for frequent (on a sprint base) and clear insights to the projects, offering: comparison between the total real cost expected and the initial budget, quantitative measurement of the team performance, evaluation of adherence of milestones to the plans, a strategy definition in case of an off-track project.

Thanks to this quantitative evaluation, the mid and top management can obtain a clear insight into the project and is able to actively define strategies insuring every project is successful.

2.1 Slick® Processes and Metrics

Slick® is based on a light version of the Agile scrum [3] methodology and is suitable for managing any kind of innovative project from a technical point of view.

Even if experienced scrum masters are able to evaluate the progress and to predict the future evolution of a project using "custom" and personal metrics; managers, CFOs, CEOs, company owners, start-uppers, and investors need to clearly know how much budget has been burning, how many resources are required to complete the project, how the team is performing, or if the budget is enough to complete the project.

Slick® answered all these question in a standard, clear, easily understandable fashion. Thanks to unique concepts like: *nominal point value, time tracking, fluid resources allocation*, partial and dynamical roles, at the end of each sprint Slick® provide three crucial metrics:

1. CB: Cost to Budget—the overall comparison between the predicted cost at the end of the project and the budget set at the beginning.
2. TE: Team overall Effort—the real team effort compared to the effort expected nominally.
3. EB: Effort to Budget—the comparison between the total amount of points assigned to the backlog during the scrum meetings and the budget set by the manager.

These three metrics are ratios that, in a perfectly on-track project, all set to 1. Values greater than 1 describe difficulties in keeping the project on-track.

The core relation between the three indexes is (the demonstration is beyond the objective of this paper):

$$CB = EB * TE. \tag{1}$$

To analyse CB in (1) for different values of EB and TE we use the four quadrants chart as in Fig. 1.

The border between the green and the warm colours area is an hyperbole placed at $CB = 1$ that is $TE * BE = 1$.

The green area corresponds to $CB < 1$. In this area the predicted cost at the end of the project will be less than the budget. This is a safe area.

The warm colours correspond to $CB > 1$. In this area the predicted cost at the end of the project will be greater than budget.

Slick® offers an interpretation of what is going wrong with the project when CB lies in the red, yellow and orange areas:

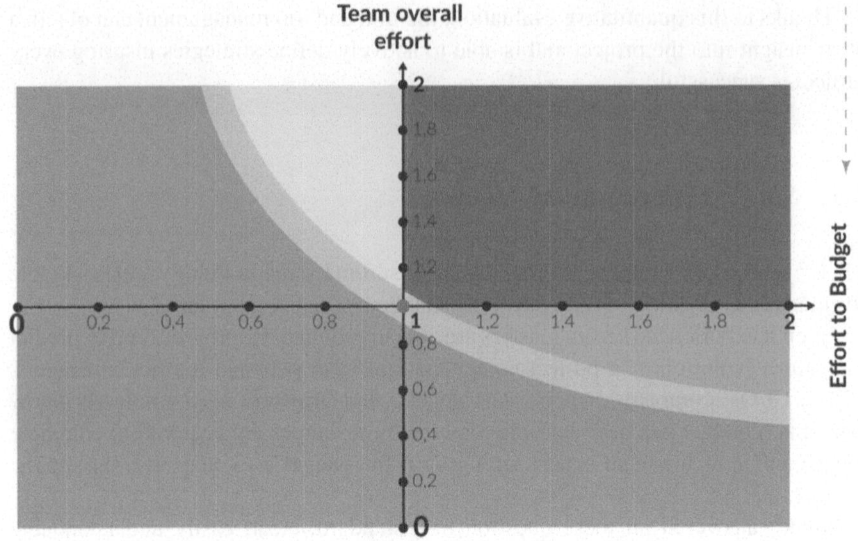

Fig. 1 Slick® chart—the two indexes EB and TE are placed on the axis. Both indexes are 1 centred that is the "normal" value for an on-track project

1. The yellow area corresponds to $EB < 1$ and $TE > 1$. This means that project is off-track because the team is under performing even if the budget is correctly set.
2. The orange area corresponds to the opposite situation in which $EB > 1$ and $TE < 1$. This means that project is off-track because the budget is underestimated even if the team is over performing.
3. The red area corresponds to the worst situation in which the project is off-track because of both the problems: budget underestimated and team underperforming.

3 Design of the Learning Experience: Slick-Farm

Slick® is a revolutionary approach to agile management that offers a unique point of view on the project progress. In order to spread this concepts to companies, R&D centres, start-ups we needed a simple, intuitive and effective way to promote the idea behind Slick®. We decided to investigate how a serious game would be effective.

The main objectives of the serious game were:

1. Demonstrate that Slick® is effective in managing innovative projects.
2. Understand the benefits of managing a project following a structured path compared to follow an unstructured one.
3. Learn how to manage uncertainty in projects using a structured methodology.
4. Identify the variables that determine the progress of a project.

5. Learn how Slick® indexes can help managers to make pivoting based on numerical evidences.
6. Develop personal project evaluation capabilities.
7. Optimise the budget usage.

We used the "lean start-up" framework to identify scientifically the MVP of the game and we applied an iterative design process as follows: design team set-up, assumption definition, first version of the prototype, field test, feedback collection, feedback analysis, improvement of the tool, field test, feedback collection, feedback analysis, improvement of the tool, final test and validation.

3.1 The First MVP

The design team identified the founding hypotheses about the game and the participants using the "I believe… " pattern [8].

Hypotheses about the game	Hypotheses about the participant
I believe applying Slick® in a simplified case is an efficient way to understand quickly the benefit of the new methodology	I believe the participants will be engaged by a serious game
I believe a project simulation will be a good approach to engage participants	I believe the participants will be able to make a comparison between the two project methodologies
I believe a comparative simulation based on a first "free" phase followed by a "slick-like" session is the most efficient way to highlight the power of Slick®	I believe the participants will feel frustrated and ineffective during the first "free" phase
I believe the two phases will be recognized as equally difficult	I believe the participants will feel confident and in control during the second "slick-like" session
I believe the game rules will be easy and fast to understand	I believe the participants will be able to use the Slick indexes to critically analyse the project progress and understand that Slick® offers disrupting opportunities

Considering these hypotheses we designed a two phases game provided though a workshop in which people are randomly divided in teams composed of 3–5 members. The objective of each phase is to create an imaginative farm completing a predefined set of "connect the dots" animal figures in 15 min. The first phase is completely "free" in terms of organization and methodology to be used. The second phase has the same objective as in the first phase and is managed using Slick®.

In both phases the team performance is evaluated at the end of the 15 min session. The session is considered successful if all the animals required to complete the farm are correctly drawn. The time budget is therefore 15 min, but no one can predict how much effort is required to complete the task, in fact is very difficult to predict how long is drawing a figure and, even more complex, how much is drawing all the figures.

During the second phase the team has the additional game sheet as in Fig. 2. This sheet guides the team in a sequence of three sprints and calculates the Slick® indexes: CB, TE and EB as defined in paragraph "Slick® processes and metrics". Thanks to this sheet, the team is expected to follow a structured strategy to plan and monitor the activities. In fact, effort estimation is as difficult as in phase 1, but in this phase the sprint by sprint Slick® analysis would help the team in evaluating the real progress and would offer quantitative elements to understand how to deal with the fixed budget of 15 min.

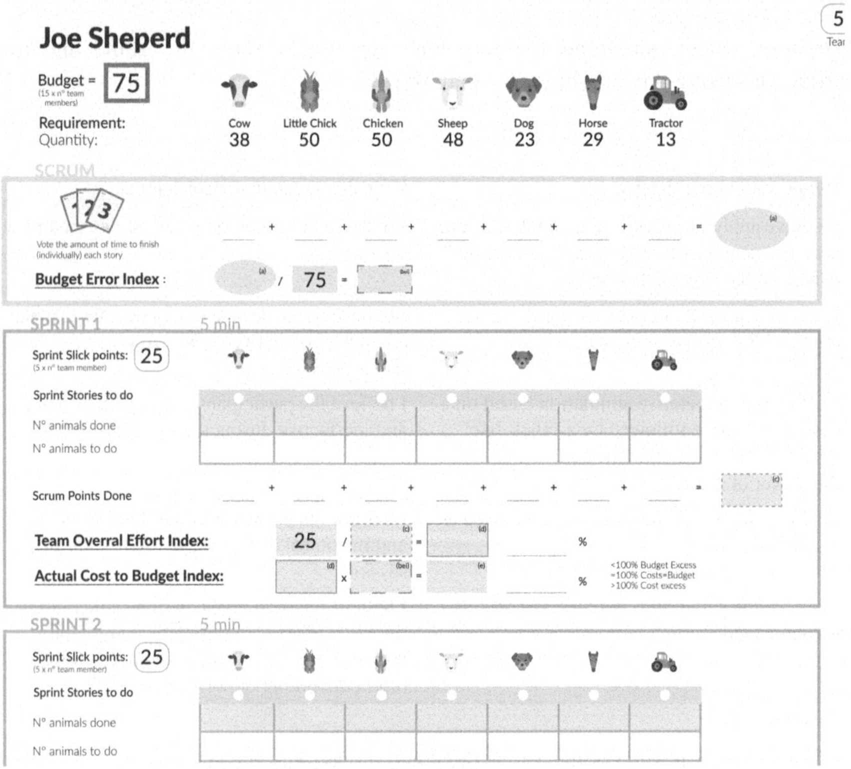

Fig. 2 Game sheet used in the second phase of the game

3.2 *First Validating Trial*

After creating a survey to collect feedback from participants, we run the first trial workshop. We had the opportunity to test the game involving 50 people from all over the world, each of them representing different professional backgrounds, experiences and agile knowledge. At the end of the trial we collected the following feedback with the aim to validate or reject the hypotheses (Tables 1, 2 and 3).

The survey highlighted that not all the initial hypotheses were confirmed, in particular we rejected the comparative design. It was clear from question 1a, 2a and

Table 1 Relevant feedback collected after 1st MVP—phase 1

Question	Answers
1a—Did you follow a specific strategy? If yes describe it	85% yes, but no strategies described
2a—Was the strategy repeatable?	12% yes
3a—When you started, did you know how much time it was needed to make all the animals requested?	92% no
4a—Was the initial budget adequate? If not, when did you realize it?	80% yes, but all the teams failed
5a—How much do you think you had the budget under control?	40% no 41% medium control 19% discrete control
6a—Was this phase hard?	73% yes
7a—Did you ever feel frustrated?	60% no

Table 2 Relevant feedback collected after 1st MVP—phase 2

Question	Answers
1b—Did you follow a specific strategy? If yes describe it	95% yes
2b—Was the strategy repeatable?	95% yes
3b—When you started, did you know how much time it was needed to make all the animals requested?	11% no, 89% yes
4b—Was the initial budget adequate? If not, when did you realize it?	50% yes
5b—How much do you think you had the budget under control?	76% full control
6b—Was this phase hard?	43% yes
7b—Did you ever feel frustrated?	60% no

Table 3 Relevant general feedback related to 1st MVP

Question	Answers
8—What did you learn from this game?	70% control and organization
9—Were the game instructions clear?	>90% yes

4a that teams never realized that phase 1 was unstructured and off-control. On the other hand was clear to the teams that phase 2 offered a better way of controlling the project (question 1b, 2b, 5b, 8).

The questions 3b and 4b compared to the 3a and 4a told us a very important thing: the teams took advantage from the phase 1 to better estimate phase 2. We were not satisfied by answer 4b, we expected a more clear understanding of the budget. Summarizing, we confirmed hypotheses 1, 2, 5, 6, 9, 15 and we rejected hypotheses 3, 4, 7, 8.

3.3 *Second MVP*

Considering the feedback we decided to drop the idea of comparative phases and focus on a unique Slick phase, we also decided to change the task, branching out from a repetitive set of equal short tasks in a small set of slower and varied tasks.

We easily identified Lego animals as an opportunity but we immediately pivoted this MVP before making any trial because the animals were too difficult to create.

3.4 *Third and Final MVP*

We finally decided to keep the "connect the dot" approach, to remove the "free" phase 1 and include some degree of variability in the task. We created 5 different farm scenarios (the field, the market, the ranch, the cowshed, the cowboy shop) each composed by 20–30 different objects to be drawn, cut and glued on a paper. In this MVP all the tasks are different, without any kind of repetition.

We also redesign the game-sheet to offer a clearer and more immediate insight into the budget control offered by Slick®. With this objective in mind we added to the game sheet the following elements: the slick® chart, intuitive questions on how to evaluate the project: what's the project status (safe, aligned, danger), what are the causes of the project's problem (team, budget, both).

We tested the third MVP in new workshop dividing 12 people into 3 teams. From this trial we collected the following feedback (Fig. 3; Table 4).

4 Limits of the Current Version and Further Improvements

Once collected the third trial's feedback we decided that the 3rd MVP is the product we were looking for to show the most relevant Slick® benefits. In fact the feedback was fully aligned with the remaining hypotheses and with all the learning objectives.

We are aware that this version is not able to promote all the possible benefits offered by Slick®, in fact the following concept are missing:

Fig. 3 Game sheet with added Slick® chart and clearer instructions

Table 4 Relevant feedback related to the 3rd MVP

Question	Answers
1—Did you follow a specific strategy? If yes describe it	100% yes
2—Was the strategy repeatable?	100% yes
3—When you started, did you know how much time it was needed to make all the animals requested?	100% no
4—Was the initial budget adequate? If not, when did you realize it?	100% no 66% after 1st sprint, 34% after 2nd sprint
5—How much do you think you had the budget under control?	100% yes
6—What did you learn from this game?	90% control and organization
7—Were the game instructions clear?	90% yes

1. Ability to deal with budget changes and requirements revisions.
2. Possibility to have planning poker session in every scrum meeting.
3. Ability to ask for new budget to the farmer.
4. Extend the project beyond the 3rd sprint.

In order to keep the instructions clear and simple, we prefer not to introduce new concepts to this version of the game, indeed we are planning a new workshop, containing the additional concepts, which can be provided after the base version.

We are also redesigning the game sheet to provide a "white label"/empty sheet useful to participants as a template for real-life, simple, custom projects.

5 Conclusions

Slick® is a disruptive agile methodology based on Agile Scrum that offers clear and quantitative insight on any kind of innovative project, even to non-technical managers responsible of the economic performance of the project. Slick® is a low footprint set of processes any company can apply to obtain monitoring signals.

In order to show the benefit of Slick® to companies and teams we designed the serious game Slick-farm which is able to practically demonstrate how to successfully deal with uncertain projects. We used the lean start-up methodology to define the final form of the game going through three different MVP and using feedback collected in trial workshops to pivot the wrong hypothesis.

Acknowledgements We thank Marta Zuliani for the effort in producing the graphical content and motivating the design team with constant new ideas and graphical sketches. We thank IdeaTech team members for the enthusiasm during the brain storming sessions and the time spent pre-validating the tasks.

References

1. Boehm BW (1981) Software engineering economics. Prentice-Hall, Englewood Cliffs, NJ
2. Abrahamsson P, Salo O, Ronkainen J, Warsta J (2002) Agile software development methods: review and analysis. arXiv preprint arXiv: 1709.08439
3. Larman C, Basili VR (2003) Iterative and incremental developments. A brief history. IEEE Comput 36:47–56
4. Singh A (2012) Agile: analysis of its problems and their solutions. IJCA Proc Int Conf Recent Adv Future Trends Inf Technol 32–35. https://research.ijcaonline.org/irafit/number6/irafit1047.pdf
5. Lopez-Martinez J, Juarez-Ramirez R, Huertas C, Jimenez S, Guerra-Garcia C (2016) Problems in the adoption of agile-scrum methodologies: a systematic literature review. In: 2016 4th international conference in software engineering research and innovation, pp 141–148. https://academic.microsoft.com/paper/2398354586
6. Isaacs M (2017) The #NoEstimates debate: an unbiased look at the origins, arguments, and thought leaders behind the movement. https://techbeacon.com/app-devtesting/noestimates-debate-unbiased-look-origins-arguments-thought-leaders-behindmovement
7. Manifesto for agile software development (n.d.) https://agilemanifesto.org/
8. Ries E (2018) The startup way. Verlag Franz Vahlen GmbH

Standardized Work Framework Applied to ETO Context

Sergio Ferigo, Francesca Tiraboschi, Monica Rossi and Matteo Consagra

Abstract Customer satisfaction is one of the main factor leading to customer loyalty and therefore continuous business relationship [1]. Customer satisfaction is correlated to the perceived quality of service, that is meant both as freedom from deficiencies and as ability to meet customer needs [2]. This dualism indicates that customers choose their business partners not only for their ability to standardize the provided products and services guaranteeing stable quality standards, but also for their awareness and flexibility in implementing customized solutions. The aim of this paper is to investigate how standardization, accomplished with Standardized Work (SW), is applicable to low-volume-high-mix context as Engineering-to-Order (ETO) ones and which benefits, if any, could be achieved. The study is developed by defining an application framework from literature first to high-volume-low-mix context, and further developed and refined with empirical research to fit ETO context. Empirical research has been supported and applied to FPZ Group.

Keywords Engineering-to-order · Low-volume-high-mix · Standardized work · Framework · Lean thinking

S. Ferigo · F. Tiraboschi (✉)
FPZ S.p.A, via F.lli Cervi 16, 20863 Concorezzo, Monza Brianza, Italy
e-mail: francesca.tiraboschi@fpz.com

S. Ferigo
e-mail: sergio.ferigo@fpz.com

M. Rossi
Department of Management, Economics and Industrial Engineering, Via Raffaele Lambruschini, 4b, 20156 Milan, Italy
e-mail: monica.rossi@polimi.it

M. Consagra
Istituto Lean Management S.r.l, Via Col D'Echele 15, 36100 Vicenza, Italy
e-mail: mconsagra@istitutolean.it

M. Rossi et al. (eds.), *Proceedings of the 6th European Lean Educator Conference*, Lecture Notes in Networks and Systems 122,
https://doi.org/10.1007/978-3-030-41429-0_40

1 Introduction

Customer satisfaction is one of the primary factors leading to customer loyalty and continuation of relationship [1]. Satisfaction concept is strictly related to quality one. Quality can be conceived as consisting of two elements: freedom from deficiencies and meeting customer needs [2]. This dual nature of quality requires both standardization and customization. Standardization provides a mean to service reliability and freedom from defects. Its main goal is to keep every process under control as to minimize defects and reworks while maximizing efficiency. As direct consequence, costs and time are reduced. On the other hand, customization improves the probability of meeting customer needs, developing solutions that are tailored to clients' requests. The main problem is that standardization and customization require different organizational resources to accomplish. Standardization calls for a strict process while customization requires flexibility and innovation [3].

Among the different typologies of production systems, ETO is the best known for allowing the highest level of customization by designing every product from scratch, based on customer inputs. On the other side, standardization can be addressed through the topic of SW, that consists in a procedure defining the best method an operator has to follow to perform a process: in this way the task is realized avoiding wastes while maximizing performances [4, 5].

The increasing popularity of ETO production systems leads this paper focalizing on how SW can be applied in those contexts in order to understand if any benefit can be achieved as in traditional ones.

2 Standardized Work Framework in Traditional Context

The only evidence in literature of SW application in ETO context concerns construction field. However, the higher availability of publications in *traditional context* as well as the fact that the reference context remains the industrial one, lead to consider the investigation on how the SW is applied in *traditional contexts* as the starting point of this paper.

For *traditional context* it is considered the high-volume-low-mix context, characterized by a production stability, in terms of quantity and type of product and well as process. Literature review regarding SW and *traditional context* brought to the framework for SW application here proposed (Table 1). This framework has been confirmed by its application for implementing SW in *Effepizeta Blower* division (Fig. 1), characterized by a *traditional context* production, with the aim to move the production towards the logic of cellular manufacturing.

Table 1 Standardized work framework for *traditional context*

Step 1: Product/process identification and takt time calculation
What to do: identify a small number of product families [6], calculate Takt Time (TT) Why: to avoid spreading the analysis on a too high number of item codes, and to identify the customer demand, through the TT
Step 2: Observation
What to do: observe the process and design it with tools such VSM, Production flow charts, Process flow charts, Material flow charts, Spaghetti charts, Makigami [7, 8] Why: to understand how process is organized, current status and problems to focus on
Step 3: Assess current situation as starting point
What to do: define the main activities performed in the process, collect data for cycle time definition through Process Study Sheets (PSS) and video recording [7–10] Why: to "take a picture" of the actual condition of the considered process. This condition represents what is the best situation achievable at the moment
Step 4: Analyse current situation
What to do: calculate how many resources are needed in the process defining the Full Time Equivalent (FTE), check if TT is respected through the Yamazumi chart [7, 9] Why: to define the necessary machinery and workforce in the process to meet TT
Step 5: Kaizen activities
What to do: focus on the three Ks, that are Kaizen on information and material flow and process, Kaizen on equipment, Kaizen on layout in order to improve the process (i.e. to meet the TT target, to decrease quality issues, to decrease setup time) [7, 11] Why: to improve performances of the process, starting from what observed in step 2 and 3
Step 6: Standardized work definition
What to do: define the SW through its main tools, that are Standardized Work Sheet, Standardized Work Combination Chart, Work Instruction [7–11] Why: to establish the most efficient way to perform a job in the production process
Step 7: Training
What to do: define Job Instruction as last step of SW definition, train people adopting the Training Within Industry (TWI) method [12] Why: to train the maximum number of operators in performing the job following the established standards, and therefore in improving the SW
Step 8: Evaluate
What to do: Define KPIs that represent the process performances, with specific targets, share KPIs with the people involved in the process [13] Why: to evaluate the production performances after SW implementation, to identify opportunities for further improvements, to ensure sustainability in SW process, to strengthen people involvement in the SW sustainability and improvement

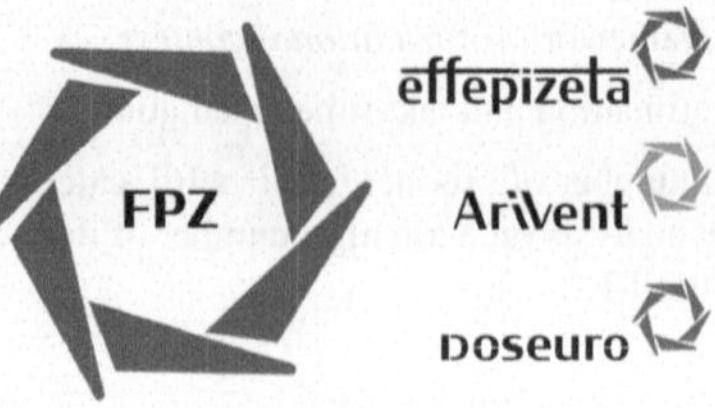

Fig. 1 *FPZ* Group divisions: *Effepizeta*, *Arivent*, *Doseuro*

Table 2 Literature review results

Title, abstract, keywords	Findings	And	Limit to subjarea	Limit to language	Findings
Standardized-work	305	–	ENGI	English	102
Standardized-work	305	Engineering-to-order	–	–	1
Standardized-work	305	Customization	–	–	2
Standardized-work	305	High-mix-low-volume	–	–	0
Standardized-work	305	One-of-a-kind	–	–	0
Standardized-work	305	Construction	–	English	43
Lean-thinking	7844	Engineering-to-order	ENGI	English	31

3 Standardized Work Framework in Non-traditional Context

3.1 *Literature Review Results*

Non-traditional context is used to make reference to those production systems featured by low-volume-high-mix as in case of ETO. Table 2 represents the results for literature review finalized to identify previous analysis of SW studies in ETO context, through Scopus and Google Scholar database search; papers focused on ETO production and customization/standardized work are rear (3 cases in total). This highlights the innovation of this case study.

3.2 *Application of Traditional Context Framework to ETO*

The above mentioned SW framework is applied on the production system of Arivent. It is an Italian company belonging to FPZ Group (Fig. 1), counting 164 employees, with a 2018 revenue of 40 M€. Arivent produces industrial fans based on clients specification, in an integrated process. The production process starts with design phase, continuous with internal metal sheet automatic cut, internal welding and assembly of the components, painting of final products and testing. Arivent process is strongly based on human work content.

The variety of product features among which clients can choose as well as the low number of pieces per order, make Arivent fit the ETO context. The case study is performed by following the SW framework guidelines step by step.

Step 1: product/process identification and takt time calculation. On the contrary of traditional context application, here it is stressed the necessity to focus not on product categories, neither on every type of product, because the production variability typical of an ETO system would make the analysis too fragmented. An advice is to identify common features the can group Arivent products in few families.

In this case, fan accommodation (four types) and fan dimensions (three types) are taken as reference features through which define twelve products families. Takt time is calculated in order to define customer demand.

Step 2: observation. As suggested by the framework, this step is performed through Gemba Walk in the production shopfloor. Information collected has been represented with VSM and Spaghetti Chart (Fig. 2a, b). The analysis revealed elevated movements of the operators due to the bad disposition of different workstations, stock of material and equipment, and high WIP amount distributed without a clear rule in the available space. While defining the VSM, critical resources have been assessed: unique equipment and human capabilities have been signed. This leads

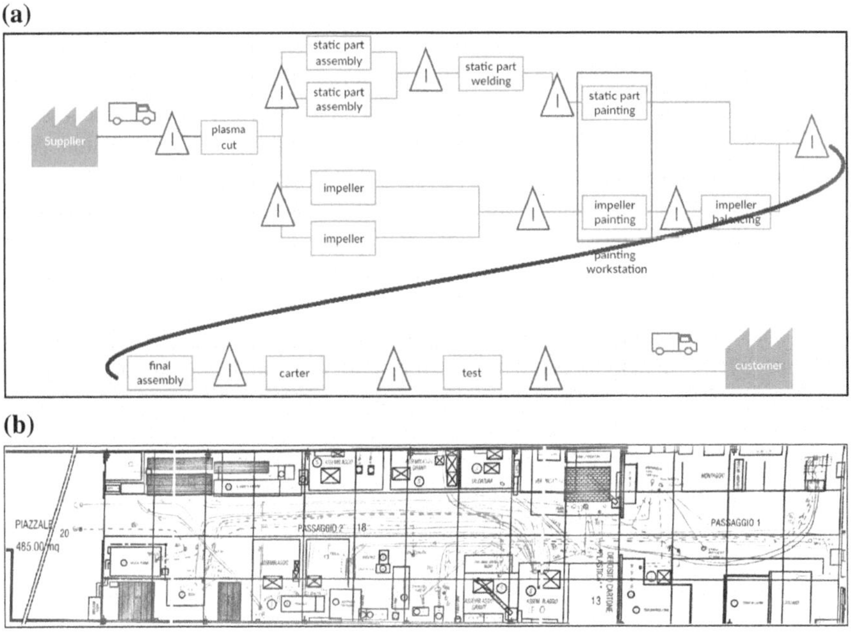

Fig. 2 **a** VSM material flow—simplified version. **b** Spaghetti chart

Table 3 Fans product families production volume percentages

		Accommodation				Total (%)
		4 (%)	8 (%)	9S (%)	12 (%)	
Size	S (<=600)	24	0.80	9.80	1	35.6
	M (<=1000)	20.80	2	20.50	3.20	46.5
	L (>1000)	1.30	2.40	6.80	1.90	12.4
	Total	46.1	5.2	37.1	6.1	

to the need to define a simple skill matrix: which activities could every operator perform, and how many activities could be performed by more than one operator.

Step 3: assess current situation as starting point. On the contrary of *traditional context*, even though the process for producing an industrial fan is always the same (same VSM for different product families), CT is different for every product family. For this reason, the analysis focuses on CT calculated as weighted means depending on the production percentages of the cited features (Table 3). Variability of CT is considered through standard deviation parameter. Moreover, the study considered also engineering phase, being an active part of the supply chain for ETO systems.

The CT collection was collected thanks to the help of operators, who were asked to register the time needed to complete a specific task on a task list provided together with the work order for a specific fan. Time measurements were taken on fans representing the majority of Arivent production, following Pareto rule.

Step 4: analyse current situation. The analysis of the process CT and lead time through VSM highlighted the unbalance of the process: this is possible to check by considering the high amount of WIP places between workstations against the logic of one piece flow, and by building the Yamazumi chart of the AS-IS situation, underling that work content among workers was unfairly distributed (Fig. 3).

FTE calculation reveals that the actual workforce in R&D office is not sufficient to cover the workload of the designing phase, while in the shopfloor the actual number of operators exceeds the necessary one, confirming the need to balance activities.

Step 5: Kaizen activities. Kaizen activities are based on previous analyses. The main issues were found for people wasting time in "walking" and "moving" material as shown by the Spaghetti chart, and in terms of line balancing as shown by the Yamazumi AS-IS. The line unbalance does not permit to have a continuous flow of material in production line, creating WIP accumulations in the line. Thus the line balancing issue becomes the main focus of Kaizen activity.

K on flow and processes: dealing with line balancing, a tool that can be very useful is the skill matrix [8]. A skill matrix is a table that clearly and visibly illustrates the skills and competences held by individuals within a team [14]. The skill matrix defined for Arivent needs, brought to a new Yamazumi TO-BE (Fig. 4), and therefore to the request for skill improvements.

K on equipment: not considered relevant in this context, being Arivent a mainly human work content based production.

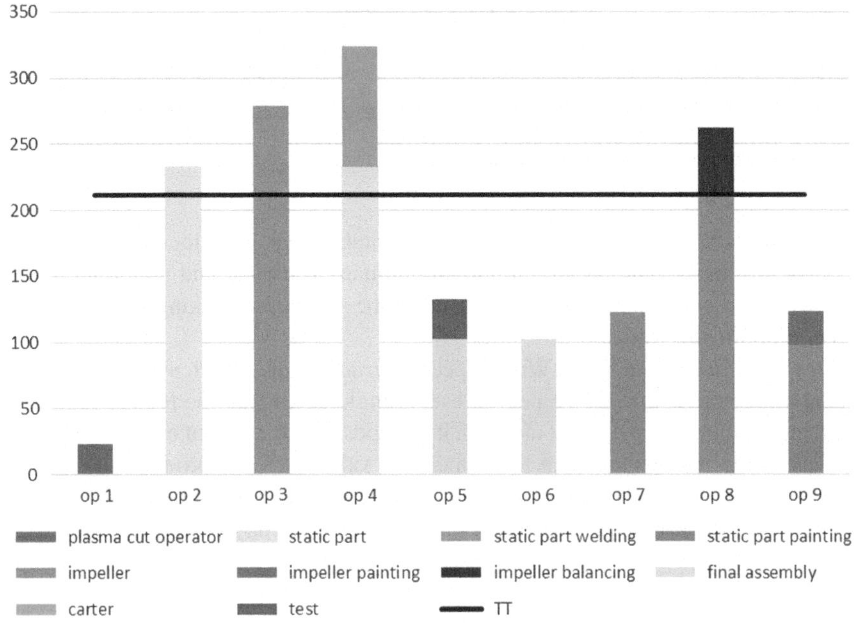

Fig. 3 Work content study and Yamazumi AS-IS

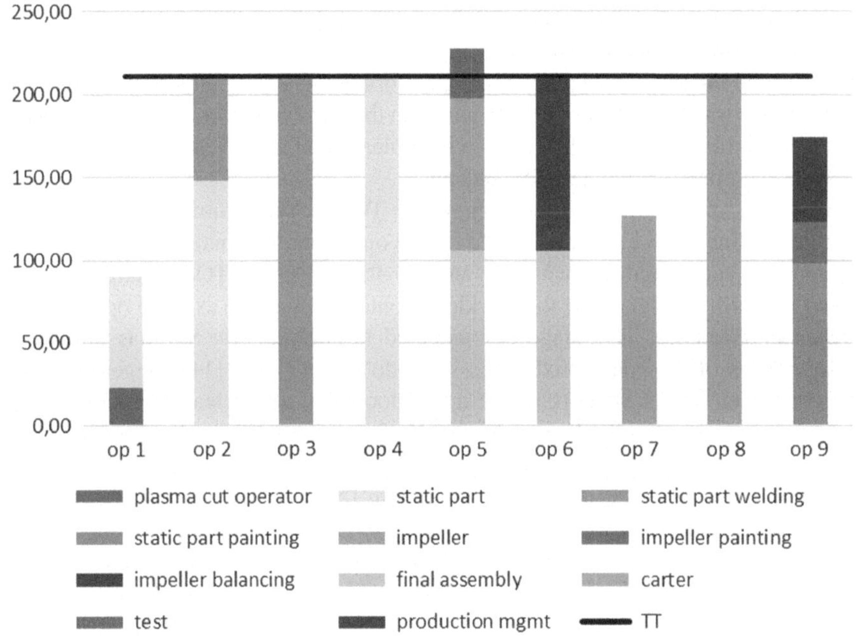

Fig. 4 Yamazumi TO-BE

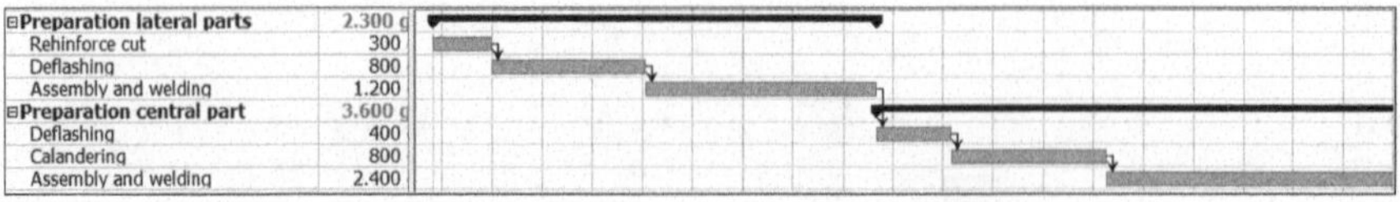

Fig. 5 Standardized work combination chart

K on layout: a project to boost 5S implementation was implemented in Arivent, in order to redesign the layout minimizing human "walking" and material "transportation" and promoting material flow and, at the same time, making more efficient and ergonomic the workspace.

Step 6: Standardized Work. With respect to a *traditional context*, what is important to underline here is that CT respect of every single task is not the focus of the SW implementation. Being Arivent an ETO, it is impossible to say that every single task will always be completed with the same time value for each product, due to the high variability in product range and clients' specifications: based on features of the product to be realized, the same activity in the VSM can occupy more or less time with respect to other products. So, the most important part of this SW implementation is that the operator follows the Standardized Work Combination Chart sequence (Fig. 5), allowing the easiest, safest and most efficient way of performing a job. Figure 5 shows a selected part of the sequence of activities performed by an operator for a specific process step (static part welding); in the Gantt chart the red horizontal bars show the specific work content (in seconds) for an examined product.

Step 7: Training. As next step after implementation of SW, it assumes high importance to train the maximum number of operators in respecting the established standards. In this way it will be possible to maintain the SW even when there are absences or employee turnover [9, 15]. This step is implemented in Arivent by explaining to the involved operators the new SW sequence.

Moreover, as stated, in order to achieve the TO-BE Yamazumi, a yearly training plan was defined. It was defined based on the operators skill matrix, that identified the gap between current known skills and TO-BE skills. The TO-BE skills allow a higher flexibility in balancing the workload content among the existing operators.

Step 8: Evaluate. This stage is conducted to analyse how SW has improved performances of the considered process. To do this, [7] and [11] suggest the use of Cell Kaizen Target Sheet (CKTS); it is a document where lean metrics, or Key Performance Indicators (KPIs) are identified before and after Kaizen activities implementation. In Arivent, Kaizen on process and on layout brought to a saving of meters walked by operator: it was possible to move from 1500 to 700 m/fan, saving on average 40 min for each fan to be produced, and therefore impacting positively on the KPI Work-order Lead Time, calculating the working days from client's confirmation order to shipment. Another visual KPI defined is the ability to respect the WIP spaces defined with the new layout: no more WIP products are allowed outside the defined areas, otherwise it means that neither the K on layout nor the balancing have been respected.

4 Conclusions

From the performed analysis on Arivent case study, it is possible to demonstrate that SW framework guidelines are applicable in ETO contexts [16]. Companies operating in this field can be incentivized and motivated in approaching lean thinking, without assuming that lean management is suitable only for production in series. An important finding is the different focus of SW in ETO with respect to traditional context: traditional contexts are hinged on respecting cycle time (CT) of each job element, whereas ETO context is more about respecting the sequence of job elements, since work content (and CTs) is highly variable.

The main limit of this paper is the fact that the study is performed only in one industrial sector, regarding fluid motion. It would be interesting to follow the same methodology in other ETO sectors to compare the results, identifying if there are critical factors typical of ETO context thus checking whether and how the framework adapts to different industrial realities. Moreover, the ETO company chosen for the case study shows variability mainly in CT, while the production process remains always the same. It would be interesting to analyse how the framework evolves in ETO companies featured by high variability both in CT and in production processes.

References

1. Rust RT, Chung TS (2006) Marketing models of service and relationships. Mark Sci 25(6):560–580
2. Juran JM, Szemere G (1989) La Perfezione Possibile (Juran on Planning for Quality). IPSOA scuola d'impresa
3. McCutcheon DM, Raturi AS, Meredith JR (1994) The customization-responsiveness squeeze. Sloan Manag Rev 35(2):89
4. Mĺkva M, Prajová V, Yakimovich B, Korshunov A, Tyurin I (2016) Standardization-one of the tools of continuous improvement. Procedia Eng 149:329–332
5. Pereira A et al (2005) Reconfigurable standardized work in a lean company—a case study. Procedia CIRP 52:239–244
6. Rother M, Harris R (2001) Creating continuous flow : an action guide for managers, engineers and production associates. Lean Enterprise Institute, USA
7. Jaffar A, Halim NHA, Yusoff N (2012) Effective data collection and analysis for efficient implementation of standardized work (SW). J Mech Eng 9(1):45–78
8. Bragança S, Costa E (2015) An application of the lean production tool standard work. J Teknol 76(1):47–53
9. Chan CO, Tay HL (2018) Combining lean tools application in Kaizen: a field study on the printing industry. Int J Product Perform Manag 67(1):45–65
10. Mor RS, Bhardwaj A, Singh S, Sachdeva A (2018) Productivity gains through standardization-of-work in a manufacturing company. J Manuf Technol Manag
11. Halim NHA et al (2015) Standardized work in TPS production line. J Teknol 76(6):73–78
12. Piatkowski M (2012) Lean transformation solutions 1–33
13. Lu J, Yang T (2015) Implementing lean standard work to solve a low work- in-process buffer problem in a highly automated manufacturing environment. Int J Prod Res 53(8):2285–2305
14. What is a skills matrix?—ability6. [Online]. Available: https://ability6.com/what-is-a-skills-matrix/. Accessed: 05 Mar 2019

15. Fin JC, Vidor G, Cecconello I, Machado VDC (2017) Improvement based on standardized work: an implementation case study. Braz J Oper Prod Manag 14(3):388
16. Standardized work: from mass production to ETO. Roberta Curti. Available: https://www.politesi.polimi.it/

Maritime Design Process Improvement Through a Lean Transformation

Brendan P. Sullivan, Monica Rossi, Lucia Ramundo and Sergio Terzi

Abstract The maritime industry has been working to apply unique solutions capable of improving design and development performance to ensure competitiveness [1–4]. Over the past several decades, the industry has faced changes related to the increasing complexity of the dynamic global market. Activities to meet these challenges have been carried out in the marine space (e.g. aquaculture, renewable energy, environmental monitoring, accident response and clean up), however, design methodologies have remained rather unchanged, regardless of demand for the incorporation of new technologies and materials. Literature and multiple case studies have been utilized to present a design methodology and its preliminary results, which aim to help transform the design process for the extension of serviceability and reduction of design time. This paper describes an innovative maritime design framework, based on the comprehensive usage of design methodologies and advanced technology integration in vessel design.

Keywords Lean design · Maritime design · Collaborative design · Case study · Design methodology

---

B. P. Sullivan (✉) · M. Rossi · L. Ramundo · S. Terzi
Department of Management, Economics and Industrial Engineering, Politecnico di Milano, Via Lambruschini, 4/B, 20156 Milan, MI, Italy
e-mail: brendan.sullivan@polimi.it

M. Rossi
e-mail: monica.rossi@polimi.it

L. Ramundo
e-mail: lucia.ramundo@polimi.it

S. Terzi
e-mail: sergio.terzi@polimi.it

M. Rossi et al. (eds.), *Proceedings of the 6th European Lean Educator Conference*, Lecture Notes in Networks and Systems 122,
https://doi.org/10.1007/978-3-030-41429-0_41

1 Introduction

The maritime industry has throughout history been a vital aspect to economic development, creating wealth and expanding influence through international and interregional trade. The continuing diversification of economies, the flow of knowledge, resources, goods and services have become more vital than ever owed to technological and sociocultural forces [5–7]. These continued advancements require that vessel designers develop new approaches capable of creating value-added services, management, and designs.

Like other industries it has become increasingly important for companies to improve their processes to maintain competitiveness, prosperity, and survival. Recently European maritime industry has been dealing with increased competition, due to ports becoming strategic poles of distribution for maritime traffic to/from new emerging countries. To facilitate and increase the competitiveness of the European maritime sector, lessons learned from the naval and specialized vessel industry have served as a key element in future development. This includes the incorporation of new technologies, materials and optimization processes into the engineering and design practices of the greater industry [7, 8]. At the same time, the European Maritime sector must diversify their business offerings and be capable of integrating these advanced technologies into design processes to be competitive in the global market. To date, the industry has been working to implement and apply many of these advancements in their engineering and development processes [2]. However further work remains beyond application alone to identify and develop value-added solutions, products and services that can accommodate changing demands in the sector.

In order to address the challenge of implementing lean in maritime design, both academic literature and industrial feedback are utilized to reduce the challenge of implementing lean for designer and engineers during the conceptual, preliminary, contract and detailed vessel design phases. This paper introduces a unique lean method for process improvement that was developed alongside industry partners for Small-Medium-Enterprises (SMEs), to leverage technological solutions and overcome barriers that inhibit the implementation of lean, such as: (1) the isolated implementation of lean tools in design processes, (2) the articulation of requirements for specialty and one-of-a-kind vessels between designers and stakeholders, and (3) overcoming entrenched habits amongst vessel designers.

2 Lean

The original definition of lean thinking aims at "creating more value, defined from the customer's perspective, while consuming fewer resources" [9]. Since emerging in the early 1990s from the Japanese automotive sector, leans ability to facilitate improvement in project management, simultaneous development, teamwork, and communication has made it well suited for applications in design and development.

Since its introduction, lean has evolved into an approach that considers entire enterprises with emphasis on organizational behavior, work flow, design, and production processes through continual improvement. Expanding from the traditional production sectors, to industries involved in advanced and complex design and development processes such as the space industry (JPL, Team-X and ESA) [10–12]. However, the possibility to improve the overall performance of complex system development through the proper adoption of best practices and innovative methods has been identified as critical area for several sectors including maritime.

2.1 Lean Product Development

Lean Product Development (LPD) is the application of lean thinking and principles to product development (PD) projects, in an attempt to support companies willing to implement lean thinking, tools and techniques to improve the overall performance of their processes [9, 13– 15]. Through emphasis on waste reduction and process improvement Lean has successfully demonstrated that when considered in early-stage decisions, lifecycle value and product performance (quality, reliability, safety, etc.) are improved.

In recognizing the dispersion of the concept, when being applied to PD several considerations must be understood. First, PD is different from a repetitive production process, in that the former has information as its main input and output, while the latter mainly deals with materials [15, 16]. Second, PD is characterized by a higher level of variability and uncertainty compared to the production process of large-scale manufacturing companies and this could prevent an effective delivery of value and identification of "wastes" [17]. As such this consideration is critical for the management of projects, where the complexity of the system being developed is high and cost as well as delivery time are critical [18–20]. The literature showed that in cases of successful LPD implementation there are at least three important advantages provided that include a/an:

- Improved development performance [21, 22].
- Improved efficiency in manufacturing and processes [21].
- Structure for establishing a lean enterprise [9, 23].

2.2 Lean Product and Process Development

Lean Product and Process Development (Lean PPD) considers the design and development of engineering products to facilitate the integration of manufacturing knowledge during the conceptual design stage to increase designer awareness of manufacturing constraints and manufacturing capabilities [24, 25]. To prevent diminished value, it is necessary to analyze and specify requirements with the stakeholders at the very beginning of the project, thereby allowing for rapid alternative generation. In considering the flow of information, it is necessary for strong coordination so that interrelated information flows can be brought together. This allows for the transfer of knowledge in a harmonic manner, whereby analysis, simulation, prototypes, and test information are available to not only those performing them, but all who are involved in the design process.

Similarly, lean design is the application of lean tools and principles, for the elimination waste and non-value added activities in design related processes [26]. In application, this approach aims to fulfil stakeholder needs through the exploration of alternatives to determine how requirements can most effectively be accommodated to deliver the highest level of value. Through the consideration of design alterations, information flow, and value generation, lean reduces waste by minimizing the amount of time before information gets used, the time spent inspecting information for conformance to requirements, the time spent reworking information to achieve conformance, and the time spent moving information from one design contributor to the next [27, 28]. The difference in these considerations is how the process describes the critical design aspects and properties.

3 Maritime Design

Vessels are inherently complex one-of-a-kind products that are built by highly skilled persons according to specific requirements set by the stakeholders and many times require unique solutions that are drawn upon from previous or existing designs [29, 30]. This practice allows for a designs process that is quick and straight forward to apply. Managing the specific requirements, stakeholder interests (builders, cargo owners, customers, ports operators, classification societies, environmental matters, comfort of the crew/passengers, etc.), optimization criteria and technical feasibility. In addition when designing a "standard" vessel, this approach reduces the risk for failure and ease for verifying compliance to national and international legal regulations [31, 32]. Despite the obvious benefits presented by this approach, it can be difficult to incorporate new innovative solutions due to the restrictiveness of the design space [33].

3.1 Lean in the Maritime Industry

In the maritime industry lean is a growing area within engineering and manufacturing of maritime vessels due to cost, size and complexity. As in many industries, vessel design has many Non-Value-Added Activities (referred throughout this deliverable as WASTE or "MUDA") which increase costs, compromises quality and can propagate delays in the delivery schedule, which all in effect provide no positive contribution for the stakeholders. With origins of the concept generally contributed to Japanese shipyards which reached peak profitability and competitiveness far before the more well-known auto industry [1, 34]. Despite this the number of publications on lean are limited, with few published cases that utilize lean principles in an organized manner during design. This can be contributed to the novelty and the general restrictive nature of the concept, particularly related to the processes of organization being used in each yard [35].

Lean construction/shipbuilding is one of the more researched areas of lean in relation to the maritime industry, representing an adaptation of lean manufacturing principles and practices to construction and shipbuilding through advanced focus on planning. It is critical to note that these approaches and principles are not learned from the Japanese ship industry which as stated by Koenig could have served as the initial knowledge flow of lean in the industry [1].

Shipbuilding can therefore be viewed as a process that extends beyond the typical design phases, incorporating the requirements selection, stakeholder interests (builders, cargo owners, customers, ports operators, classification societies, environmental matters, comfort of the crew/passengers, etc.), optimization criteria and technical feasibility. The increased phases of construction mean it is necessary for design information to be well organized and defined so labor and materials can establish to reduce inventories. According to the description of the concept, lean construction tries to manage and improve shipbuilding through advanced planning, and emphasis on concurrent/continuous improvements throughout the entire systems life [30, 35, 36].

4 Research Approach

The research covered in this paper utilized a systems development approach due to its ability to facilitate a close examination of the lean design process in the maritime industry and specific lean tool implications. Whereby product design literature was applied to develop a comprehensive methodological foundation that included elements of lean systems engineering, and lean design. The System Development Methodology is classified as an applied method that allows for planning, creating and testing of systems through an integrated multi-methodological research approach (Fig. 1). Theory Building is the central tenant of this process and aims to improve the practical understanding of a problem space.

Through the identification of existing practices as well as the incorporation of best practices identified through literature, a methodology was constructed based on the unique needs of the maritime cases.

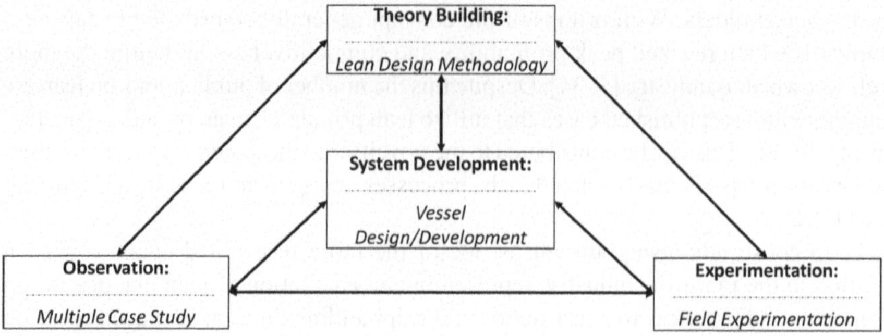

Fig. 1 Approach for system development [37]

4.1 Research Design

The research presented draws upon the findings of multiple case studies conducted in collaboration with three vessel producers due to its ability to provide for an in-depth investigation and ability to leverage different data collection techniques to provide a comprehensive overview of the problem under investigation. In order to acquire relevant and quality data to test the research hypothesis it was necessary for the project size and sea related sector and offshore and coastal activities of the vessels to be similar as shown in Table 1 [38]. Further, the business type and production approach were specified, while design process similarity was desired the cases were not selected based on this criterion.

5 Case in Maritime Design

Most shipyards and those involved in vessel design (managers, designers, engineers) are acutely aware of the need for innovations and continuous development. Frequent product introductions, upgrades, facelifts etc. are always on top of the agenda for shipyards and the maritime industry respectively. Even though some level of standardization has been achieved, the level of customization is high compared to other branches such as automotive. This makes design and production processes difficult to manage since every vessels typically requires new and unique solutions.

Table 1 Business case characteristics

Case	Business	Vessel classification
#1	SME	MCA CAT 2 (>60 nautical miles from safe haven)
#2	SME	MCA CAT 2 (>60 nautical miles from safe haven)
#3	SME	MCA CAT 3 (>20 nautical miles from safe haven)

Table 2 Maritime design process maturity

Case	Organization	Process	Knowledge management
#1	(Mature) 67.8%	(Intermediate) 58.9%	(Intermediate) 55.7%
#2	(Mature) 74.3%	(Mature) 65.3%	(Mature) 75.7%
#3	(Mature) 69.8%	(Intermediate) 56.1%	(Intermediate) 55.2%
Average	(Mature) 70.6%	(Intermediate) 60.1%	(Mature) 62.2%

5.1 Design Process Maturity Analysis

Interviews were conducted using a reference questionnaire, comprised of 33 questions identified through literature and best practices. In each case, the technical director was interviewed, lasting roughly 3 h. The questions were scored using a Likert scale, so the respondent could choose their company's process level. The lowest levels scored with 1 correspond to a poor practice, while the highest-level scored correspond to a best practice. The score for each area was calculated using an additive scale (summing the single scores of the questions describing the area) then normalized in % [39].

Based on the maturity level established by the CLIMB assessment method, efficacy of the practice was calculated as being either: Chaos (0–20), Low 24–40%, Intermediate 41–60%), Mature (61–80%), or Best Practice (81–100%) [39, 40].

- Organization: how people are involved in daily activities: coordination of people and activities, roles of engineers and designers, skills/expertise of practitioners.
- Process: how vessels are developed, based on four areas: design rules (Methods), control mechanisms (Process Management), how decisions are made (Decision Making), and competitors and customers involvement (Activities and Value).
- Knowledge Management: how the case creates, shares, represents and re-uses knowledge. Including how knowledge is formalized, shared and integrated into the design process.

The design process maturity in each case (Table 2) illustrates that in each case while scoring relatively high there are still obvious deficiencies, particularly relating to: Process and Knowledge Management which have the lower average score.

5.2 Waste Analysis

Improper or incomplete designs are one of the biggest causes of waste due to the negative effect these design failures can have during later stages [15]. For this reason, it was important early on to determine the types of waste present in maritime design. To do this, activities that failed to generate value were categorized in three ways: (1) Waste of Time (WoT), (2) Waste of Knowledge (WoK), or (3) Waste of Resources (WoR). According to this classification 10 wastes of a possible 33 proposed by Rossi et al. [13] were selected to be measured, as shown in Table 3.

The identified wastes were measured in each case through the waste identification method developed by Rossi et al. [13], and is based on the FMEA (Failure Modes and Effects Analysis). Through this approach 5 engineers/designers in each business were asked to evaluate the Probability (P), Severity (S), Detectability (D), and Avoidability (A) of the afore mentioned wastes in their design processes.

Post waste evaluation of the Priority Index of Intervention (P * S * D * A = PII) for each waste was calculated individually, and the average PII was calculated. In the following histograms the average PII value for each waste is presented (Fig. 2—Left) and the average PII value of each waste class (Fig. 2—Right), the height of the bar (y axis) corresponds to the value of PII and the x axis the waste or class being measured. The black line on (Fig. 2—Right) compare the average SME waste value in traditional industries against those in design [13].

Table 3 Business case characteristics [13]

Waste	Description
#1 Change priorities	Time making changes, reviews, modifications or reworks
#2 Reworks	Projects (of parts of projects) re-done due to incorrect, incomplete, unreliable data, information, specifications
#3 Unneeded functions	Functionalities and features that are not necessary or required
#4 Knowledge retrieval	Difficulty to retrieve and access previous knowledge
#5 Project leftovers	Projects are realized that never reach the market
#6 Data transcoding	Information and data are inserted and coded manually
#7 Missing authorization	Projects are delayed due to authorizations challenges
#8 Defect	Defective products are introduced; requiring modifications
#9 No knowledge reuse	Projects, developed in the past, are redone due to previous knowledge not being retrieved
#10 Over engineered	Over-engineering; cost increase without value added

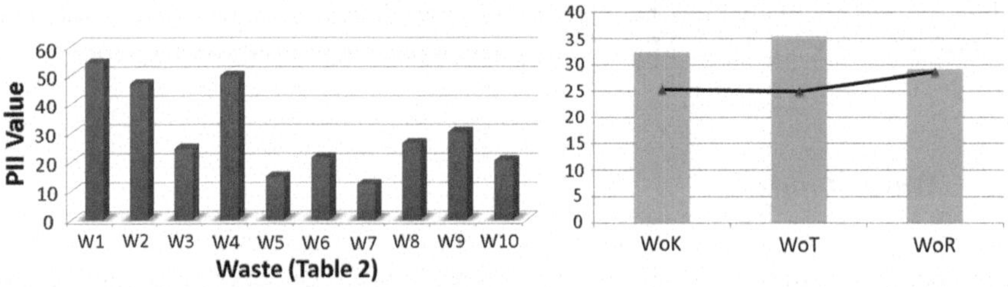

Fig. 2 Waste PII in maritime design

5.3 Barriers to Lean Implementation

Based on the process maturity and wastes identified it has been determined that there is a relationship between how mature a process is and how present a waste is. Amongst the maritime cases analyzed it is therefore necessary for the methodology supporting lean transformation to be able to consider and address multiple areas of the organization simultaneously, particularly through the better leveraging of knowledge and improvement on processes to reduce changes [41]. Through the analyses performed and interviews conducted three distinct barriers to lean implementation were identified: (1) the isolated implementation of lean tools in design processes, (2) the articulation of requirements for specialty and one-of-a-kind production (OKP) vessels between designers and stakeholders, and (3) overcoming entrenched habits amongst vessel designers.

The first barrier represents within the maritime industry a general lack of methodologies capable of support the integration and implementation of lean into the design process. In each case, it was found that at some point over the past 5-years an effort had been undertaken to implement lean or a lean tool. In each case the effort was ultimately abandoned and attributed to there being little to no cohesion between multiple lean efforts, and the inability to train each person involved how best to leverage the capabilities of the tool. This is potentially caused by the limited number of solutions that consider the integration of supporting tools according to the unique needs of specific industries or organizations.

This general lack of how to integrate lean into one's process is further explained in the difficulty of articulating requirements for on-of-a-kind vessels where changes in customer requirements lead to the recreation of designs, drawings and documentation. In discussions surrounding past projects there are instances where the customer is paying for the convenience to change their mind at any point during development. While this ability to satisfy customer expectations is valuable, the output should be able to be generated in the most efficient manner possible. All improvements relating to value require that there be some action taken towards the associated cost or the function, so that requisites can be transmitted effectively and, in a value, positive manner.

In discussing the third barrier, which is overcoming extant habits it is understood that the maritime industry is risk averse, as careful considerations are always made to ensure the safety and reliability of vessels. Despite the obvious benefits presented by this, it can be difficult to incorporate new innovative solutions due to the restrictiveness of the design space and the entrenched educational aspects that have been taught. This third barrier could in part explain why in review of the available literature there very little information is available for integrating lean, and methods to facilitate a lean transformation.

6 Design Process Improvement Through Lean Transformation

The maritime industry is one of the most recent industries that has been working to apply unique solutions capable of improving their development performances. However, based on the cases covered in this paper there is not only a need to overcome the identified barriers, but also provide an approach that overcomes the principles wastes and would facilitate process maturity. For this to be realized the lean transformation of the design process must have a wider scope than previous lean efforts and involve all levels of the organization. This comprehensive approach is necessary due to there being a critical distinction between mass production and OKP, as is the case in many vessel design efforts.

Considering this in the development of the methodology it was determined to be necessary to draw a distinction between the creation of a methodology that dictates the technical procedures/considerations for vessel design and the managing process of the design that describes how tools and approaches should be chosen to successfully implement lean. In addressing the latter, this Lean Transformation Framework (LTF) is presented to combat delays and barriers being faced when implementing lean in vessel design.

Like product development which deals with information and processes, involving a multitude of choices and iterations that are used for problem solving and decision-making activities [42]. The maritime industry is confronted with the management of iterations and complex development, the LTF aims at improving development performances by promoting a stronger relationship between the designer and customer, and by facilitating an improved approach for utilization of past designs and vessel data [43]. To facilitate the optimization of this initiative, an evaluation of diffused strategies was undertaken.

Through a holistic question-based approach the methodology offers a customizable set of tasks and steps that can be easily modified to facilitate implementation of lean in various maritime design process or organizations.

- What problem is trying to be solved?
- What is the value for the stakeholder?
- How can the organization improve to better deliver the identified value?
- What skills, attributes, or characteristics are necessary to engage in these processes to best deliver value to the stakeholder?
- Is the current management system capable of supporting this way of working, and what management system is?
- What considerations and values/goals will be used to guide the lean transformation of the organization? The set of foundational things such as how to carry oneself, "attitude".

6.1 *Lean Transformation Facets*

The LTF focuses on value creation through the elimination of non-value adding activities through the identification and implementation of tools that are well suited to vessel design. Therefore, one should identify first the right customer value the product should fulfil. Further, any activities that add no value for the customer should be identified and eliminated from the process.

6.1.1 Purpose

In considering ways to address 'Purpose' there are several tools and approaches, which have been identified in the state of art and practice could be included. One such approach that is well suited to meet this area based on a review of the literature is the improvement of customer involvement in the design process, this relates to the identification customer requirements through increased user involvement.

User involvement allows for the identification and accurate definition of system requirements to maximize value. Clear calculation of the 'purpose' addresses customer needs and attributes of the stakeholders regarding the product/system, by articulating what the customer wants to have in the vessel and the functions they want it to perform in a specific environment. Needs in this context are a wish list based on the expectations that stakeholders have for the system. In determining and considering this list, it is critical to recognize that a wide range of stakeholders can affect the product development process, a successful design must consider all those stakeholders involved: designers, engineers, shipyards and suppliers. Through greater customer involvement the following approaches were found to be well suited for cost-minimization, and increased design efficiency.

- Requirement Engineering.
- Customer Value Integration and Definition.

Recognizing that the maritime development process is a sequence of steps and activities that are employed to conceive, design, and successfully deploy a vessel. To succeed in these efforts, designs must maximize customer value, while ensuring careful conformity to regulations and safety concerns.

6.2 *Process*

'Process' is the manner in how work can be done and organized to fulfil desired goals and objectives related to the delivery of value for the customer. Referring not only to how a process should be performed, but also how the work can be continuously improved. This includes implementing:

- Set-Based-Concurrent-Engineering or Concurrent Engineering.
- Knowledge Based Engineering.
- DfX.

Each implementable process must fulfil the overarching business needs and is intended to promote efficiency without creating or establishing non-value (wasteful) efforts, allowing for the use of fewer/less resources. These approaches when integrated in the design process can be adapted for to ensure customer needs are met and allow for the integration of user data for future improvement.

6.3 People

In considering ways to address 'People' it is critical for the purpose and process of the transformation to be well understood by all employees. This verifies that the necessary skills, and attributes, can be offered during every stage of design process so that designers and engineers have the appropriate and necessary skills to tackle problems effectively utilizing product usage information and experience to generate and evaluate alternative design concepts in a collaborative and efficient manner.

- Skill Training and Engagement.
- Knowledge Sharing.

Through training and knowledge sharing in the vessel design process communication facilitates a pull event where teams can visualize risk and opportunities to improve and enhance design efforts. 'Serious Games' have also been identified as a successful approach, which allows players to assume different roles and engage them in simple and complicated decision-making processes. This process of gamification also makes it possible for a safe and entertaining environment to be established, where players are openly encouraged to investigate processes without fear of interfering in actual design.

6.4 Basic Thinking

Basic thinking considers the values/goals that will be used to guide the lean transformation of the organization. Based on the business needs value delivery efforts, it is possible to continually improve the design and development process, in a sustainable and environmentally responsible manner [44]. In considering the values/goals that can be used to guide the lean transformation of the following approaches were identified:

- 5-step methodology for continuous improvement.
- Fostering an Innovative Culture.

These approaches serve to integrate all parts of the transformation and facilitate the improvement of the design process (no matter if it is lean or not) through a lean perspective.

6.5 *Leadership and Management*

The leadership aspect represents the fifth part of the transformation and represents the soft skills, behaviors and tools to be utilized by leaders and everyone within the organization so that they can meet the objectives of the process improvement effort in an effective and collaborative manner. It has been found that the way peoples engage and interact with one another directly impacts the amount of respect, empowerment, and support in the organization. In involving people, the leadership can directly increase employee-driven participation (strongly related to Fostering an Innovative Culture).

In pursuing these aims every program/project (the entire transformation and smaller projects) should use a Chief Engineer role to lead and integrate the program from start to finish. Through the utilization of a chief engineer, leaders are responsible for defining, translating and communicating the purpose, strategy and goals throughout the organization to everyone, ensuring that a clearly crafted message is developed. Which provides flexibility for those involved to determine how to best accomplish and measure them. Failure to provide such a collaborative approach has limited efficacy and staying power if the right people are not supported at the right time.

7 Concluding Remarks and Future Work

This paper which part of a greater combined research effort is representing preliminary and ongoing work, aiming to introduce a lean transformation methodology that is and easily implementable in the maritime industry and through a series of questions can be tailored to unique needs while overcoming barriers that challenge lean implementation in vessel design. Through the implementation of the LTF for design process improvement it is expected that designers and engineers will have a more comprehensive view of the design, in order to reduce wastes in the process.

This methodology seeks to extend the advantages of lean to a new area of the industry to improve the value of SME vessel offerings. Emphasizing the objective of reducing design time through the reduction of the most significant wastes measured during the observation phase of this research we sought to improve employee skills and increase the synchronicity of processes. The current state of the implementation has demonstrated strong potential with sea trials in two cases underway. While work is ongoing, several benefits have been detected, specifically a faster pace for design variations of vessels, and the diffusion of skills and knowledge to support efforts.

However, there are further refinements and details expected to be made for the method to be suitable for the greater Maritime Sector. This will involve testing to verify its efficacy and ability to be applied to the design of different vessel types. Additional studies will be needed to understand if the use of the methodology and best practices are context dependent. Additional data related to the value (benefit) of unique lean tools is being collected and will then begin to integrate new sensor data into the design processes for each vessel. The KPI data has not yet been reported. Through complete implementation in the three cases (from concept design to vessel delivery) the methodology will be validated to confirm the objectives set forth in the project. This will allow for the ultimate determination if the methodology needs to be adjusted to better serve the purpose of supporting such a challenging industry.

Acknowledgements The content of this paper is based on the LINCOLN project (EU H2020 727982) that is currently under development (www.lincolnproject.eu). We wish to express our deep gratitude to all LINCOLN partners.

References

1. Koenig PC, Narita H, Baba K (2002) Lean production in the Japanese shipbuilding industry? J Ship Prod 18(3):167–174
2. Milanovic V (2016) Application of set-based concurrent engineering to shipbuilding projects. MS Engineering and ICT. Norwegian University of Science and Technology
3. Liker J, Lamb T (2001) Lean Shipbuilding. In: Ship Production Symposium, Ypsilanti, MI
4. Radovic I, Macclaren BJ (2004) Lean six sigma in shipbuilding. In: European shipbuilding and repair conference, London, 2 November 2004
5. Corbett JJ, Winebrake J (2008) The impacts of globalisation on international maritime transport activity. In: Global forum on transport and environment in a globalising world
6. Markit IHS (2016) Five trends shaping the global maritime industry
7. Thanopoulou H, Strandenes SP (2017) A theoretical framework for analysing long-term uncertainty in shipping. Case Stud Transp Policy 5(2)
8. Luglietti R, Wurst S, Sassanelli C, Terzi S, Martín CM (2018) Towards the definition of specialized vessels' model. In: 2018 IEEE international conference on engineering, technology and innovation (ICE/ITMC)
9. Womack JP, Jones DT (1996) Lean thinking: banish waste and create wealth in your corporation. Simon & Schuster, New York
10. Whalen M (2014) Low cost, big return. JPL - Universe, Pasadena
11. León HCM, Farris JA (2011) Lean product development research: current state and future directions. EMJ Eng Manag J 23(1):29–51
12. Welo T (2011) On the application of lean principles in product development: a commentary on models and practices. Int J Prod Dev 13(4):316
13. Rossi M et al (2011) Proposal of a method to systematically identify wastes in new product development process. In: 2011 17th international conference on concurrent enterprising, (Ice), pp 1–9
14. Oehmen J, Rebentisch E (2010) Waste in lean product development. Lean.Mit.Edu 19(1): 471–473
15. Morgan JM, Liker JK (2006) The Toyota product development system: integrating people, process and technology, vol 24

16. Krishnan V, Ulrich KT (2001) Product development decisions: a review of the literature. Manage Sci. 47(1):1–21
17. Browning TR (2003) On customer value and improvement in product development processes. Syst Eng 6(1):49–61
18. Rebentisch E, Rhodes DH, Murman E (2004) Lean systems engineering: research initiatives in support of a new paradigm. Syst Eng 4(1)
19. Chase JP (2001) Value creation in the product development process. M.Sc. thesis, pp 1–133
20. Oppenheim BW (2011) Lean enablers for system engineering. Syst Eng 14(3):305–326
21. Liker JK, Morgan J (2011) Lean product development as a system: a case study of body and stamping development at ford. EMJ Eng Manag J 23(1):16–28
22. Rossi M, Morgan J, Shook J (2017) Lean product and process development. In: The Routledge companion to lean management
23. Garcia P, Drogosz J (2007) Lean engineering—best practice in the automotive industry. SAE Tech Pap
24. Khan M et al (2013) Towards lean product and process development. Int J Comput Integr Manuf 26(12):1105–1116
25. Kerga E, Taisch M, Terzi S, Bessega W, Rosso A (2014) Set-based concurrent engineering innovation roadmap (SBCE IR): a case on adiabatic humidification system. Int J Des Creativity Innovation 2(4):224–255
26. Mascitelli R (2004) The lean design guidebook: everything your product development team needs to slash manufacturing cost
27. Oppenheim BW (2004) Lean product development flow. Syst Eng 7(4):352–376
28. Browning TR et al (2002) Adding value in product development by creating information and reducing risk. IEEE Trans Eng Manag 49(4):443–458
29. Longva KK (2009) Warehouse management in a lean shipbuilding perspective—an exploratory case study of Ulstein Verft. Høgskolen Molde
30. Salem O, Solomon J, Genaidy A, Minkarah I (2006) Lean construction: from theory to implementation. J Manag Eng 22(4)
31. Hamann R, Peschmann J (2013) Goal-based standards and risk-based design. Sh Technol Res 60(2):46–56
32. Molland AF (2008) The maritime engineering reference book, 1st edn. Butterworth-Heinemann
33. Papanikolaou A (2009) Risk-based ship design: methods, tools and applications. Springer
34. Liker JK, Lamb T (2002) What is lean ship construction and repair. J Sh Prod 18(3):121–142
35. Phogat S (2013) An introduction to applicability of lean shipbuilding. Int J Latest Res Sci Technol 2(6):85–89
36. Dugnas K, Oterhals O (2008) State-of-the-art shipbuilding: towards unique and integrated lean production systems. In: Proceedings of IGLC16: 16th annual conference of the international group for lean construction, 6411, pp 321–331
37. Nunamaker J, Chen M, Purdin T (1991) Systems development in Information systems research. J Manag Inf Syst 7:89–106
38. DG for Maritime Affairs and Fisheries (2009) The role of maritime clusters and development executive summary ECO
39. Rossi M, Terzi S (2017) CLIMB: maturity assessment model for design and engineering processes. Int J Prod Lifecycle Manag 10(1)
40. Rossi M, Terzi S, Taisch M (2014) Engineering and design best practices in new product development: an empirical research. Procedia CIRP 21:2.
41. Nonaka I (1994) A dynamic theory of organizational knowledge creation. Organ Sci 5(1)
42. Shook J (2018) The lean transformation framework. The Lean Enterprise Institute. https://www.lean.org/WhatsLean/TransformationFramework.cfm
43. Sullivan BP, Rossi M, Terzi S (2018) A customizable lean design methodology for maritime. In: 5th IFIP WG 5.1 international conference, Turin, Italy. https://doi.org/0.1007/978-3-030-01614-2_47
44. Rossi M, Taisch M, Terzi S (2012) Lean product development: a five-steps methodology for continuous improvement. In: 2012 18th international conference on engineering, technology and innovation, ICE 2012—conference proceedings. https://doi.org/10.1109/ICE.2012.6297704

The manufacturer's authorised representative in the EU is Springer
Nature Customer Service Centre GmbH, Europaplatz 3, 69115 Heidelberg,
Germany. If you have any concerns regarding our products, please
contact ProductSafety@springernature.com

Printed and bound by CPI Group (UK) Ltd, Croydon, CR0 4YY
05/10/2026
02228121-0019